Environmental Pollu
Medicinal Plants

EXPLORING MEDICINAL PLANTS
Series Editor
Azamal Husen
Wolaita Sodo University, Ethiopia

Medicinal plants render a rich source of bioactive compounds used in drug formulation and development; they play a key role in both traditional and indigenous health systems. As the demand for herbal medicines increases worldwide, supply is declining, as most of the harvest is derived from naturally growing vegetation. Considering global interests and covering several important aspects associated with medicinal plants, the Exploring Medicinal Plants series comprises volumes valuable to academia, practitioners, and researchers interested in medicinal plants. Topics provide information on a range of subjects including diversity, conservation, propagation, cultivation, physiology, molecular biology, growth response under extreme environments, handling, storage, bioactive compounds, secondary metabolites, extraction, therapeutics, mode of action, and healthcare practices.

Led by Azamal Husen, PhD, this series is directed toward a broad range of researchers and professionals, consisting of topical books exploring information related to medicinal plants. It includes edited volumes, reference works, and textbooks available for individual print and electronic purchase.

Traditional Herbal Therapy for the Human Immune System, *Azamal Husen*

Environmental Pollution and Medicinal Plants, *Azamal Husen*

For more information on this book series, visit www.routledge.com/Exploring-Medicinal-Plants/book-series/CRCEMP

Environmental Pollution and Medicinal Plants

Edited by
Azamal Husen

CRC Press
Taylor & Francis Group
Boca Raton   London   New York

CRC Press is an imprint of the
Taylor & Francis Group, an **informa** business

First edition published 2022
by CRC Press
6000 Broken Sound Parkway NW, Suite 300, Boca Raton, FL 33487-2742

and by CRC Press
4 Park Square, Milton Park, Abingdon, Oxon, OX14 4RN

CRC Press is an imprint of Taylor & Francis Group, LLC

© 2022 Taylor & Francis Group, LLC

Library of Congress Cataloging-in-Publication Data
Names: Husen, Azamal, editor.
Title: Environmental pollution and medicinal plants: impact and adaptation / edited by Azamal Husen.
Description: First edition. | Boca Raton: CRC Press, 2022. |
Includes bibliographical references and index. | Summary: "Environmental pollution mainly caused by anthropogenic activities can damage medicinal plants and alter their growth, gas exchange parameters, biochemical attributes, bioactive substances, antioxidant activity, and gene expression. This book presents information on impacts of environmental pollution on the performance of medicinal plants at various levels including damage detection, adaptation, tolerance, physiological and molecular responses. Discusses trends and responses of medicinal plant tolerance and adaptation to environmental pollution. Focuses on secondary metabolites, phytochemicals, and bioactive compounds associated with medicinal plants growing in contamination conditions"– Provided by publisher.
Identifiers: LCCN 2021053253 (print) | LCCN 2021053254 (ebook) |
ISBN 9781032014845 (hardback) | ISBN 9781032014944 (paperback) |
ISBN 9781003178866 (ebook)
Subjects: LCSH: Medicinal plants–Effect of pollution on.
Classification: LCC SB293 .E58 2022 (print) |
LCC SB293 (ebook) | DDC 633.8/8–dc23/eng/20220218
LC record available at https://lccn.loc.gov/2021053253
LC ebook record available at https://lccn.loc.gov/2021053254

ISBN: 978-1-03-201484-5 (hbk)
ISBN: 978-1-03-201494-4 (pbk)
ISBN: 978-1-00-317886-6 (ebk)

DOI: 10.1201/9781003178866

Typeset in Times
by Newgen Publishing UK

Dedication

To my mentor, Professor (Dr) Muhammad Iqbal, in recognition of his outstanding scientific contributions

Contents

Preface

Health is wealth. In consonance with this proverb, humans have always been in search of various types of medicine to eliminate illnesses. Plants are the natural source of these medicines, mainly due to their secondary metabolites, and have long been in use as medicine in the crude extract form. They are also used to isolate the bioactive compounds in modern medicine and herbal medicine systems. Thus, they play an important role in the development, synthesis, and formulation of new drugs. Right now, numerous plant-based drugs are available in the market, making a remarkable contribution to disease management. However, research on medicinal plants is mostly restricted to relatively small areas, limited species, and basic investigations and descriptions. Medicinal plants and their parts contain chemically active constituents and produce physiological responses when used in the treatment of various ailments.

Environmental pollution, mainly caused by anthropogenic activities, is one of the most serious global challenges and has a negative impact on human health and plant growth/production. It is now realized that the continuously increasing environmental pollution can cause the accumulation of a number of impurities in medicinal plants that are used for healing purposes. As many herbal materials have medicinal properties and are used as medicinal remedies, they must be subjected to rigorous assessments according to material quality standards. Currently, there is no scientific basis to assure that the quantity and quality of medicinal bioactive substances in plants growing in polluted conditions is unchanged. Compared to herbs which used to grow earlier in a clean environment. Exposure to environmental pollution, comprising mainly of sulphur dioxide, elevated ozone, UV irradiation, heavy metal, industrial waste, and contaminated water, induces acute or chronic injury, depending on the type and concentration of pollutants, exposure duration, geographical location, season, and plant species. Pollutants affect the growth rate, gas exchange parameters, biochemical attributes, bioactive substances, antioxidant activity, and gene expression in plants. The impact of environmental pollution depends on the balance between the production of reactive oxygen species and the scavenging mechanisms through both enzymatic and metabolic antioxidants. Attention needs to be focused not only on obtaining new bioactive compounds for new herbal drug preparation, but also on ensuring high standards of quality and chemical purity of medicinal plants growing under polluted conditions. Various biotechnological approaches have been used to increase the production and extraction of secondary metabolites of medicinal plants at the industrial level. Further, the 'omics' approaches have enabled researchers to identify the genetics underlying medicinal plant responses to adaptation mechanisms under a changed environment.

Taken together, the present book covers a wide range of topics as mentioned above, and discusses the latest trends and responses of medicinal plants in relation to tolerance and adaptation to environmental pollution. Additionally, future challenges and opportunities have been addressed. I hope this book will cater to the needs of graduate students as a textbook, of researchers and practitioners as a reference book, and will also inspire industrialists and policymakers. With great pleasure, I extend my sincere thanks to all the contributors for their timely response, and the excellent, up-to-date contributions. I am thankful to Ms Randy Brehm, Dr Julia Tanner, and all associates at Taylor & Francis Group, LLC/ CRC Press for their sustained cooperation. Finally, my special thanks goes to Shagufta, Zaara, Mehwish, and Huzaifa for their time, immediate, extended support and incentive to put everything together. I shall be happy to receive comments and criticism, if any, from subject experts and general readers of this book.

Azamal Husen
Wolaita, Ethiopia

Editor

Professor Azamal Husen (BSc from Shri Murli Manohar Town Post Graduate College, Ballia, UP, MSc from Hamdard University, New Delhi, and PhD from Forest Research Institute, Dehra Dun, India) is a foreign delegate at Wolaita Sodo University, Wolaita, Ethiopia. He has served the University of Gondar, Ethiopia, as a full professor of biology, and worked there as the coordinator of the MSc program and as the head of the Department of Biology. He was a visiting faculty member of the Forest Research Institute, and the Doon College of Agriculture and Forest at Dehra Dun, India. He has more than 20 years' experience in teaching, research, and administration.

Dr Husen specializes in biogenic nanomaterial fabrication and application, plant responses to nanomaterials, plant adaptation to harsh environments at the physiological, biochemical, and molecular levels, herbal medicine, and clonal propagation for improvement of tree species. He has conducted several research projects sponsored by various funding agencies, including the World Bank, the Indian Council of Agriculture Research (ICAR), the Indian Council of Forest Research Education (ICFRE), and the Japan Bank for International Cooperation (JBIC).

He has published more than 150 research papers, review articles, and book chapters, edited books of international repute, presented papers at several conferences, and produced more than a dozen manuals and monographs. He has received four fellowships from India and a recognition award from the University of Gondar, Ethiopia, for excellent teaching, research, and community service. An active organizer of seminars/conferences and an efficient evaluator of research projects and book proposals, Dr Husen has been on the editorial board and the panel of reviewers of several reputed journals published by Elsevier, Frontiers Media SA, Taylor & Francis, Springer Nature, RSC, Oxford University Press, Sciendo, The Royal Society, CSIRO, PLOS, and John Wiley & Sons. He is on the advisory board of Cambridge Scholars Publishing, UK. He is a Fellow of the Plantae group of the American Society of Plant Biologists, and a member of the International Society of Root Research, Asian Council of Science Editors, and INPST.

Dr Husen is Editor-in-Chief of the *American Journal of Plant Physiology*. He is also working as Series Editor of 'Exploring Medicinal Plants', published by Taylor & Francis Group, USA; 'Plant Biology, Sustainability, and Climate Change', published by Elsevier Inc., USA; and 'Smart Nanomaterials Technology', published by Springer Nature Singapore Pte Ltd. Singapore.

Contributors

Mahendra L. Ahire
Department of Botany
Yashavantrao Chavan Institute of
  Science
Satara, Maharashtra, India
and
Life Science Laboratory
Rayat Institute of Research and
  Development
Satara, Maharashtra, India

Iftikhar Ahmad
Department of Environmental
  Sciences
COMSATS University Islamabad
Vehari-Campus, Vehari, Pakistan

Arvind Arya
Noida Institute of Engineering and
  Technology
Greater Noida, Uttar Pradesh, India

Archana (Joshi) Bachheti
Department of Environmental Science
Graphic Era University
Dehradun, India

Rakesh Kumar Bachheti
Centre of Excellence in Nanotechnology
Addis Ababa Science and Technology
  University
Addis Ababa, Ethiopia

Piyush Bhalla
Chemistry and Bioprospecting Division
Forest Research Institute
Dehradun, India

Kiran Chauhan
Chemistry and Bioprospecting Division
Forest Research Institute
Dehradun, India

Jaykumar J. Chavan
Department of Biotechnology
Yashavantrao Chavan Institute of
  Science
Satara, Maharashtra, India
Department of Botany
and
Yashavantrao Chavan Institute of
  Science
Satara, Maharashtra, India
and
Life Science Laboratory
Rayat Institute of Research and
  Development
Satara, Maharashtra, India

Anuj Choudhary
Department of Botany
Punjab Agricultural University
Ludhiana, India

Deepti
Department of Environmental
  Science
Graphic Era University
Dehradun, India

Allah Ditta
Department of Environmental
  Sciences
Shaheed Benazir Bhutto University
  Sheringal
Upper Dir, Khyber Pakhtunkhwa,
  Pakistan
and
School of Biological Sciences
The University of Western Australia
Perth, Australia

Himani Gautam
Department of Entomology
Dr. Y.S Parmar University of
 Horticulture and Forestry
Nauni, Solan, Himachal Pradesh, India

Swati T. Gurme
Department of Biotechnology
Yashavantrao Chavan Institute of
 Science
Satara, Maharashtra, India

Azamal Husen
Wolaita Sodo University
Wolaita, Ethiopia

Muhammad Imtiaz
Soil and Environmental Biotechnology
 Division
National Institute for Biotechnology and
 Genetic Engineering
Faisalabad, Pakistan

Amin Ullah Jan
Department of Biotechnology
Shaheed Benazir Bhutto University
 Sheringal, Dir (U)
Khyber Pakhtunkhwa, Pakistan

Shipra Jha
Amity Institute of Biotechnology
Amity University
Uttar Pradesh, Noida, India

Madhavi Joshi
Department of BioSciences
Himachal Pradesh University
Shimla, Himachal Pradesh, India

Rohit Kashyap
SR Institute of Management and
 Technology
Lucknow, Uttar Pradesh, India

Gurleen Kaur
California Baptist University
Riverside, California, USA

Harmanjot Kaur
Department of Botany
Punjab Agricultural University
Ludhiana, India

Aryan Khattri
Faculty of Biotechnology
Institute of Bio-Sciences and Technology
Shri Ramswaroop Memorial University
UP, India

Antul Kumar
Department of Botany
Punjab Agricultural University
Ludhiana, India

Ritesh Kumar
Department of Agronomy
Kansas State University
Manhattan, Kansas, USA

Xiaomin Li
School of Space and Environment
Beihang University
Haidian, Beijing, PR China

Sajid Mehmood
Guangdong Provincial Key Laboratory
 for Radionuclides Pollution Control
 and Resources
School of Environmental Science and
 Engineering
Guangzhou University
Guangzhou, China
School of Civil Engineering
Guangzhou University
Guangzhou, PR China

Sahil Mehta
International Centre for Genetic
 Engineering and Biotechnology
Aruna Asaf Ali Marg
New Delhi, India
and
School of Agricultural Sciences
K.R. Mangalam University
Sohna Rural, Haryana, India

V.K. Mishra
Department of Biotechnology
Devbhoomi Uttarakhand University
Manduwala, Dehra Dun, Uttarakhand,
    India

Pankaj S. Mundada
Department of Biotechnology
Yashavantrao Chavan Institute of
    Science
Satara, Maharashtra, India

Saumya Pandey
Department of Biotechnology
IILM-College of Engineering and
    Technology
Greater Noida, Uttar Pradesh, India

Rahat Parveen
Faculty of Biotechnology
Institute of Bio-Sciences and
    Technology
Shri Ramswaroop Memorial University
UP, India

Paras Porwal
Amity Institute of Biotechnology
Amity University
Lucknow, Uttar Pradesh, India

Shakeelur Rahman
Prakriti Bachao Foundation
Bariatu, Ranchi, Jharkhand, India

Nisha Rani
Department of BioSciences
Himachal Pradesh University
Shimla, Himachal Pradesh, India.

Rahul Rawat
SR Institute of Management and
    Technology
Lucknow, Uttar Pradesh, India

Muhammad Rizwan
Institute of Soil Science
PMAS-Arid Agriculture University
Rawalpindi, Pakistan

Muhammad Shahid Rizwan
Cholistan Institute of Desert Studies
The Islamia University of Bahawalpur
Pakistan

Anand Sagar
Department of BioSciences
Himachal Pradesh University
Shimla, Himachal Pradesh, India.

Prachi Sao
Faculty of Biotechnology
Institute of Bio-Sciences and
    Technology
Shri Ramswaroop Memorial University
Uttar Pradesh, India

Hardeep Rai Sharma
Institute of Environmental Studies
Kurukshetra University
Kurukshetra, Haryana, India

Radhika Sharma
Department of Soil Science
Dr. Y.S Parmar University of
    Horticulture and Forestry
Nauni, Solan, Himachal Pradesh, India

Shubhra Sharma
Faculty of Biotechnology
Institute of Bio-Sciences and
    Technology
Shri Ramswaroop Memorial
    University
UP, India

Mandeep Singh
Department of Agronomy
Punjab Agricultural University
Ludhiana, India

Pooja Singh
SR Institute of Management and
   Technology
Lucknow, Uttar Pradesh, India

Sachidanand Singh
Faculty of Biotechnology
Institute of Bio-Sciences and
   Technology
Shri Ramswaroop Memorial University
Uttar Pradesh, India
and
Department of Biotechnology
Vignan's Foundation for Science
Technology and Research (Deemed to
   be University)
Vadlamudi, Guntur, Andhra Pradesh,
   India

Ghulam Sarwar Soomro
School of Energy and Environmental
   Engineering
University of Science and Technology
Beijing, PR China

Gurparsad Singh Suri
California Baptist University
Riverside, California, USA

Neha Tiwari
Faculty of Biotechnology
Institute of Bio-Sciences and
   Technology
Shri Ramswaroop Memorial University
UP, India

Naseer Ullah
Environmental Chemistry Laboratory
Department of Environmental Science
   and Engineering
School of Space and Environment
Beihang University
Beijing, PR China

1 Potential Impacts of Environmental Pollution on the Growth and Metabolism of Medicinal Plants
An Overview

Nisha Rani, Madhavi Joshi, Anand Sagar,
Hardeep Rai Sharma

CONTENTS

1.1 INTRODUCTION

Since time immemorial medicinal plants have been used in the healthcare system. These are living factories for the production of a massive amount of secondary metabolites (SMs) that form the basis not only of modern pharmaceuticals but also of herbal remedies of traditional medicines which involve the use of active ingredients of plant extracts. The World Health Organization (WHO 2005) has estimated that 80 per cent of the world's population living in any region relies on forms of medicinal plant. Different chemical constituents and their associated biochemical properties and biological activities not only improve human health via food additives and pharmaceuticals but have large-scale use in various sectors, such as nutraceuticals, perfumes, cosmetics, phytochemicals, and agrochemicals.

DOI: 10.1201/9781003178866-1

1

According to studies conducted by the Indian Council of Forestry Research and Education (ICFRE), about 6,000 plant species have been codified and used under the natural and herbal medical system (Ved and Goraya 2007). About 350 species in 18 Indian states including the northeast are threatened and listed in the code red category of the International Union for Conservation of Nature (IUCN). Most of these threatened species in the wild are reaching the stage of 'no return', meaning that they are on the brink of extinction. Such an alarming condition has constrained pharmaceutical industries to use substitutes which have low efficiency. Increasing biotic, as well as abiotic, stresses, along with excessive and destructive harvesting, are the main causes of concern. The extensive degradation of climate is considered a cause behind the deteriorating resilience of the rootstocks of medicinal plants. Disappearing water springs, invasive attacks of alien species, drying of alpine lakes, retreating glaciers, and effects of glacier melting have further aggravated the situation. Despite the production of SMs by plants as protective measures to various abiotic as well as biotic factors caused by degrading and changing environmental conditions, their quality, as well as quantity to be used in medicinal purposes, is deteriorating. In this situation, there is an urgent need to assess the impacts of various environmental pollutants and changing climatic conditions on the growth and metabolism of medicinally important plants.

1.2 IMPORTANCE OF MEDICINAL PLANTS

Since prehistoric periods medicinal plants have been in use for various healthcare systems. 'A medicinal plant is any plant which, in one or more of its organs/parts, contains substances that can be used for therapeutic purposes, or which are precursors for chemo-pharmaceutical semi synthesis' (Yudharaj et al. 2016). Herbs are mainly indicated as medicinal, but local individuals also used shrubs, trees, epiphytes, vines, lianas, and ferns.

Medicinal plants are considered rich resources of ingredients and have played an important role in the development of human cultures around the whole world. The Indian subcontinent is a home for mega-biodiversity hot spots and is known for its rich deposit of medicinal and aromatic plants. The raw materials for the production of drugs and perfumes are largely collected from Indian forests. For thousands of years, plant-based medicines have been part of traditional healthcare and there is an increasing interest in plants as agents to fight various diseases.

The importance of medicinal plants for healing as well as for curing human diseases lies in their phytochemical constituents. These are naturally present in the plants' parts, such as roots, stems, leaves, and fruits. Phytoconstituents play a big role in defence mechanisms and protect plants against the stressing environment and from various diseases. They are further categorized as primary and secondary compounds. Primary constituents are chlorophyll, proteins, and common sugars, while secondary compounds include terpenoid, alkaloids and phenolic compounds (Wadood et al. 2016).

Aloe vera, tulsi, neem, turmeric, and mint are examples of commonly used medicinal plants for home remedies in many regions of the country. Besides having medicinal properties, herbs are also used in pest control, natural dye, food, tea, perfume, and so on.

Medicinal plants/herbs can be classified into various categories according to their nature, usage, and active constituents. There are five major types of herb according to their active constituents: astringents (tannins), aromatic (volatile oils), bitter (phenol compounds, saponins, and alkaloids), nutritive (foodstuffs), and mucilaginous (polysaccharides). Examples include *Aegle marmelos* (alkaloids, terpenoids, saponins), *Acorus calamus* (alkaloids), *Carica papaya* (terpenoids. saponins, tannins), and *Aloe vera* (vitamins A, C, and E, carotenoids) (Yudharaj et al. 2016). Details of some important medicinal plants and their uses are mentioned in Table 1.1.

Phytoconstituents are the main components of medicinal plants and their presence is considered safe for curing many diseases with no or minimal side effects. This has led to the worldwide popularity of herbal treatment. The herbs that have medicinal properties offer rational means for the cure of many internal diseases, which are difficult to treat even through allopathy.

TABLE 1.1
Some Important Medicinal Plants and Their Uses

Name of Plant (scientific)	Family	Common Name	Uses/Properties
Aloe barbadensis	Liliaceae	Aloe vera	Used during sunburn, and for pimples.
Azadirachta indica	Meliaceae	Neem	Anti-inflammatory, antipyretic, and antimalarial.
Bergenia ciliata	Saxifragaceae	Pashan bheda, Patharchatta	Antipyretic, anticancerous, to stop bleeding, piles, and pulmonary infections.
Curcuma longa	Zingiberaceae	Haldi	Healing of wounds, and as antiseptic.
Erigeron alpines	Asteraceae	Bashakar	Used for curing of colds and coughs and rheumatism.
Gentianella moorcroftiana	Gentianaceae	Tikta	Fever, cough, cold, nausea, and gastric.
Geranium nepalense	Geraniaceae	Ratanjot, Bhanda	Ulcers, renal diseases, and haemorrhoids.
Melia azedarach	Meliaceae	Drek, Mahanimba	Joint pain, skin ailments, and to cure piles.
Mentha sylvestris	Lamiaceae	Pudina	Indigestion, as juice in diarrhoea.
Moringa oleifera	Moringaceae	Sehjan	Stomachache, swelling, asthma and cough.
Murraya koenigii	Rutaceae	Gandala, Karripata	Stimulant, antidysentric, anti-helminthic, and analgesic.
Ocimum sanctum	Lamiaceae	Tulsi	Cough and cold, reducing stress.
Tinospora cordifolia	Menispermaceae	Giloye	Tonic, and joint pain.
Vitex negundo	Verbenaceae	Chinese chaste tree	Headache, cataract and swelling of joints.
Zanthoxylum alatum	Rutaceae	Tirmira	Teeth cleaning, fever, scabies, skin diseases, and cholera.

1.3 DIFFERENT ENVIRONMENTAL PROBLEMS FACING THE WORLD

In general, environment means the combination of all the external influences to which an organism is exposed and it comprises four main components: the lithosphere, hydrosphere, atmosphere, and biosphere. The environment is changing constantly due to increasing population, pollution, industrialization, and the unsustainable use of natural resources. According to United Nations 2017 population estimates, the world population of 7.6 billion will increase to about 9.8 billion by 2050. The population explosion is responsible for more urbanization, migration, and natural resources exploitation. Also, industries release various types of effluents into air, water, and soil, causing pollution. The transformation of forest land to agricultural land causes deforestation and accelerates soil erosion. Industrial air emissions, forest fires, incineration of wastes and crop residues, and use of fossil fuels release different gases into the atmosphere, thereby changing its composition and causing air pollution. Greenhouse gases including carbon dioxide, methane, NOx, and CFCs strongly absorb infrared radiation in the atmosphere and effectively block a large portion of the earth's emitted radiation. The net result is heating of the earth's surface, a phenomenon called the greenhouse effect and a process known as global warming that ultimately leads to global climate change. Secondary pollutants along with primary pollutants initiate different photochemical reactions in the atmosphere, which cause acid rain, photochemical smog, and depletion of the ozone layer.

In recent years, personal care products and pharmaceuticals have emerged as a possible threat to the human population and to aquatic ecosystems. These chemicals once utilized by humans and livestock enter the environment, thereby causing a negative effect on non-target organisms in the aquatic bodies. Antibiotics' effects on biological functions may cause a change in nutrient cycling, organic matter degradation, nitrogen transformation, sulphate reduction, and methanogenesis (Grenni et al. 2018). According to UNESCO (2017), despite improvement in some regions, water pollution is on the rise globally with eutrophication as the most prevalent water problem. In developing countries more than 80 per cent of sewage is discharged untreated, polluting lakes, rivers, and coastal areas. Elevated temperature, destruction of habitats, plastic debris, overfishing, pollution, waste dumping, and ocean acidification are the main threats to marine environments. Due to improper waste disposal, about five trillion plastic fragments weighing more than 260,000 tonnes are afloat over the ocean surface of the world (Eriksen et al. 2014). Due to acidification and ocean warming, more than half of warm-water coral reefs are predicted to be at high to critical risk by 2030 (IOC-UNESCO and UNEP, 2016). Sea level rise, melting of ice sheets and glaciers, ocean acidification, mass extinction of species, the spread of diseases, extreme weather events, and change in agricultural productivity are some of the consequences of climate change. The Greenland ice sheet in the Arctic poses the greatest threat to sea levels because melting land ice is the leading cause of rising sea levels. As per the World Economic Forum's Global Risks Report (2018), natural disasters, extreme weather events, loss of biodiversity, ecosystem collapse, water crises, soil, water and air pollution, as well as a failure of mitigation and adaptation measures of climate change, are the most pressing threats that are predicted to have the biggest impact in the next ten years. Very recently scientists have warned that big lakes are facing degradation and loss of ice, depletion of resources (water and food), destruction of habitats and ecosystems, loss of species, and accelerating pollution, due to rapid warming (Jenny et al. 2020).

Many researchers believe that biodiversity loss is more than an environmental emergency. Biodiversity loss undermines the ability of ecosystems to work properly and efficiently and thus weakens nature's capacity to sustain a healthy environment (Roe 2019). The report of the Intergovernmental Science-Policy Platform on Biodiversity and Ecosystem Services (Diaz 2019) warned that the declining rate of nature is exceptional in human history, with an accelerating speed of species extinctions. The report further mentioned that the global species extinction rate is already at least 10–100 times higher than the average rate over the past 10 million years. With global number of cases 422 million and death toll surpassing 5.8 million as of third week of February 2022 (WHO, 2022), the COVID-19 pandemic is certainly one of the major threats that humankind have faced in modern times. The COVID-19 pandemic has created a global emergency crisis in terms of socio-economic and environmental challenges. The pandemic has changed the qualitative and quantitative aspects of wastes and their management, causing a serious threat to sanitation workers (Mallapur 2020). Improper management of different biomedical wastes like masks, gloves, and personal protective equipment (PPE) may further increase the risk of virus transmission. Phelps Bondaroff and Cooke (2020) estimated that about 1.56 billion masks would get entry to our oceans in 2020, which would amount to between 4,680 and 6,240 tonnes of plastic pollution. All these environmental problems directly or indirectly affect the flora and fauna in their respective area of concern. It is the collective responsibility of all, from individual citizens to governments, policymakers, scientists, world leaders, and communities across regions and countries, to reduce these environmental problems by working together to make a better world for all.

1.4 EFFECT OF AIR POLLUTION ON THE GROWTH AND METABOLISM OF MEDICINAL PLANTS

The response and sensitivities of medicinal plants to various environmental stresses vary among species. The residence time of air pollutants and their physiochemical properties largely depend upon the climatic conditions of the area. Rapid industrialization and increased anthropogenic activities into the natural ecosystem have brought huge quantities of pollutants into the atmosphere as well as into the land and water ecosystem. These include oxides of sulphur, nitrogen, and carbon, soot particles, various organic and inorganic chemical compounds, radioactive isotopes, and toxic heavy metals. Air pollution is aesthetically unfriendly and poses serious hazards to biota (Agbaire and Akporhonor 2014). The air pollutants through dry and wet deposition make their way into the plant ecosystem and get incorporated through metabolic pathways. They affect plants directly via leaves and indirectly through soil acidification. The physiological damage is more pronounced in some species, with discernible harm to leaves on exposure to various pollutants (Liu and Ding 2008). The synergistic effects of these pollutants are even more harmful when they are present in varying concentrations. The deposition of oxides of nitrogen, sulphur, carbon, and trace elements on leaves affects the physiological behaviour of some plant species (Van Wittenberghe et al. 2013). Reduced photosynthetic activity growth and yield, leaf injury, stomata damage, senescence, and reduced membrane permeability were observed by Tiwari et al. (2006) in some sensitive plant species.

The air pollutants released from municipal solid waste gets deposited on medicinal foliage, which significantly reduces the chlorophyll content by up to 60 per cent and ascorbic acid up to 30 per cent, as studied in *Morinda lucida* and *Chromolaena odorata* collected from Aba-Eku, Ajakanga, Awotan, and Lapite dumpsites in Ibadan, Nigeria (Falusi et al. 2016). This also has major effects on the physiological and biochemical characteristics of plants' growth and metabolism, i.e. periodicity, phenology, fruiting, flowering, leaf wax, biomass production, seed germination, and seed growth (Iqbal et al. 2015).

The air pollutants (SO_2, CO, NOx) trigger the rapid induction of reactive oxygen species (ROS) which can pose a threat to cells and can perform as an indicator to activate the ascorbate-glutathione cycle as a stress response pathway (Mittler et al. 2004; Foyer et al. 2005). Increased levels of these pollutants result in a net reduction of chlorophyll and carotenoid by up to 70–80 per cent, as compared to the control, in leaves of strawberry plants (Muneer et al. 2014).

The phytotoxicity of these pollutants largely depends upon the concentration of air pollutants as low levels of these can be useful to plants – for instance, SO_2 at low levels alleviates nutrient deficiencies and is a structural component of amino acid, proteins, vitamins, and chlorophyll (De-Kok 1990; Maugh 1979). It not only affects carbohydrate metabolism but also enhances nitrogen fixation and nodule development. However, exposure to high doses can lead to necrosis, leaf chlorosis, growth inhibition, and ultimately the death of plants (Agrawal et al. 2003). The effects of some atmospheric pollutants on the production of SMs of medicinal plants are depicted in Table 1.2.

1.5 IMPACTS OF CLIMATE CHANGE ON PROPERTIES OF MEDICINAL PLANTS

Climate change, as defined by various environmental organizations and governmental agencies (IPCC 2007), is a measure of significant changes in climate, i.e. temperature, precipitation, or wind, over a prolonged period, for decades or even longer. Global warming refers to a rise in atmospheric temperature that can contribute to change in global climate patterns. This climate change over time may occur due to natural unpredictability or as a result of anthropogenic activities. According to the United Nations Framework Convention on Climate Change (UNFCC) definition, climate change is a change in climate, attributable directly or indirectly to human activity, that changes atmospheric composition. The Intergovernmental Panel on Climate Change (IPCC) projected a global average temperature rise of 4.2°C towards the end of the twenty-first century.

Considering the importance of medicinal plants, there is the utmost need to understand the pressure and impacts arising from climate change that are already visible at

TABLE 1.2
Effect of Air Pollutants on Secondary Metabolites (SMs) of Plants

Plant species	Family	Uses	Effected SM	References
		Ozone (O_3)		
Capsicum baccatum	Solanaceae	Potential antioxidant and anti-inflammatory compounds	Capsaicin, dihydrocapsaicin, carotenids, and phenolics	Bortolin et al. (2016)
Salvia officinalis	Lamiaceae	Pain relief, antioxidant and anti-inflammatory compounds	Caffeic acid, catechinic acid, gallic acid, and rosmarinic acid	Pellegrini et al. (2015)
Melissa officinalis	Lamiaceae	Health-promoting and antidiabetic properties	Phenolics, anthocyanins, and tannins	Tonelli et al. (2015)
Pueraria thomsonii	Fabaceae	Treat cardiovascular diseases	Increased puerarin, and ABA	Sun et al. (2011)
Hypericum perforatum	Hypericaceae	Antidepressant properties	Hypericin	Xu et al. (2011)
Ginkgo biloba	Ginkgoaceae	Mental health conditions, Alzheimer's disease, and fatigue	Tannins, quercetinaglycon, keampferolaglycon, isorhamnetin, and bilobalide	He et al. (2009); Huang et al. (2010)
Betula pendula	Betulaceae	Analgesic, antiseptic, anti-inflammatory, astringent	Platyfylloside, dehydrosalidroside betuloside, salidroside, hyperoside, papyriferic acid, and hyperoside	Lavola et al. (1994); Saleem et al. (2001)
Loblolly pine	Pinaceae	Antiseptic, diuretic, rubefacient and vermifuge	Tannins	Jordan et al. (1991)
		Carbon dioxide (CO_2)		
Zingiber officinale	Zingiberaceae	Analgesic properties	Flavonoids, and phenolics	Ghasemzadeh et al. (2010)
Ginkgo biloba	Ginkgoaceae	Mental health conditions, Alzheimer's disease, and fatigue	Tannins, quercetinaglycon, keampferolaglycon, isorhamnetin, bilobalide	He et al. (2009); Huang et al. (2010)

(continued)

TABLE 1.2 (Continued)
Effect of Air Pollutants on Secondary Metabolites (SMs) of Plants

Plant species	Family	Uses	Effected SM	References
Quercus ilicifolia	Fagaceae	Treating gynaecological problems	Tannins, phenolics	Stiling and Cornelissen (2007)
Papaver setigerum	Papaveraceae	Treat asthma, stomach illnesses	Codeine, morphine, papaverine and Noscapine	Ziska et al. (2008)
Hypericum perforatum	Hypericaceae	Antidepressant	Hypericin, pseudohypericin, and hyperforin	Zobayed and Saxena (2004) Mosaleeyanon et al. (2005),
Brassica oleracea var. *italica* Plenck	Brassicaceae	Antioxidant and anticancer potential	Methylsulfinylalkyl glucosinolates, glucoraphanin, and glucoiberin	Schonhof et al. (2007)
Panax ginseng	Araliaceae	Maintain homeostasis of the body and antifatigue	Increase in flavonoids, and phenolics	Ali et al. (2005)
Pseudotsuga menziesii	Pinaceae	Used as a poultice to treat cuts, burns, wounds and other skin ailments	Reduction in monoterpenes	Snow et al. (2003)
Sulfur dioxide (SO₂) and Hydrogen sulfide (H₂S)				
Nicotiana tabacum	Solanaceae	Inflammatory diseases	Increased in total phenolics contents, flavonols, carotenoids, and anthocyanins	Montesinos-Pereira et al. (2016)
Brassica oleracea Acid rain	Brassicaceae	Cardiotonic and stomachic	Mediate nicotine biosynthesis	Chen et al. (2016)
Pereskia bleo	Cactaceae	Detoxification, inflammatory diseases such as dermatitis, hypertension, diabetes, stomach ache, treatment of cancer and rheumatism	Increases in proline and antioxidant levels	Sulandjari and Dewi (2020)

all levels of species survival and conservation. Like other living components, medicinal plants are not immune to the changing climatic conditions and noticeable effects have been observed on the lifecycle and distribution of vegetation across the world. The medicinal plants that are endemic to a particular geographical region or ecosystem are more vulnerable to changes in climate due to anthropogenic activities (Neilson et al. 2005). The impacts are so intense that they will result in the loss of some keystone species of aromatic and medicinal plants from our biodiversity mega spots around the world.

Food plants are valuable for the production of carbohydrates, sugars, proteins, and various vitamins, while medicinal plants owe their importance to secondary metabolites (SMs). These are not essential for the normal growth of plants but are synthesized to protect and adapt the plants to various biotic as well as abiotic stresses present in the environment. These metabolites and their associated biological activities form the basis for many herbal remedies and commercial pharmaceutical drugs available to modern as well as ancient medicinal systems. The pharmaceutical and food industries use these biochemicals and their derivatives to improve human health. These mainly include polyphenols, alkaloids, glycosides, terpenes, flavonoids, coumarins, and tannins, etc. as discussed in Section 2.

Global warming has resulted in many abiotic pressures like flooding, salinity, drought, and extreme temperatures. These factors drastically affect the growth and development of plant species as they have specific temperature requirements (Wani and Sah 2014). As observed in recent years, the intensity of heatwaves is very high; even for a short duration, a rise in temperature of 5°C above the normal has exhibited a relatively larger impact on medicinal plant productivity.

The production and accumulation of SMs in medicinal plants depend upon various cultural conditions and are worst affected by temperature fluctuation (Gershenzon 1984; Gleadow and Woodrow 2002; Ballhorn et al. 2011). Elevated temperature reduced the accumulation of carotenoids (β-carotene) in species of the *Brassicaceae* family and ginsenoside in cell cultures of *Panax quinquefolius* (Jochum et al. 2007). Unpredictable events, like abnormally hot summers in Germany and Poland, have prevented seeding in chamomile (*Matricaria recutita*), and extreme events of flooding in Hungary have led to a reduction of harvests in fennel (*Foeniculum vulgare*) and anise (*Pimpinella anisum*) (Pompe et al. 2008). Climate change also changes the taste and effectiveness of medicinal properties of some Arctic plants (Gore 2006).

Studies conducted on the accumulation of SMs in response to varying temperatures of lower range (12–18°C) and higher of 30°C in somatic embryos of *Eleutherococcus senticosus* showed that a rise in temperature reduced the build-up (flavonoids, phenolics, and total eleutheroside) while a lower temperature favours the accumulation of eleutheroside E (Shohael et al. 2006). The temperature range between 17 and 25°C is usually considered optimal for the growth, production, and accumulation of SMs in the cells of medicinal plants. Murthy et al. 2014 optimized the synthesis of ginsenosides (major constituents of ginseng, with antitumour and antioxidant properties) in the cell cultures of *P. Ginseng* at 25±1°C. The effect of climate change on both cultivated and wild medicinal plants is very significant. The drought stress increases the concentration of bioactive compounds in a variety of species (Selmar et al. 2013 and Al-Gabbies et al. 2015) and includes essential oils and terpenes, simple and complex phenolic compounds, alkaloids, and glucosinolates. The active metabolites in wild plants such as *Vitellaria paradoxa* are found to be higher in drier areas (Maranz and Wiesman 2004). Water-stressed conditions favour the production of certain chemicals as compared to low-temperature conditions, as seen in *Rehmannia glutinosa* (Gaertn.), one of the 50 fundamental herbs used in traditional Chinese medicine as shown by Chung et al. 2006. However, as a consequence of global warming, if drought is accompanied by an increase in temperature the beneficial effect on the concentration of SMs could be counteracted. Both these stresses may increase the concentration of bio-compounds but also lead to a significant reduction in biomass, as observed in American ginseng. This increased unsustainability of the harvest would result in severe economic loss, affecting people's subsistence. Under drought and temperature stress, an increased concentration of SMs as a protecting solute may include toxic metabolites, as observed by Kirk et al. (2010) where the pyrrolizidine alkaloids in *Senecio* species are responsible for liver damage. However, *Withania somnifera*, the ashwagandha or Indian ginseng, exhibited an increase in key bioactive compound (steroidal lactones; withanolide) accumulation in foliage to cold stress response (Mir et al. 2015; Kumar et al. 2012).

The most apparent effect of climate change as suggested by evidence on vegetation is a shift in phenology, i.e. the timing of life cycle events in plants mainly in relation to climate and distribution (IPCC 2007). In response to changing weather patterns and temperature, early flowering and shifts in geographical ranges in some medicinal and aromatic plants have been reported. Although in the first instance this seems of little importance, it presents great challenges to survival. Studies by Walther et al. (2002) indicates the emergence of an early spring since 1960, and this accelerated onset of spring has caused perceptible changes in phonological events such as the timing of plant budburst, first leafing and flowering, fruit and seed dispersal, etc. Temperature-driven early blooming can be damaging for areas prone to chilly spells late in the spring. If these two events are followed by one another within a few days, those early buds or fruits freeze, potentially affecting or killing the yield of some economically beneficial plants (Zobayed et al. 2005). Similar results have been observed in the medicinal plant bloodroot (*Sanguinaria Canadensis*) belonging to the family *Papaveraceae* (Shea 2008). As data from the Nature Calendar indicates, many medicinal plants such as horse chestnut (*Aesculus hippocastanum*, *Hippocastanaceae*) and hawthorn (*Crataegus monogyna; C. laevigata, Rosaceae*) started blooming earlier in the UK (Cavaliere 2009). Such variations in geographic range and phenology, and unpredictable shifts in species qualities, could threaten the usability of these plants as medicine.

The changes in weather patterns have severely impacted seasonal lifecycle timing as well as geographical ranges for most plants, which could eventually threaten some wild medicinal plant populations in their habitat. To add to the list, severe climate events have commenced, affecting the harvesting and production of various SMs in medicinal plants all around the world.

India, where the annual monsoon largely controls the climate, appears to be experiencing increasingly severe and irregular rainfall. Increased rainfall events and their intensity, as well as frequency of flooding, have increased since the 1980s (Goswami et al. 2006). All such events result in heavy loss of vegetation.

However, particular and inconsistent weather patterns cannot definitively be blamed on global warming and climate change, but the negative effects of some recent floods, storms, and droughts on herbal crops reveal the threats that increased extreme weather could pose to the availability and supply of important medicinal plants.

1.6 EFFECT OF SALINITY ON THE GROWTH AND METABOLISM OF MEDICINAL PLANTS

Salinity is a major global concern and a significant abiotic stress that hampers growth as well as the development of plants and also plays the most important role in determining the ecological distribution of medicinal plant species. Salinization of soil and water is a major reason for medicinal plants encountering this stress. In reality, approximately 40 per cent of the earth's land area suffers from these problems (Vriezen et al. 2007). There has been an estimated 10 per cent annual increase in the salinized areas for diverse reasons, including more surface evaporation followed by less rainfall, weathering of rocks, poor agriculture and irrigational practices. It has been predicted that, by the year 2050, more than 50 per cent of the arable land will be salinized (Jamil et al. 2011; Mondal and Kaur 2017).

Saline soils suppress plant growth by disturbing osmotic effects, toxic solutes, and sodicity. In addition to decreasing the yield of most crops, salinity also affects soil's physicochemical properties (seed germination, plant growth, reproductive development, oxidative stress, water and nutrient uptake) and the ecological balance of the area (Akbarimoghaddam et al. 2011; Srivastava and Kumar 2015).

Germination of seed is one of the most salt-sensitive stages, and increasing salinity severely inhibits growth either by destroying the embryo or by disturbing soil potential so that water uptake is hampered (Sosa et al. 2005; Banerjee and Aryadeep 2017). Germination inhibition was observed by Belaqziz et al. 2009 and Ramin 2005 in *Thymus maroccanus* Ball. and in *Ocimum basilicum* L., respectively.

The next most susceptible stage in the life cycle of plants is the seedling stage. Salinity has severely impacted the growth and development in seedling growth, as reported in *Thymus maroccanus* (Ramin 2005), chamomile, and sweet marjoram (Ali et al. 2007). Chauhan and Kumar (2014) showed in their research on *Citronella java* plants that high salinity levels lead to significant reduction in the number of tillers. Salinity stresses affect the metabolism of plants, leading to suspended cell division, enlarging and injuring hypocotyls, slow or less mobilization of reserve foods, etc. Detrimental and damaging effects of salt can be seen at the whole-plant level. Reductions in seed production and yield components per plant were observed in the milk thistle, *Trachyspermum ammi*, and *Cuminum cyminum* L. (Ashraf and Orooj 2006; Ghavami and Ramin 2008).

Under salt stress, a plant also synthesizes various reactive oxygen scavenging enzymes, SMs and phytohormones as protective mechanisms to minimize the biological loss caused (Laloi et al. 2004; Spoel and Dong 2008). In an experiment conducted by Gengmao et al. 2014 on *Salvia miltiorrhiza*, catalase (CAT) activity was increased under salt stress. In *Mentha pulegium* peroxidase (POD) activity was found to be increased with increasing NaCl concentration (Oueslati et al. 2010). Increased superoxide dismutase (SOD) activity assists plants to resist the potential oxidative damage produced by salinity stress (Gengmao et al. 2014).

Similarly, SMs that play a vital role in providing defence to plants against pathogens, grazing animals and diverse environmental stresses were also found to increase under stressed conditions (Bennett and Wallsgrove 1994; Akula and Ravishankar 2011). Essential oil yield was found to decrease under salt stress in *Trachyspermum ammi* (Ashraf and Orooj, 2006). Salt stress decreased essential oil production in *Mentha piperita* (Tabatabaie and Nazari, 2007), pennyroyal, peppermint, and apple mint (Aziz et al, 2008), and basil (Said-Al and Mahmoud, 2010). Effects of increased salinity on medicinal plants are shown in Table 1.3.

1.7   EFFECT OF HEAVY METALS ON THE GROWTH AND METABOLISM OF MEDICINAL PLANTS

Heavy metals are always present in the environment as they are found as a majority in dispersed form in rock formations and pedogenesis. Due to anthropogenic activities, the concentration of numerous heavy metals is also increasing in certain ecosystems of the world. Many of the heavy metals are considered to be an essential factor for plant growth. Some heavy metals, like Cu and Zn, either serve as a cofactor or many times as activators of enzyme reactions (Mildvan 1970). But high concentrations of heavy metals are associated with toxic/harmful effects on the growth, development, and metabolism of medicinal plants (Rai and Mehrotra 2008). This can alter seed germination potential and reduction in root and shoot length, leaf area, and biomass production (Street et al. 2007).

Heavy metals also act as a barrier to metabolic processes, thereby causing disturbances in protein structure by blocking the functional groups of important cellular molecules, malfunctioning the formation of bonds between metals and sulfhydryl groups, affecting the integrity of the plasma membrane, damaging the function of essential metals in biomolecules like pigments or enzymes, and resulting in the disturbing of vital activities like photosynthesis, respiration, and enzymatic actions. A higher concentration of heavy metals results in the increased production of reactive oxygen species (ROS), leading to oxidative stress by upsetting the balance of antioxidants and pro-oxidants within the plant cells (Ali et al. 2013).

The net photosynthetic rate was found to be decreased under the toxic effects of lead on *Davidia involucrata*. In *Vigna radiata*, antioxidant enzyme activity was found to decrease under cobalt stress, and a decrease in protein content, plant sugar, and amino acid was also recorded. Heavy metal stress shows varying degrees of secondary metabolite response. Chromium (Cr) stress induced the production of eugenol, an important essential oil found in *Ocimum tenuiflorum* (Rai et al. 2004). Increased production of diosgenin was reported in *Dioscorea bulbifera* L. under nickel (Ni) stress (Narula et al. 2005). Under arsenic stress in *Artemisia annua* L. the increased production of artemisinin was reported (Rai et al. 2011).

In contrast to all the negative and growth-inhibiting impacts of heavy metal toxicity, enhanced plant yield has also been reported in some medicinal plants under the influence of Zn, Co, Pb, and Ni in *Matricaria chamomilla, Mentha arvensis,* and *Stevia rebaudiana* (Misra 1992; Kartosentono et al. 2002; Das et al. 2005; Grejtovsky et al. 2006). Some of the adverse effects of heavy metals shown on medicinal plants have been compiled in Table 1.4.

TABLE 1.3

Effects of Increased Salinity on Medicinal Plants

Plant species	Family	Salt conc. used	Impact on plant	References
Foeniculum vulgare	Apiaceae	25–75 mM (NaCl)	Significant reduction in fresh and dry weight, chl a, b and β carotenoid content of seedlings.	Nourimand et al. 2012
Ocimum tenuiflorum	Lamiaceae	25–125 mM (NaCl)	Decreased eugenol content.	Rastogi et al. 2019
A. annua	Asteraceae	100 ml of 100 mM NaCl	Artemisinin content increased.	Vashishth et al. 2018
Schizonepeta tenuifolia	Lamiaceae	25, 50, 75, 100 mM (NaCl)	Plant height, dry biomass reduced by salt treatments except 25 mM NaCl. Contents of antioxidants, declined severely at 75 and 100 mM.	Zhou et al. 2018
Mentha pulegium	Labiatae	25, 50, 75 and 100 mM. (NaCl)	Dry weight of the whole plant decreased along with salinity stress, total protein, POD activity increased with increasing concentration.	Oueslati et al. 2010
Coriandrum sativum	Apiaceae	0, 25, 50 and 75 mM (NaCl)	Linalool yield in the fruits increased at high salt conc. Total phenolic content decreased with increasing salinity.	Neffati et al. 2008
Thymus maroccanus	Lamiaceae	0–200 mM (NaCl)	Increased salinity caused a reduction in seed germination.	Belaqzi et al. 2009
Citronella java	Cardiopteridaceae	0–20 d S/m concentration of salts of chloride, sulphate, and bicarbonate (6:3:1) of sodium and calcium (4:1)	All treatments significantly reduced the number of tillers.	Chauhan and Kumar 2014
Ocimum basilicum	Lamiaceae	0, 1, 3, 6, 9, 12, and 15 dS/m (NaCl)	Germination reduced and abnormal seedling development observed at salt conc. ≥ 12 dS/m.	Ramin 2005

(continued)

TABLE 1.3 (Continued)
Effects of Increased Salinity on Medicinal Plants

Plant species	Family	Salt conc. used	Impact on plant	References
Aloe vera	Liliaceae	0, 2, 4, 6, and 8 ds/m (NaCl)	Number of leaves, height of plant, number of sprouts, leaf, root and plant weight, Vitamin C, and total soluble solids (TSS) were affected.	Moghbeli et al. 2012
Chamomilla recutita L.	Asteraceae	0, 6, 12, and 18 dSm -1(NaCl)	Reduced dry matter yield with increasing salt conc.	Ghanavati and Sengul 2010
Salvia miltiorrhiza	Lamiaceae	0, 25, 50, 75, and 100 mM (NaCl)	Decreased the Chl-b, significantly decreased the superoxide dismutase (SOD) activity, the catalase activity in the leaves increased markedly.	Gengmao et al. 2014
Trachyspermum ammi	Apiaceae	0, 40, 80, and 120 mM L^{-1} (NaCl)	Reduction in fresh and dry weight of roots and shoots as well as seed yield.	Ashraf and Orooj 2006
Achillea fragratissima	Asteraceae	700, 2000, and 4000 ppm (NaCl)	Highest conc. depressed plant height, fresh and dry weight, crude protein, total ash, essential oil and potassium percentage.	Abd EL-Azim and Ahmed 2009
Thymus vulgaris and *Thymus daenensis*	Lamiaceae	0, 30, 60, and 90 mM NaCl	Plant dry matter, K$^+$, Ca^{2+} contents decreased at highest salinity while, Na$^+$, Cinnamic acid, Gallic acid content increased.	Bistgani et al. 2019

TABLE 1.4

Effect of Heavy Metals on Growth and Metabolism of Medicinal Plants

Heavy metal	Medicinal Plant species	Family	Effects on plants	References
Cd	Zea mays	Poaceae	Shoot and root growth is inhibited (10^{-4} M)	Wang et al. 2007
	Allium sativum	Amaryllidaceae	Seedlings growth reduction (10^{-3}–10^{-2} M)	Jiang et al. 2001
	Foeniculum vulgare	Apiaceae	Root growth significantly reduced (10 mg/L)	Jeliazkova and Craker 2003
	Trigonella foenum-graecum	Fabaceae	Diosgenin levels increased (300 µM $CdCl_2$) growth remained unaffected	De and De 2011
Cr	Allium cepa	Amaryllidaceae	Germination process inhibited, plant biomass reduced (>150 mg/L)	Nematshahi et al. 2012
	Ocimum tenuiflorum	Lamiaceae	Increased eugenol levels (3.83% more at 100 µM)	Rai et al. 2004
Mn	Mentha spicata	Lamiaceae	Chl a, carotenoid contents decreased (15.75 µM) and anthocyanin content increased	Asrar 2005
Zn	Bowiea volubilis	Hyacinthaceae	50 mg/L inhibited shoot/root length	Street 2007
Cu	Merwilla natalensis	Hyacinthaceae	Seed germination unaffected, shoot, root and seedlings fresh weight decreased (50 mg/L)	Street 2007
As	Polygonum convolvulus	Polygonaceae	Decrease in biomass and seed, mortality	Kjaer and Elmegaard 1996
	Brassica napus	Brassicaceae	Reduced growth, wilting with chlorosis	Cox et al. 1996.
	Brassica juncea	Brassicaceae	Reduced seed germination, shoot-root length, protein and chlorophyll content	Chaturvedi 2006
	Allium cepa	Amaryllidaceae	Increases chlorophyll-a and -b content in leaves	Miteva and Merakchiyska 2002
Ni	Artemisia annua	Asteraceae	Artemisinin production increased at 0–4500 µg L^{-1}	Rai et al. 2011
	Hypericum perforatum	Hypericaceae	Decreased production of hypericin and pseudohypericin	Murch et al. 2003
	Dioscorea bulbifera	Dioscoreaceae	Stimulated production of diosgenin	Narula et al. 2005
Pb	Avena sativa	Poacae	CO_2 fixation affected due to enzyme inhibition	Moustakas et al. 1994
	Davidia involucrata	Davidiaceae	Inhibition of net photosynthetic rate	Yang et al. 2020

1.8   EFFORTS TO CONSERVE MEDICINAL PLANTS

Medicinal plants are comprised of great biodiversity on earth and are under threat due to various developmental activities. Therefore, conservation is the need of the hour and its goals support sustainable development, i.e. using and protecting biological resources in such a way as not to diminish genetic diversity, species diversity, and important habitats, as well as ecosystems. Among various methods available, in-situ and ex-situ conservation are the two main methods. In this biotechnological era, there are various methods and techniques available through which we can preserve and maintain biodiversity. However, conserving the ecosystem and plant diversity along with their natural habitat is a great challenge. Efforts are being made by national and international governments and agencies toward sustainable use and conservation, toward reduction in environmental pollution, and to curtail greenhouse gas emissions, in order to mitigate adverse effects of a changing climate on biodiversity.

1.9   CONCLUSIONS

Air pollution, acid rain, global warming-induced climate change, salinity, and various heavy metals certainly affect the properties, growth, and metabolism of medicinal plants. Any change in a particular medicinal plant's properties will affect its usage and role in pharmaceutical and traditional medicine and also impact the people associated with them directly and indirectly. Reduction in yield will cause stress on their demand and price which may severely affect their survival. Sincere efforts are needed to avoid their over-exploitation and to protect them from environmental factors affecting their growth and metabolism.

REFERENCES

Abd EL-Azim, W.M., and S.T.H. Ahmed. 2009. Effect of salinity and cutting date on growth and chemical constituents of *Achillea fragratissima* Forssk, under Ras Sudr conditions. *Research Journal of Agriculture and Biological Sciences* 5 (6): 1121–1129.

Agbaire, P.O., and E.E. Akporhonor. 2014. The effects of air pollution on plants around the vicinity of the Delta Steel Company, Ovwian-Aladja, Delta State, Nigeria. *IOSR Journal of Environmental Science, Toxicology and Food Technology* 8: 61–65.

Agrawal, M., B. Singha, M. Rajputa, F. Marshall, and J.N.B. Bell. 2003. Effect of air pollution on peri-urban agriculture: a case study. *Environmental Pollution* 126: 323–329.

Akbarimoghaddam, H., M. Galavi, A. Ghanbari, and N. Panjehkeh. 2011. Salinity effects on seed germination and seedling growth of bread wheat cultivars. *Trakia Journal of Sciences* 9: 43–50.

Akula, R., and G.A. Ravishankar. 2011. Influence of abiotic stress signals on secondary metabolites in plants. *Plant Signaling and Behaviour* 6(11): 1720–1731.

Al-Gabbies A., M. Kleinwächter, and D. Selmar. 2015. Influencing the contents of secondary metabolites in spice and medicinal plants by deliberately applying drought stress during their cultivation. *Jordan Journal of Biological Sciences* 8: 1–10.

Ali M, E., Hahn, and K. Paek. 2005. CO(2)-induced total phenolics in suspension cultures of *Panax ginseng* C. A. Mayer roots: role of antioxidants and enzymes. *Plant Physiology and Biochemistry* 43: 449–457.

Ali, R.M, H.M. Abbas, and R.K. Kamal. 2007. The effects of treatment with polyamines on dry matter, oil and flavonoid contents in salinity stressed chamomile and sweet marjoram. *Plant Soil and Environment* 53: 529–543.

Ali. H., E. Khan, and M.A. Sajad. 2013. Phytoremediation of heavy metals – concepts and applications. *Chemosphere* 91: 869–881.

Ashraf, M., and A. Orooj. 2006. Salt stress effects on growth, ion accumulation and seed oil concentration in an arid zone traditional medicinal plant ajwain (*Trachyspermum ammi* [L.] Sprague). *Journal of Arid Environments* 64: 209–220.

Asrar Z., R.A. Khavari-Nejad, and H. Heidari. 2005. Excess manganese effects on pigments of *Mentha spicata* at flowering stage. *Archives of Agronomy and Soil Science* 51(1): 101–107

Aziz E.E., H. Al-Amier, and L.E. Craker. 2008. Influence of salt stress on growth and essential oil production in peppermint, pennyroyal, and apple mint. *Journal of Herbs, Spices and Medicinal Plants* 14(1 & 2): 77–87.

Ballhorn, D.J., S. Kautz, M. Jensen, S. Schmitt, M. Heil, and A.D. Hegeman. 2011. Genetic and environmental interactions determine plant defences against herbivores. *Journal of Chemical Ecology* 99: 313–326.

Banerjee, A., and A. Roychoudhury. 2017. Effect of salinity stress on growth and physiology of medicinal plants. *In:* Medicinal Plants and Environmental Challenges. Eds. Mansour, G., and A. Varma, pp. 177–188. Springer International Publishing. ISBN 978-3-319-68716-2 ISBN 978-3-319-68717-9 (eBook) https://doi.org/10.1007/978-3-319-68717-9

Belaqziz, R., R. Abderrahmane, and A. Abbad. 2009. Salt stress effects on germination, growth and essential oil content of an endemic thyme species in Morocco (*Thymus maroccanus* Ball.). *Journal of Applied Science Research* 5(7): 858–863.

Bennett, R.N., and R.M. Wallsgrove. 1994. Secondary metabolites in plant defence mechanisms. *New Phytologist* 127: 617–633.

Bistgani, Z.E., M. Hashemi, M. DaCosta, L. Craker, F. Maggi, and M.R. Morshedloo. 2019. Effect of salinity stress on the physiological characteristics, phenolic compounds and antioxidant activity of *Thymus vulgaris* L. and *Thymus daenesis* Celak. *Industrial Crops and Products* 135: 311–320.

Bortolin, R., F. Caregnato, et al. 2016. Chronic ozone exposure alters the secondary metabolite profile, antioxidant potential, anti-inflammatory property, and quality of red pepper fruit from *Capsicum baccatum*. *Ecotoxicology and Environmental Safety* 129:16–24.

Cavaliere, C. 2009. The effects of climate change on medicinal and aromatic plants. *HerbalGram* 81: 44–57.

Chaturvedi, I. 2006. Effects of arsenic concentrations and forms on growth and arsenic uptake and accumulation by Indian mustard (*Brassica juncea* L.) genotypes. *Journal of Central European Agriculture* 7(1): 31–40.

Chauhan, N., and D. Kumar. 2014. Effect of salinity stress on growth performance of *Citronella java*. *International Journal of Geology, Agriculture and Environmental Science* 2: 11–14

Chen, X., Q. Chen, X. Zhang, R. Li, Y. Jia, A.A. Ef, A. Jia, L. Hu, and X. Hu. 2016. Hydrogen sulphide mediates nicotine biosynthesis in tobacco (*Nicotiana tabacum*) under high temperature conditions. *Plant Physiology and Biochemistry* 104: 174–179. DOI:10.1016/j.plaphy.2016.02.033.

Chung, I.M., J.J. Kim, J.D. Lim, C.Y. Yu, S.H. Kim, and S.J. Hahn. 2006. Comparison of resveratrol, SOD activity, phenolic compounds and free amino acids in *Rehmannia glutinosa* under temperature and water stress. *Environmental and Experimental Botany* 56: 44–53.

Cox, M., S.P.F. Bell, and J.L. Kovar. 1996. Differential tolerance of canola to arsenic when grown hydroponically or in soil. *Journal of Plant Nutrition* 19(12): 1599–1610.

Das, K., R. Dang, T.N. Shivananda, and P. Sur. 2005. Interaction between phosphorus and zinc on the biomass yield and yield attributes of the medicinal plant Stevia (*Stevia rebaudiana*). *The Scientific World Journal* 10: 390–395.

De, D., and B. De. 2011. Elicitation of diosgenin production in *Trigonella foenum-graecum* L. seedlings by heavy metals and signaling molecules. *Acta Physiologiae Plantarum* 33: 1585–1590.

De-Kok, L.J. 1990. Sulfur metabolism in plants exposed to atmospheric sulfur. *In*: Sulfur Nutrition and Sulfur Assimilation in Higher Plants. Eds. Rennenberg, M., C. Brunold, L.J. De- Kok, and I. Stulen, p. 111e130. SPB Academic Publishing, The Hague.

Díaz, S., J. Settele, E. Brondízio, H. Ngo, M. Guèze, et al. 2019. Summary for policymakers of the global assessment report on biodiversity and ecosystem services of the Intergovernmental Science-Policy Platform on Biodiversity and Ecosystem Services. Available at: https://uwe-repository.worktribe.com/output/1493508 (accessed on 07.07.2021). DOI: 10.1126/science.205.4404.383.

Eriksen, M., L.C.M. Lebreton, and H.S. Carson. 2014. Plastic pollution in the world's oceans: more than 5 trillion plastic pieces weighing over 250,000 tons afloat at sea. *PLOS ONE* 9(12): e111913.

Falusi, B.A., O.A Odedokun, A. Abubakar, and A. Agoh. 2016. Effects of dumpsites air pollution on the ascorbic acid and chlorophyll contents of medicinal plants. *Cogent Environmental Science* 2:1 1170585, DOI: 10.1080/23311843.2016.1170585

Foyer, C.H., and G. Noctor. 2005. Oxidant and antioxidant signalling in plants: a re-evaluation of the concept of oxidative stress in a physiological context. *Plant Cell and Environment* 28:1056–1071.

Gengmao Z., S. Quanmei, H. Yu, L. Shihui, and W. Changhai. The physiological and biochemical responses of a medicinal plant (*Salvia miltiorrhiza* L.) to stress caused by various concentrations of NaCl. *PLOS ONE* 9(2): e89624. doi:10.1371/journal.pone.0089624

Gershenzon, J. 1984. Changes in the levels of plant secondary metabolites under water and nutrient stress. *Recent Advances in Phytochemistry* 18: 273–320.

Ghanavati M., and S. Sengul. 2010. Salinity effect on the germination and some chemical components of *Chamomilla recutita* L. *Asian Journal of Chemistry* 22: 859–866.

Ghasemzadeh, A., H. Jaafar, and A. Rahmat. 2010. Elevated carbon dioxide increases contents of flavonoids and phenolic compounds, and antioxidant activities in Malaysian young ginger (*Zingiber officinale* Roscoe.) varieties. *Molecules* 15: 7907–7922.

Ghavami, A., and A. Ramin. 2008. Grain yield and active substances of milk thistle as affected by soil salinity. *Communications in Soil Science and Plant Analysis* 39: 2608–2618.

Gleadow, R.M., and I.E. Woodrow. 2002. Defense chemistry of cyanogenic Eucalyptus cladocalyx seedlings is affected by water supply. *Tree Physiology* 22: 939–945.

Gore, A. 2006. An Inconvenient Truth: The Planetary Emergency of Global Warming and What We Can Do About It. New York: Rodale Press, p. 325.

Goswami, B.N., V. Venugopal , D. Sengupta, M.S. Madhusoodanan, and P.K. Xavier. 2006. Increasing trend of extreme rain events over India in a warming environment. *Science* 314: 1442–1445.

Grejtovsky, A., K. Markusova, and A. Eliasova. 2006. The response of chamomile (*Matricaria chamomilla* L.) plants to soil zinc supply. *Plant Soil Environment* 52: 1–7.

Grenni, P., V. Ancona, and A.B. Caracciolo. 2018. Ecological effects of antibiotics on natural ecosystems: A review. *Microchemical Journal* 136: 25–39.

He, X.Y., W. Huang, W. Chen, T. Dong, C.B. Liu, Z.J. Chen, S. Xu, and Y.N. Ruan. 2009. Changes of main secondary metabolites in leaves of *Ginkgo biloba* in response to ozone fumigation. *Journal of Environmental Sciences* 21: 199–203.

Huang, W., X. He, C. Liu, et al. 2010. Effects of elevated carbon dioxide and ozone on foliar flavonoids of Ginkgo biloba. *Advanced Materials Research* 113: 165–169.

Intergovernmental Panel on Climate Change (IPCC). 2007. Climate Change: Synthesis Report November 2007.

IOC-UNESCO and UNEP. 2016. Large Marine Ecosystems: Status and Trends, Summary for Policy Makers. United Nations Environment Programme (UNEP), Nairobi.

Iqbal, M.Z., M. Shafig, S. Qamar Zaidi, and M. Athar. 2015. Effect of automobile pollution on chlorophyll content of roadside urban trees. *Global Journal of Environmental Science and Management* 1: 283–296.

Jamil, A., S. Riaz, M. Ashraf, and M.R. Foolad. 2011. Gene expression profiling of plants under salt stress. *Critical Reviews in Plant Sciences* 30: 435–458.

Jeliazkova, E.A., and L.E. Craker. 2003. Seed germination of some medicinal and aromatic plants in heavy metal environment. *Journal of Herbs, Spices & Medicinal Plants* 10(2): 105–112. http://dx.doi.org/10.1300/J044v10n02_12

Jenny, J.P., O. Anneville, F. Arnaud, Y. Baulaz, D. Bouffard, et al. 2020. Scientists' warning to humanity: rapid degradation of the world's large lakes. *Journal of Great Lakes Research* 46: 686–702.

Jiang, W., D. Liu, and W. Hou. 2001. Hyperaccumulation of cadmium by roots, bulbs and shoots of garlic. *Bioresource Technology* 76(1): 9–13.

Jochum, G.M., K.W. Mudge, and R.B. Thomas. 2007. Elevated temperatures increase leaf senescence and root secondary metabolite concentrations in the understory herb *Panax quinquefolius* (Araliaceae). *American Journal of Botany* 94: 819–826.

Jordan, D., T. Green, and A. Chappelka. 1991. Response of total tannins and phenolics on Loblolly pine foliage exposed to ozone and acid rain. *Journal of Chemical Ecology* 17: 505–513.

Kartosentono, S., S. Suryawati, G. Indrayanto, and N.C. Zaini. 2002. Accumulation of Cd2+ and Pb2+ in the suspension cultures of *Agave amaniensis* and *Costus speciosus* and the determination of the culture's growth and phytosteroid content. *Biotechnology Letters* 24: 687–690.

Kirk, H., K. Vrieling, E. van der Meijden, and P.G.L. Klinkhamer. 2010. Species by environment interactions affect pyrrolizidine alkaloid expression in *Senecio jacobaea*, *Senecio aquaticus*, and their hybrids. *Journal of Chemical Ecology* 36: 378–387.

Kjær, C., and N. Elmegaard. 1996. Effects of copper sulfate on black bindweed (*Polygonum convolvulus* L). *Ecotoxicology and Environmental Safety* 33(2): 110–117.

Kumar, A., E. Abrol, S. Koul, and D. Vyas. 2012. Seasonal low temperature plays an important role in increasing metabolic content of secondary metabolites in *Withania somnifera* (L.) Dunal and affects the time of harvesting. *Acta Physiologiae Plantarum* 34: 2027–2031.

Laloi, C., K. Appel, and A. Danon. 2004. Reactive oxygen signaling: The latest news. *Current Opinion in Plant Biology* 7: 323–328.

Lavola, A., R. Julkunen-Tiitto, and E. Pakkonen, 1994. Does ozone stress change the primary or secondary metabolites of Birch (*Betula pendul* Roth.)? *New Phytologist* 126: 637–642.

Liu, Y. J., and H. Ding. 2008. Variation in air pollution tolerance index of plant near a steel factory: implication for landscape-plant species selection for industrial areas. *WSEAS Transactions on Environment and Development* 4: 24–32.

Mallapur, C. 2020. Sanitation workers at risk from discarded medical waste related to COVID-19. IndiaSpend. Available at: www.indiaspend.com/sanitation-workersat-risk-from-discarded-medical-waste-related-tocovid-19/

Maranz, S., and Z. Wiesman. 2004. Influence of climate on the tocopherol content of shea butter. *Journal of Agricultural and Food Chemistry* 52: 2934–2937.

Maugh, T.H. 1979. SO$_2$ pollution may be good for plants. *Science* 27: 205(4404): 383.

Mildvan, A.S. 1970. Metal in enzymes catalysis. *In*: The Enzymes, vol 11. Ed. Boyer, D.D., pp. 445–536. Academic Press, London.

Mir, B.A., S.A. Mir, J. Khazir, L.B. Tonfack, D.A. Cowan, D. Vyas, and S. Koul. 2015. Cold stress affects antioxidative response and accumulation of medicinally important withanolides in *Withania somnifera* (L.) Dunal. *Industrial Crops and Products* 74: 1008–1016.

Misra, A. 1992. Effect of zinc stress in Japanese mint as related to growth, photosynthesis, chlorophyll content and secondary plant products – the monoterpenes. *Photosynthetica* 26: 225–234.

Miteva, E., and M. Merakchiyska. 2002. Response of chloroplasts and photosynthetic mechanism of bean plants to excess arsenic in soil. *Bulgarian Journal of Agricultural Science* 8: 151–156.

Mittler, R., S. Vanderauwera, M. Gollery, and F. Van Breusegem. 2004. Reactive oxygen gene network of plants. *Trends in Plant Science* 9(10): 490–498.

Moghbeli E, S. Fathollahi, H. Salari, G. Ahmadi, et al. 2012. Effects of salinity stress on growth and yield of *Aloe vera* L. *Journal of Medicinal Plants Research* 6: 3272–3277.

Mondal, H.K., and Kaur, H. 2017. Effect of salt stress on medicinal plants and its amelioration by plant growth promoting microbes. *International Journal of Bio-resource and Stress Management* 8(2): 316–326.

Montesinos-Pereira, D., Y. Barrameda-Medina, N. Baenas, D.A. Moreno, E. Sanchez-Rodriguez, B. Blasco, and J.M. Ruiz. 2016. Evaluation of hydrogen sulfide supply to biostimulate the nutritive and phytochemical quality and the antioxidant capacity of Cabbage (*Brassica oleracea* L. 'Bronco'). *Journal of Applied Botany and Food Quality* 89. doi:10.5073/JABFQ.2016.089.038

Mosaleeyanon, K., S.M.A., Zobayed, F. Afreen, and T. Kozai. 2005. Relationships between net photosynthetic rate and secondary metabolite contents in St. John's wort. *Plant Science* 169: 523–531.

Moustakas, M., T. Lanaras, L. Symeonidis, and S. Karataglis. 1994. Growth and some photosynthetic characteristics of field grown *Avena sativa* under copper and lead stress. *Photosynthetica* 30(3): 389–396.

Muneer, S., T.H. Kim, B.C. Choi, B.S. Lee, J.H. Lee. 2014. Effect of CO, NO$_x$ and SO$_2$ on ROS production, photosynthesis and ascorbate-glutathione pathway to induce *Fragaria* x *annasa* as a hyperaccumulator. *Redox Biology* 2: 91–98.

Murch, S.J., K. Haq, H.P.V Rupasinghe, and P.K. Saxena. 2003. Nickel contamination affects growth and secondary metabolite composition of St. John's wort (*Hypericum perforatum* L.). *Environmental and Experimental Botany* 49: 251–257.

Murthy, H. Niranjana, M.I. Georgiev, Y.S. Kim, C.S. Jeong, S.J. Kim, S.Y. Park, and K.Y. Paek. 2014. Ginsenosides: prospective for sustainable biotechnological production. *Applied Microbiology Biotechnology* 98(14): 6243–54. doi: 10.1007/s00253-014-5801-9.

Narula, A., A. Kumar, and P.S. Srivastava. 2005. Abiotic metal stress enhances diosgenin yield in *Dioscorea bulbifera* L. cultures. *Plant Cell Reports* 24: 250–254.

Neffati, M., and B. Marzouk. 2008. Changes in essential oil and fatty acid composition in coriander (*Coriandrum sativum* L.) leaves under saline conditions. *Industrial Crops and Products* 28: 137–142.

Neilson, R.P., L.F. Pitelka, A.M. Solomon, et al. 2005. Forecasting regional to global plant migration in response to climate change. *BioSciences* 55(9): 749–759.

Nematshahi, N., M. Lahouti, and A. Ganjeali. 2012. Accumulation of chromium and its effect on growth of (*Allium cepa* cv. Hybrid). *European Journal of Experimental Biology* 2: 969– 974.

Nourimand, M.D., S. Mohsenzadeh, and J.A. Teixeira da Silva. 2012. Physiological responses of fennel seedling to four environmental stresses. *Iranian Journal of Science & Technology* A1: 37–46.

Oueslati, S., N. Karray-Bouraoui, H. Attia, M. Rabhi, K. Riadh, and M. Lachaal. 2010. Physiological and antioxidant responses of *Mentha pulegium* (Pennyroyal R.,) to salt stress. *Acta Physiologiae Plantarum* 32: 289–296.

Pellegrini, E., A. Francini, G. Lorenzini, et al. 2015. Ecophysiological and antioxidant traits of *Salvia officinalis* under ozone stress. *Environmental Science and Pollution Research* 22: 13083–13093.

Phelps Bondaroff, T., and S. Cooke. 2020. Masks on the Beach: The impact of COVID-19 on marine plastic pollution. *OceansAsia.* Available at: https://oceansasia.org/wp-content/uploads/2020/12/Marine-Plastic-Pollution-FINAL-1.pdf

Pompe, S., J. Hanspach, F. Badeck, S. Klotz, W. Thuiller, and I. Kuhn. 2008. Climate and land use change impacts on plant distributions in Germany. *Biology Letters* 4: 564–567.

Rai, R., S. Pandey, and S.P. Rai. 2011. Arsenic induced changes in morphological, physiological and biochemical attributes and artemisinin biosynthesis in *Artemisia annua*, an antimalarial plant. *Ecotoxicology* 20: 1900–1913.

Rai, V., and S. Mehrotra. 2008. Chromium-induced changes in ultramorphology and secondary metabolites of *Phyllanthus amarus* Schum & Thonn. – an hepatoprotective plant. *Environmental Monitoring and Assessment* 147: 307–315.

Rai, V., P. Vajpayee, S.N. Singh, and S. Mehrotra. 2004. Effect of chromium accumulation on photosynthetic pigments, oxidative stress defense system, nitrate reduction, proline level and eugenol content of *Ocimum tenuiflorum* L. *Plant Science* 167: 1159–1169.

Ramin, A.A. 2005. Effects of salinity and temperature on germination and seedling establishment of sweet basil (*Ocimum basilicum* L.). *Journal of Herbs, Spices and Medicinal Plants* 11: 81–90.

Rastogi, S., S. Shah, R. Kumar, D. Vashisth, M.D. Akhtar, and A. Kumar. 2019. *Ocimum metabolomics* in response to abiotic stresses: Cold, flood, drought and salinity. *PLOS ONE* 14(2): 0210903.

Roe, D. 2019. Biodiversity loss – more than an environmental emergency. *The Lancet Planetary Health* 3(7): e287–e289.

Said-Al, A.H., and A.A. Mahmoud. 2010. Effect of zinc and/or iron foliar application on growth and essential oil of sweet basil (*Ocimum basilicum* L.) under salt stress. *Ozean Journal of Applied Science* 3(1): 97–111.

Saleem, A., J. Loponen, K. Pihlaja, and E. Oksanen. 2001. Effects of long-term open-field ozone exposure on leaf phenolics of European silver birch (*Betula pendula* Roth). *Journal of Chemical Ecology* 27: 1049–1062.

Schonhof, I., H. Klaring, A. Krumbein, et al. 2007. Interaction between atmospheric CO_2 and glucosinolates in broccoli. *Journal of Chemical Ecology* 33: 105–114. Doi:10.1007/s10886-006-9202-0

Selmar, D., and M. Kleinwächter. 2013. Influencing the product quality by deliberately applying drought stress during the cultivation of medicinal plants. *Industrial Crops and Products* 42: 558–566.

Shea, J. 2008. Apple growers hopeful after freeze. *Times-News.* April 6.

Shohael, A.M., M.B. Ali, K.W.Yu, E.J. Hahn, R. Islam, and K.Y. Paek. 2006. Effect of light on oxidative stress, secondary metabolites and induction of antioxidant enzymes in *Eleutherococcus senticosus* somatic embryos in bioreactor. *Process Biochemistry* 41: 1179–1185.

Snow M., R. Bard, D. Olszyk, et al 2003. Monoterpenes levels in needles of Douglas fir exposed to elevated CO_2 and temperature. *Physiologia Plantarum* 117: 352–358.

Sosa, L., A. Llanes, H. Reinoso, M. Reginato, and V. Luna. 2005. Osmotic and specific ion effect on the germination of *Prospis strombulifera*. *Annals of Botany* 96: 261–267.

Spoel, S.H., and X. Dong. 2008. Making sense of hormone crosstalk during plant immune response. *Cell Host & Microbe* 3: 348–351.

Srivastava, P., and R. Kumar. 2015. Soil salinity: a serious environmental issue and plant growth promoting bacteria as one of the tools for its alleviation. *Saudi Journal of Biological Sciences* 22: 123–131.

Stiling, P., and T. Cornelissen. 2007. How does elevated carbon dioxide (CO_2) affect plant–herbivore interactions? A field experiment and meta-analysis of CO_2-mediated changes on plant chemistry and herbivore performance. *Global Change Biology* 13(9): 1823–1842. doi:10.1111/j.1365- 2486.2007.01392.x

Street, R.A., M.G. Kulkarni, W.A. Stirk, C. Southway, and J. Staden Van. 2007. Toxicity of metal elements on germination and seedling growth of widely used medicinal plants belonging to Hyacinthaceae. *The Bulletin of Environmental Contamination and Toxicology* 79: 371–376.

Sulandjari, and W.S. Dewi. 2018. Effects of intermittent acid rain on proline and antioxidant content on medicinal plant 'Pereskia bleo'. *IOP Conf. Ser.: Earth Environmental Science* 129 012020.

Sun, L., H. Su, Y. Zhu et al. 2012. Involvement of abscisic acid in ozone-induced puerarin production of Pueraria thomsnii Benth. suspension cell cultures. *Plant Cell Reports* 31: 179–185.

Tabatabaie S.J., and J. Nazari. 2007. Influence of nutrient concentration and NaCl salinity on growth, photosynthesis and essential oil content of peppermint and lemon verbena. *Turkish Journal of Agriculture* 31: 245–523.

Tiwari, S., M. Agrawal, and F.M. Marshall. 2006. Evaluation of ambient air pollution impact on carrot plants at a sub urban site using open top chambers. *Environmental Monitoring and Assessment* 119: 15–30.

Tonelli, M., E.D. Pellegrini, F. Angiolillo, M. Petersen, C. Nali, L. Pistelli, and G. Lorenzini. 2015. Ozone-elicited secondary metabolites in shoot cultures of *Melissa officinalis* L. *Plant Cell, Tissue and Organ Culture* 120: 617–629.

UNESCO 2017. World Water Assessment Programme. Facts and Figures. Available at: www.unesco.org/new/en/natural-sciences/environment/water/wwap/facts-and-figures/all-facts-wwdr3/fact-15-water-pollution/ (accessed on 07.07.2021)

Van Wittenberghe, S., L. Alonso, J. Verrelst, I. Hermans, J. Delegido, F. Veroustraete, R. Samson. 2013. Upward and downward solar-induced chlorophyll fluorescence yield indices of four tree species as indicators of traffic pollution in Valencia. *Environmental Pollution* 173: 29–37.

Vashisth, D., R. Kumar, S. Rastogi, V.K. Patel, A. Kalra, M.M. Gupta, et al. 2018. Transcriptome changes induced by abiotic stresses in *Artemisia annua*. *Scientific Reports* 8: 3423.

Ved, D.K., and G.S. Goraya. 2007. Demand and Supply of Medicinal Plants in India. NMPB, New Delhi & FRLHT, Bangalore, India.

Vriezen, J.A.C., F.J. Bruijn, and K. Nusslein. 2007. Responses of *Rhizobia* to desiccation in relation to osmotic stress, oxygen, and temperature. *Applied and Environmental Microbiology* 73: 3451–3459.

Wadood, A., M. Ghufran, S.B. Jamal, M. Naeem, A. Khan, and R. Ghaffar. 2013. Phytochemical analysis of medicinal plants occurring in local area of Mardan. *Biochemistry and Analytical Biochemistry* 2(4): 1–4.

Walther, G.R., E. Post, and P. Convey. 2002. Ecological responses to recent climate change. *Nature* 416: 389–395.

Wang, M., J. Zou, X. Duan, W. Jiang, and D. Liu. 2007. Cadmium accumulation and its effects on metal uptake in maize (*Zea mays* L.). *Bioresource Technology* 98(1): 82–88.

Wani, S.H., and S.K. Sah. 2014. Biotechnology and abiotic stress tolerance in rice. *Journal of Rice Research* 2: e105

World Economic Forum. The Global Risks Report (2018). 13th Edition. World Economic Forum Geneva. Available at: www3.weforum.org/docs/WEF_GRR18_Report.pdf (accessed on 16.05.2021)

World Health Organization (WHO). 2022. Weekly Epidemiological Update on COVID-19, 22 February 2022, Edition 80. Available at: www.who.int/publications/m/item/weekly-epidemiological-update-on-covid-19-22-february-2022 (accessed on 03.03.2022)

World Health Organization (WHO). 2005. Regulation, evaluation and safety monitoring of herbal medicines: Summary report of the global survey on national policy on traditional medicine and complementary/alternative medicine and regulation of herbal medicines (ISBN 92 4 159323 7).

Xu, M., B. Yang, J. Dong, et al. 2011. Enhancing hypericin production of *Hypericum perforatum* cell suspension culture by ozone exposure. *Biotechnology Progress* 27(4): 1101–1106.

Yang, Y., L. Zhang, X. Huang, Y. Zhou, Q. Quan, Y. Li, et al. 2020 Response of photosynthesis to different concentrations of heavy metals in *Davidia involucrata*. *PLOS ONE* 15(3): e0228563. https://doi.org/10.1371/journal.pone.0228563

Yudharaj, P., R. Jasmine Priyadarshini, E. Ashok Naik, M. Shankar, R. Sowjanya, and B. Sireesha. 2016. Importance and uses of medicinal plants – an overview. *International Journal of Preclinical & Pharmaceutical Research* 7(2): 67–73.

Zhou, Y., N. Tang, L. Huang, Y. Zhao, X. Tang, and K. Wang. 2018. Effects of salt stress on plant growth, antioxidant capacity, glandular trichome density, and volatile exudates of *Schizonepeta tenuifolia* Briq. *International Journal of Molecular Science* 19(1): 252. DOI: 10.3390/ijms19010252

Ziska, L., S. Panicker, and H. Wojno. 2008. Recent and projected increases in atmospheric carbon dioxide and the potential impacts on growth and alkaloid production in wild poppy (*Papaver setigerum* DC.). *Climate Change* 91: 395–403. doi:10.1007/s10584-008-9418-9

Zobayed, S., and P. Saxena. 2004. Production of St. John's wort plants under controlled environment for maximizing biomass and secondary metabolites. *In Vitro Cellular & Developmental Biology* 40: 108–114.

Zobayed, S.M.A., F. Afreen, and T. Kozai. 2005. Temperature stress can alter the photosynthetic efficiency and secondary metabolite concentrations in St. John's wort. *Plant Physiology and Biochemistry* 43: 977–984.

2 Growth and Development of Medicinal Plants, and Production of Secondary Metabolites under Ozone Pollution

Deepti, Archana (Joshi) Bachheti, Piyush Bhalla, Rakesh Kumar Bachheti, Azamal Husen

CONTENTS

2.1 INTRODUCTION

Life on Earth is not possible without its environment or without interacting with other species or ecosystems, which are very significant units of life on Earth. The life cycle involves the assembling and disassembling of various atoms from the environment and returning them to the environment. The Industrial Revolution played a major role in the change of atmospheric gases and the presence of particulate matter and their effects on the natural environment, leading to environmental pollution. Environmental pollution can be defined as unwanted or unfavourable changes or alterations of our surroundings, mainly as a by-product of human action through direct or indirect effects of change in energy patterns, radiation levels, chemical and physical constitution, and abundance of organisms.

Our atmosphere is made up of various layers, such as the troposphere, stratosphere, mesosphere, thermosphere, and exosphere, and, as it is a blanket of air, various gases are found in different concentrations, such as nitrogen, oxygen, carbon dioxide,

DOI: 10.1201/9781003178866-2

25

helium, argon, and ozone, etc. Ozone is found in the stratosphere layer of the atmosphere and helps us by protecting us from harmful radiation coming from the sun. Slow thinning or depletion of the stratosphere ozone layer, due to the reaction of ozone and chlorofluorocarbon gases, is a major environmental problem because it increases the ultraviolet radiation that reaches the surface of the Earth. This radiation is related to numerous harmful effects in human, animal, plant, and natural ecosystems. However, technically, ozone in the troposphere is considered a greenhouse gas which may also contribute to climate change (NASA 2018). Ozone is formed by chemical reactions amid the precursors of ozone in the atmosphere; hence the actual sources of ozone are the sources of precursors leading to the formation of ozone. In the presence of high temperature, sunlight, and light winds, when precursor emissions react the unhealthy ozone levels rise. Common sources of ozone are:

- Fuel combustion and evaporation allied with cars, buses, trucks, and equipment, small stationary sources like print shops, gasoline dispensing facilities, large stationary sources like factories, oil refineries, and power plants, and off-road engines in trains, aircraft, agricultural operations, construction equipment, and garden or home equipment.
- Organic compounds' evaporation by consumer products like cleaning materials, paints, or solvents.

The data presented in the chart at Figure 2.1 is taken by satellite instrument as part of the NASA ozone watch meteorological programme (Figure 2.1). The concentration of ozone is measured in Dobson units (DU). A Dobson unit is several molecules of ozone that would be required to create a layer of pure ozone 0.01 mm thick at

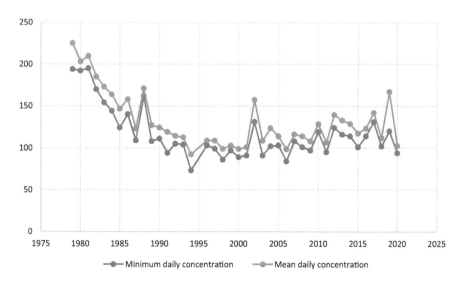

FIGURE 2.1 Variation in Stratospheric ozone concentration (adopted from Ozone watch, NASA 2018).

a temperature of 0°C and a pressure of 1 atm. The chart (Figure 2.1) shows that, from 1979 to the start of the 1990s, the concentration of stratospheric ozone in the southern hemisphere decreased by 100 DU – i.e., to a concerning 'ozone hole' level. It remained around or below 100 DU for a few decades after the 1990s and in the last few years, from 2010, it has somehow started recovering.

Primary metabolites are required for the normal growth or development of plants, while secondary metabolites are synthesized by plants to adapt, defend against hostile organisms or diseases or environment, or to adjust in the variable environmental conditions. Alkaloids, terpenes, tannins, flavonoids, polyphenols, glycosides, and coumarins are among the major secondary metabolites that are used in medicines (Bachheti et al. 2020, 2021; Beshah et al. 2020; Sharma et al. 2020). For protection against the whims of nature like flood, drought, heat, and cold, etc., and against predators and pathogens, plants have developed varied physical and chemical mechanisms during their evolution. The production of a variety of phytochemicals to face the various biotic or abiotic threats or stresses is the most thriving adaptation of plants. For instance, antioxidants are produced by plants to deal with oxidative stress. These phytochemicals are stored in various plant parts such as fruits, leaves, bark, stem, flower, or hardwood, etc. to paralyse, knock back, or kill the predator or disease-causing agents. Sometimes the production of secondary metabolites is also influenced by the collaboration of other living organisms, such as soil microbes or plants. Plant health, crop yield, and ecosystem structure or function are adversely affected by tropospheric ozone and global warming (Krupa et al. 1988; Runeckles et al. 1992; Glick et al. 1995; Holopainen et al. 2018; Ainsworth et al. 2020; Gupta et al. 2021). Many studies have been reported on medicinal plants (such as *Azadirachta indica*, *Cassia occidentalis*, *Cajanus cajan*, and *Datura innoxia*) and their interactions with climate change or air pollution (Gupta and Ghouse 1987; Husen 1997, 2020a; Husen et al. 1999; Husen and Iqbal 2004; Joshi et al. 2009; Iqbal et al. 2010a, 2010b; Adrees et al. 2016; Yadav et al. 2019; Mukherjee et al. 2019). Thus, threats to plant growth and productivity from climate change and unforeseeable environmental extremes have constantly increased. Climate change-driven effects, especially from unpredictable environmental fluctuations, can result in an increased prevalence of abiotic and biotic stresses in plants. These stresses have slowed down the global yields of the crop as well as medicinal plants (Bachheti et al. 2021; Husen 2021a). However, plant exposure to certain biostimulants, bio- and nano-fertilizers and/or hormones reduces damage caused by stress, improves the defence mechanisms involved, and also helps in disease management and nutrient-use efficiency (Singh and Husen 2020; Husen 2021b).

The tropospheric ozone is considered as a significant stressor in the natural ecosystem which is formed via a complex reaction between carbon monoxides, oxides of nitrogen, and methane and non-methane volatile organic compounds under the presence of sunlight (Kleanthous et al. 2014; Harmens et al. 2019), causing serious threat to vegetation by disturbing physiological processes or plant growth (Emberson et al. 2018; Schauberger et al. 2019; Ghosh et al. 2020; Kim et al. 2020) and yield or productivity of natural vegetation or crops (Sarkar and Agrawal 2012). An increase in the concentration of tropospheric ozone is also a cause of serious threat to life on Earth and this has been increasing over the last few decades due to anthropogenic activities and industrialization (Sharma et al. 2016; Yadav et al. 2019; Harmens et al. 2019; Agathokleous et al. 2020). Persson et al. (2003) studied leaf area reduction, premature leaf infirmity being a consequence of elevated ozone levels, leading to incorporated product loss and, according to Valkama et al. (2007), Lindroth (2010), and Li et al. (2020), in similar conditions, in many plants, the content and composition of secondary metabolites get altered. Exposure to ozone reduces plant photosynthesis and growth (Bueker et al. 2015; Li et al. 2017) and induces substantial changes in protein activity, gene expression, signalling pathways, and metabolism, even before any tissue damage can be detected (Bueker et al., 2015; Vainonen and Kangasjärvi 2015). Considering the above facts, the present chapter emphasizes the impact of ozone stress on medicinal plant growth, development, and reproductive processes. Additionally, the chapter describes the impact of ozone on natural products of medicinal plants.

2.2 OZONE UPTAKE AND QUANTIFICATION OF PLANT DESTRUCTION

In 1970, Rich et al. investigated the ozone uptake by leaves for the first time. The ozone atom accesses the leaves commonly via stomata, which is also the site for carbon dioxide uptake (Kerstiens and Lendzian 1989; Unsworth and Ormrod 1982). Later, Kerstiens and Lendzian (1989) also verified that the ozone is taken up into the leaf via the stomata. Inside the leaf, ozone reacts with constituents of the aqueous matrix associated with the cell wall to produce reactive oxygen species (ROS), resulting in the oxidation of sensitive components of the plasmalemma, and then the cytosol (Heath 1980; Chameides 1989). The destructive impacts of ozone on sensitive plant species include reduced growth and biomass production, and cell damage (Fumagalli et al. 2001). Based on the dose of ozone received by plants, it is logical to estimate the biomass loss – i.e., the time-integrated stomatal flux as stomata set up the prime route for the ozone atom that leads to plant destruction. Hence, due to the removal of ozone from various processes on distinct types of surface, the total ozone flux and stomatal ozone flux are not the same. So, it can be said that the amount of ozone access into the plant tissues – i.e., the dose or time-integrated stomatal flux (not concentration) – can significantly relate to the quantification of plant destruction

2.3 IMPACT OF OZONE ON GROWTH, DEVELOPMENT, AND REPRODUCTIVE PROCESSES

Chlorosis and necrosis are the most common foliar lesions reported at the time of ozone exposure but the rigidity of the damage in leaves was found to vary in different plant species (Kadono et al. 2006) (Figure 2.2). Ozone exposure induces early senescence while abscission layer formation decreases the size of stomata, photosynthesis, moderate biomass production via carbon availability or more directly, and also decreases translocation of fixed carbon to plants and their parts (Wilkinson et al. 2012). The most critical stage of life is that of reproductive development, and various stages of this development are sensitive to ozone (Black et al. 2000). It has been reported that exposure to ozone moderates or brings forward flowering and induces

FIGURE 2.2   Healthy (top) and ozone-injured (bottom) tulip tree (yellow poplar – used for medicinal purpose as a heart stimulant extracted from the inner bark of the root, and a tonic for treating rheumatism and dyspepsia was extracted from stem bark) foliage (adopted from US National Park Service, 2020).

pollen sterility and also ovule abortion (Wilkinson et al., 2012). According to a report of the Royal Society (2008) the time of sexual reproductive development of many plants occurs when the ozone production is highest – i.e., in warm and sunny weather – hence tropospheric ozone is the most destructive of air pollutants. Since the beginning of the Industrial Revolution, particularly over the last 60 years, the concentration level of tropospheric ozone had increased substantially and it is estimated that, if the emission rate in the northern hemisphere continues at the same level, by 2100 it will have risen by approximately 10–30 ppb. These estimated levels of ozone for the future will lead to an increment in international criteria for the protection of medicinal or crop plants, natural vegetation, human health, and the entire ecosystem.

As soon as ozone is picked up by the plants, it reacts and forms various ROS, such as singlet oxygen, hydroxyl radicals, or hydrogen peroxide (Heath 2008; Sandermann 2008; Fuhrer 2009; Feder and Shrier 1990). Hormonal levels, lipid peroxidation, and redox potential get influenced under the affluence of ROS in intercellular and apoplastic space in plant cells (Kangasjärvi et al. 2005) and this can also cause cell death at certain high levels (Overmyer et al. 2003; Heath 2008; Sandermann 2008; Fuhrer 2009). Some other studies, for instance those by Morgan et al. (2003), Fiscus et al. (2005), Ainsworth (2008), Feng et al. (2008, 2010), Booker et al. (2009), and Feng and Kobayashi (2009) concluded that the lowering of reproductive output and biomass production are among the consequences of the reduction of stomatal conductance and photosynthesis in leaves, due to chronic exposure to ozone. Nevertheless, the ratio of reduction in reproductive output due to changes in reproductive processes like pollen and ovary development, seed abortion, and flower initiation is still not known. At 70 ppb daily exposure of ozone Davison and Barnes (1998) noticed the disparate impacts on resources allotment of reproductive and vegetative organs of various species; allotment of reproductive organs of *Chenopodium album* was less while it was more in *Papaver dubium* and *Trifolium arvense*.

Floral sites, pollen, anthers, stylar and stomatal surface, seeds and fruits, and the reproductive structures can be directly damaged by ozone (Black et al. 2000) while the ability of pollen germination on stomatal surfaces can be affected by the ROS (Mumford et al. 1972; Harrison and Feder 1974; Krause et al. 1975; Feder 1981; Benoit et al. 1983; Hormaza et al. 1996). These changes in the cellular redox environment caused by ROS can result in disturbances in pollen-tube growth and guidance, which can be harmful for the fertilization of the embryo. Thus, it can be assumed that the development of viable seeds can be controlled by these downstream processes involved in disturbed reproductive growth and development caused by the influence of surface-level ozone.

2.4 IMPACT OF OZONE ON BIOCHEMICAL AND PHYSIOLOGICAL PROCESSES

Before the major harm caused to plants by increased levels of ozone, various biochemical changes occur in plants, including increment in enzyme activities that are related to the defence mechanism of plants. The production of ethylene (plant hormone) and activity of various flavonoid and phenylpropanoid pathway enzymes increase, and changes in polyamine metabolism are caused by ozone exposure. The formation of ROS from the action of apoplastic NADPH oxidase is regulated by hydrogen peroxide (H_2O_2) and superoxide (O_2-) which is produced as the result of dissolved apoplast, after the entry of ozone into the leaf through stomata (Laloi et al. 2004). Plants develop enzymatic and non-enzymatic antioxidants, to cope with such conditions – i.e., for the removal or detoxification of ROS. Glutathione, ascorbic acid, catalases, and α-tocopherol are some common examples of antioxidant defence produced by plants (Noctor and Foyer 1998). In higher plants, the central component of antioxidant defence is glutathione (Foyer et al. 1997) and it is found to be more biased towards oxidants than other oxidants like ascorbic acid (Polle 2001). Programmed cell death (PCD) can be triggered by acute exposure to ozone which exhibits the same results as in the case of plant-pathogen interactions (Kangasjärvi et al. 2005; Overmyer et al. 2005). The phytohormones such as salicylic acid, jasmonic acid, and ethylene, produced by induction of ROS at the time of ozone consolidation in the plants, help in plant defence and growth, and also to control the leaf damage caused by ozone exposure (Tamaoki et al. 2008). The activity of enzymes like peroxidase and superoxide dismutase that work to protect the cells from oxidative damage induced by H_2O_2, superoxide, and the hydroxyl radical increases. Although the gene-level variations in plant cells by ozone exposure are not known, the decline in quantum yield of photosynthesis (Barnes et al. 1990) and decrease in photosynthetic activity are caused by variation in chlorophyll-fluorescence (Agrawal et al. 1993). Out of all the air pollutants, tropospheric ozone is recognized as the most destructive for vegetation, affecting various functions and services of the ecosystem and air quality (Sitch et al. 2007; Matyssek et al. 2010; Karnosky et al. 2003; Ainsworth et al. 2012; Fuhrer et al. 2016). In the production process, it is important to maintain constant climatic conditions to appraise the attributes of raw products as it generally depends on climatic conditions and soil (Głodowska and Wozniak 2019). Based on origin, the elicitors can be divided into two, abiotic and biotic, which are involved in the ways affecting the chemical parameters, particularly those related to bioactive compounds (Radman et al. 2004). Ozone, an abiotic elicitor, is an allotrope of oxygen that can initiate changes in metabolism which lead to increment in the range of antioxidants from different chemical groups when applied in a correct way (Piechowiak et al. 2019). Xu et al. (2019) authenticated the credibility of ozone being used as an elicitor for the increment of the content of bioactive compounds. According to Jolivet et al. (2016), most of the molecular and cellular processes that harm ozone are strong oxidants. By the production of ROS, ozone alters the RuBisCO activity (Saxe 2002), early senescence, leaf chlorosis (Novak et al. 2003; Sicard et al. 2016; Waldner et al. 2007; Ferretti et al. 2018), and drop in stomatal conductance (Ainsworth et al. 2012). Ozone exposure or fumigation results in leaf yellowing, premature senescence, and sometimes abscission in *Salvia officinalis* (Pellegrini et al. 2015) and *Anthoxanthum odoratum* (Dawnay and Mills 2009). Chlorophyll degradation, disassembly of cellular integration, and reduction in photosynthetic activity are among the major cell ultrastructural and metabolic changes that have occurred due to ozone exposure (Munné-Bosch and Alegre, 2004). Membranes are the prime target of ozone as it can have a damaging impact on their integrity (Calatayud et al. 2003; Francini et al. 2007), transport capacity (Płażek et al. 2000), function (Guidi et al. 2001), and conformation (Ranieri et al. 2001).

2.5 IMPACT OF OZONE STRESS ON SECONDARY METABOLITES

Carbon dioxide, methane, and nitrous oxide are the most common greenhouse gases which lead to change in climatic conditions as well as affecting different sectors, including plants. The secondary metabolites of plants also get affected by ozone stress (Table 2.1). Recently, Ansari et al. (2021) reported that an increase in exposure to ozone can increase the concentration of secondary metabolites (tannins, lignin, saponins, and alkaloids) in the leaves and rhizome of *Costus pictus*. Similarly, alkaloids, flavonoids, terpenes, glycosides, and saponins in *Sida cordifolia* increase (Ansari et al. 2020). Another study reported that the ozone affects the biochemical quality of *Brassica campestris* (Han et al. 2020). An experiment was conducted on *Brassica campestris* for six successive days under 60 ppb ozone concentration and at different exposure conditions (0 h/day, 1 h/day, and 4 h/day). The results showed that, after the 4 h/day ozone exposure, the content of 4-methylthiobutyl-glucosinolate, total indole, and aromatic glucosinolate increases while total glucosinolate and total aliphatic glucosinolate decrease (Han et al. 2020) (Table 2.1). Other research reports showed that when *Melissa officinalis* was exposed to elevated ozone, there

TABLE 2.1
Effect of Ozone Stress on Plant Secondary Metabolite Concentration

Medicinal plant (Family)	Plant part	Medicinal uses	Major class of secondary metabolite	Name of metabolite	Environmental condition	Concentration variation of metabolites	Key references
Costus pictus (Costaceae)	Leaves and rhizome	Used as anti-diabetic mainly along with anti-bacterial, anti-cancerous, diuretic, anti-helminthic, anti-oxidant, anti-fertility, anti-glycation	Tannins, lignin, saponins and alkaloids	Corosolic acid	Elevated ozone	Increases	Ansari et al. (2021)
Glycine max (Fabaceae)	Plant	Used in the treatment of colds, fevers and headaches, insomnia, irritability, snakebite, blindness and opacity of the cornea.	Flavonoids and alkaloids	Corymboside, apigenin 6-C-glucoside 8-C-arabinoside, citrate, 2-oxoglutarate and pyrrolidine alkaloid	Elevated ozone	Decreases	Zhang et al. (2021)
				Histidine, fatty acid, lignin/ phenylpropanoids, aspartate, arginine and piperidine alkaloids		Increases	
Sida cordifolia (Malvaceae)	Plant	Used to treat asthma, tuberculosis, the common cold, flu, headaches, nasal congestion, cough and wheezing, urinary tract infections, sore mouth, and fluid retention (oedema).	Alkaloids, flavonoids, terpenes, glycosides terpenes and saponins	–	Elevated ozone	Increases	Ansari et al. (2020)

Plant	Part	Uses	Secondary metabolite	Treatment	Effect	References
Brassica campestris (Brassicaceae)	Plant	Relieves itching, skin diseases, swelling, asthma and enhances memory.	Total glucosinolate and total aliphatic glucosinolate 4-Methylthiobutyl-GLS, total indole, and aromatic GLS	-	Elevated ozone	Decreases / Increases	Han et al. (2020)
Melissa officinalis (Lamiaceae)	Shoot	Used for dementia, anxiety and central nervous system (CNS) related disorders	Phenolics and tannins	Anthocyanins	Elevated ozone	Increased	Pellegrini et al. (2011); Shakeri et al. (2016)
Capsicum baccatum (Solanaceae)	Seeds	Used for asthma and digestive problems	Alkaloids	Capsaicin Dihydrocapsaicin	Elevated ozone	Decrease No change	Bortolin et al. (2016)
Betula pendula (Betulaceae)	Leaves	Used for treatment of high blood pressure, high cholesterol, obesity, gout, kidney stones and nephritis	Flavonoid Triterpenoid Phenolics	Hyperoside Papyriferic acid Dehydrosalidroside Hyperoside and betuloside	Elevated ozone	Increase Decreased Decreased	Saleem et al. (2001); Lavola et al. (1994)
Pinus taeda (Pinaceae)	Plant	Used for treatment of kidney and bladder complaints	Tannins Phenols	-	Elevated ozone	Increase No Change	Jordan et al. (1991)
Pueraria thomsonii (Fabaceae)	Cell suspension	Used for treatment of fever, acute dysentery, diarrhoea, diabetes, and cardiovascular diseases	Sesquiterpene	Puerarin	Elevated ozone	Increase	Sun et al. (2012)

(continued)

TABLE 2.1 (Continued)
Effect of Ozone Stress on Plant Secondary Metabolite Concentration

Medicinal plant (Family)	Plant part	Medicinal uses	Major class of secondary metabolite	Name of metabolite	Environmental condition	Concentration variation of metabolites	Key references
Salvia officinalis (Lamiaceae)	Plant	Used for digestive problems, stomach pain (gastritis), diarrhoea, bloating, and heartburn	Phenolic compound	Gallic acid, catechinic acid, caffeic acid and rosmarinic acid	Ozone stress	Increase	Pellegrini et al. (2015)
Hypericum perforatum (Hypericaceae)	Cell Suspension	Overcome anxiety, mild to moderate depression, mood disorders and stress	Alkaloids	Hypericin	Elevated ozone	Increases	Xu et al. (2011)
Acer saccharum (Sapindaceae)	Plant	Used in the treatment of coughs, diarrhoea, sore eyes and blindness	Flavonoids	-	Elevated ozone	Increases	Sager et al. (2005)
Brassica napus (Brassicaceae)	Plant	Used as diuretic, anti-scurvy, anti-inflammatory of bladder and anti-goat	Glucosinolates	Aromatic glucosinolates Indolyl glucosinolates	Elevated ozone	Increases Decreases	Himanen et al. (2008)
Brassica oleracea (Brassicaceae)	Plant	Used in the treatment of gout and rheumatism, cleaning infected wounds, anthelmintic, diuretic, laxative, cardiotonic and stomachic	Terpenes	-	Elevated ozone	Decreases	Pinto et al. (2007a, 2007b)

Plant	Medicinal use	Compound	Specific compound	Condition	Effect	Reference
Gossypium hirsutum (Mallows)	Plant	Used to treat skin rashes or skin allergy	Terpenes	-	Elevated ozone	Increases	Booker (2000)
Lycopersicon esculentum (Solanaceae)	Plant	Used for treatment of burns, scalds, sunburn and skin-wash	Terpenes	Jasmonic acid (sesquiterpenes)	Elevated ozone	Increases	Zandra et al. (2006)
Phaseolus lunatus (Fabaceae)	Plant	Used in the diet of people with fevers and as astringent	Terpenes	-	Elevated ozone	Increases	Vuorinen et al. (2004)
Phaseolus vulgaris (Fabaceae)	Plant	Reduces average blood sugar, total cholesterol, and triglycerides in people with uncontrolled diabetes	Phenolics / Flavonoids	Hydroxycinnamic acid / Isoflavonoids	Elevated ozone	Decreases / Increases	Kanoun et al. (2001)
Pinus strobus (Pinaceae)	Plant	Used as antiseptic, diuretic, rubefacient and vermifuge	Phenolics / Terpenoid	- / Diterpenoids	Elevated ozone	Decreases / No effect	Shadkami et al. (2007)
Populus tremuloides (Salicaceae)	Plant	Used in treatment of rheumatism, arthritis, gout, lower back pain, urinary complaints, digestive and liver disorders, debility, anorexia, also to reduce fevers and relieve the pain of menstrual cramps	Tannins / Phenolic	Condensed tannins / Glycosides	Elevated ozone	No effect / Decreases	Lindroth et al. (2001)

was an increase in the concentration of phenolics and tannins (Pellegrini et al. 2011; Shakeri et al. 2016). In 2015, Pellegrini et al. reported an increase in the content of phenolic, gallic acid, catechinic acid, caffeic acid, and rosmarinic acid when *Salvia officinalis* was tested under ozone stress (120 ± 13 ppb for 90 days) (Pellegrini et al. 2015). In *Phaseolus vulgaris* the concentration of the phenolic increases on increasing the ozone concentration while flavonoid content decreases under similar conditions (Kanoun et al. 2001). In another study on *Populus tremuloides*, phenolic content decreases on increasing the ozone concentration whereas there was no effect on tannin content (Lindroth et al. 2001). Valkama et al. (2007) reported in a meta-analysis that the activated plant secondary compounds metabolism (increased biosynthesis and reduced turnover) elucidates improved concentrations of phenolic and terpenoid plant secondary compounds in tree foliage (Valkama et al. 2007). In response to ozone exposure, elevated levels of total phenolics in urban tree species (Gao et al. 2016), condensed tannins in *Populus tremuloides* and *Betula papyrifera* (Couture et al. 2017), and flavonoids in *Fraxinus excelsior* (Cotrozzi et al. 2018) were recorded, but this was not the case in needles of Norway spruce seedlings exposed to elevated temperature and ozone concentration (Riikonen et al. 2012).

In a single plant, the different secondary metabolites can behave differently under the same environmental conditions; for example, when ozone concentration was elevated, flavonoid (hyperoside) content of *Betula pendula* (Betulaceae) increases whereas the content of triterpenoids (papyriferic acid) and phenolics (dehydrosalidroside, hyperoside, betuloside) decreases (Lavola et al. 1994). Similarly, the tannin concentration of *Pinus taeda* (Pinaceae) increases, while there was no effect on phenol concentration when ozone concentrations increase (Jordan et al. 1991). Bortolin et al. (2016) studied the effect of ozone exposure on *Capsicum baccatum*. The results revealed that there was a decrease in the concentration of capsaicin, an active component of chilli peppers, whereas there was no effect in the concentration of dihydrocapsaicin. In *Pinus strobes* the concentration of phenolics decreases while there was no effect on terpenoid content (Shadkami et al. 2007). Other studies showed that the terpenes content in *Brassica oleracea* decreases (Pinto et al. 2007a, 2007b) while sesquiterpenes (jasmonic acid) in *Lycopersicon esculentum* (Zandra et al. 2006), terpenes in *Phaseolus lunatus* (Vuorinen et al. 2004) and *Gossypium hirsutum* (Booker 2000), and flavonoids in *Acer saccharum* (Sager et al. 2005) increase in elevated ozone concentration. Increased puerarin levels in *Pueraria thomsonii* were also the result of elevated ozone concentrations (Sun et al. 2012).

2.6 CONCLUSION

Life is possible when interacting with environmental conditions and with other species. This interaction is in terms of energy and nutrient flow, which sustain life in any ecosystem but, due to population explosion and industrialization, there have been alterations in the atmospheric conditions as well as depletion of the ozone layer which protects life from ultraviolet rays. Stratospheric ozone is called 'good ozone' whereas tropospheric ozone is called 'bad ozone' due to its most destructive nature. This chapter has demonstrated how ozone affects medicinal plant growth, development, and the reproductive process. Studies reported by several investigators have stated that the ability of pollen germination, reproductive structures, and stomata surface are directly affected by ozone. From the reported articles it has been observed that chlorophyll degradation, resulting in a reduction in photosynthetic performance and metabolic changes, has also occurred due to ozone exposure. Many medicinal plants as well as other plant species initiate defence mechanisms with the help of secondary metabolites to cope with this situation. In response to ozone exposure, some plants have shown elevated levels of total phenolics, tannins, flavonoids, etc. However, these responses varied in different plant species, mainly depending on the ozone exposure duration and other climatic factors. Overall, medicinal plants and their interaction with ozone pollution is presented in Figure 2.3.

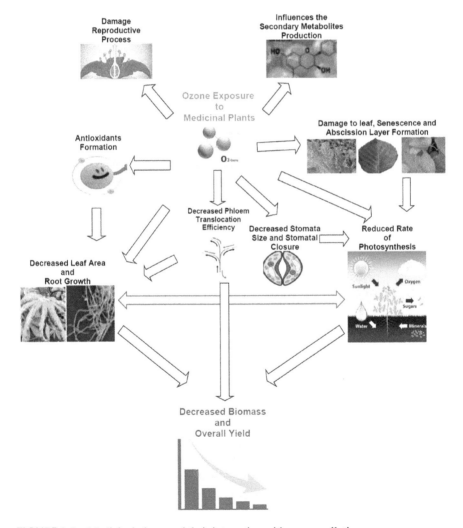

FIGURE 2.3 Medicinal plants and their interaction with ozone pollution.

REFERENCES

Adrees, M., M. Ibrahim, A.M. Shah, F. Abbas, F. Saleem, M. Rizwan, S. Hina, F. Jabeen, and S. Ali. 2016. Gaseous pollutants from brick kiln industry decreased the growth, photosynthesis, and yield of wheat (*Triticum aestivum* L.). *Environmental Monitoring and Assessment* 188(5): 267.

Agrawal, M., D.T. Krizek, S.B. Argawal, G.F. Kramer, E.H. Lee, R.M. Mirecki, and R.A. Rowland. 1993. Influence of inverse day/night temperature on ozone sensitivity and selected morphological and physiological responses of cucumber. *Journal of the American Society for Horticultural Science* 118: 649–654.

Agathokleous, E., Z. Feng, E. Oksanen, P. Sicard, Q.Wang, C. J. Saitanis, V. Araminiene, J.D. Blande, F. Hayes, V. Calatayud, M. Domingos, S.D. Veresoglou, J. Peñuelas, D. A. Wardle, A. De Marco, Z. Li, H. Harmens, X. Yuan, M. Vitale, and E. Paoletti. 2020. Ozone affects plant, insect, and soil microbial communities: a threat to terrestrial ecosystems and biodiversity. *Science Advances* 6(33): eabc1176.

Ainsworth, E.A. 2008. Rice production in a changing climate: a meta-analysis of responses to elevated carbon dioxide and elevated ozone concentration. *Global Change Biodiversity* 14: 1642–1650.

Ainsworth, E.A., P. Lemonnier and J.M. Wedow 2020. The influence of rising tropospheric carbon dioxide and ozone on plant productivity. *Plant Biology* 22 (Suppl. 1): 5–11.

Ainsworth, E.A., C.R. Yendrek, S. Sitch, W.J. Collins, and L.D. Emberson. 2012. The effects of tropospheric ozone on net primary productivity and implications for climate change. *Annual Review of Plant Biology* 63: 637–661.

Ansari, N., D.S. Yadav, M. Agrawal, and S.B. Agrawal. 2021. The impact of elevated ozone on growth, secondary metabolites, production of reactive oxygen species and antioxidant response in an anti-diabetic plant *Costus pictus*. *Functional Plant Biology* 48: 597–610.

Ansari, N., M. Agrawal, and S.B. Agrawal. 2020. An assessment of growth, floral morphology, and metabolites of a medicinal plant *Sida cordifolia* L. under the influence of elevated ozone. *Environmental Science and Pollution Research* 28: 832–845.

Bachheti, A., A. Sharma, R.K. Bachheti, A. Husen and D.P. Pandey. 2020. Plant Allelochemicals and Their Various Applications. In: Mérillon, J.M., and Ramawat, K. (eds). Co-Evolution of Secondary Metabolites. Reference Series in Phytochemistry. Springer, Cham. pp. 441–465.

Bachheti, A., Deepti, R.K. Bachheti, and A. Husen. 2021. Medicinal Plants and Their Pharmaceutical Properties Under Adverse Environmental Conditions. In: Husen, A. (ed.). Harsh Environment and Plant Resilience. Springer, Cham. pp. 457–502. https://doi.org/10.1007/978-3-030-65912-7_19

Barnes, J.D., D. Velissariou, A.W. Davison, and C.D. Holevas. 1990. Comparative ozone sensitivity of old and modern Greek cultivars of spring wheat. *New Phytologist* 116: 707–714.

Benoit, L.F., J.M. Skelly, L.D. Moore, and L.S. Dochinger. 1983. The influence of ozone on *Pinus strobus* L. pollen germination. *Canadian Journal of Forest Research – Revue Canadienne de Recherche Forestière* 13: 184–187.

Beshah, F., Y. Hunde, M. Getachew, R.K. Bachheti, A. Husen, and A. Bachheti. 2020. Ethnopharmacological, phytochemistry and other potential applications of *Dodonaea* genus: A comprehensive review. *Current Research in Biotechnology* 2: 103–119.

Black, V.J., C.R. Black, J.A. Roberts, and C.A. Stewart. 2000. Impact of ozone on the reproductive development of plants. *New Phytologist* 147: 421–447.

Booker, F.L. 2000. Influence of carbon dioxide enrichment, ozone and nitrogen fertilization on cotton (*Gossypium hirsutum* L.) leaf and root composition. *Plant, Cell and Environment* 23: 573–583.

Booker, F., R. Muntifering, M. Mcgrath, K. Burkey, D. Decoteau, E. Fiscus, W. Manning, S. Krupa, A. Chappelka, and D. Grantz. 2009. The ozone component of global change: potential effects on agricultural and horticultural plant yield, product quality and interactions with invasive species. *Journal of Integrative Plant Biology* 51: 337–351.

Bortolin, R.C., F.F. Caregnato, A.M.D. Junior, A. Zanotto-Filho, K.S. Moresco, and A. de Oliveira Rios. 2016. Chronic ozone exposure alters the secondary metabolite profile, antioxidant potential, anti-inflammatory property, and quality of red pepper fruit from *Capsicum baccatum*. *Ecotoxicology and Environmental Safety* 129: 16–24.

Bueker, P., Z. Feng, J. Uddling, A. Briolat, R. Alonso, S. Braun, S. Elvira, G. Gerosa, P.E. Karlsson, D. Le Thiec, and R. Marzuoli. 2015. New flux-based dose-response relationships for ozone for European forest tree species. *Environmental Pollution* 206: 163–174. DOI: 10.1016/j.envpol.2015.06.033

Calatayud, A., D.J. Iglesias, M. Talòn, and E. Barreno. 2003. Effects of 2-month ozone exposure in spinach leaves on photosynthesis, antioxidant systems and lipid peroxidation. *Plant Physiology and Biochemistry* 41 : 839–845.

Chameides, W.L. 1989. The chemistry of ozone deposition to plant leaves: role of ascorbic acid. *Environmental Science & Technology* 23(5): 595–600.

Cotrozzi, L., A. Campanella, E. Pellegrini, G. Lorenzini, C. Nali, and E. Paoletti. 2018. Phenylpropanoids are key players in the antioxidant defense to ozone of European ash, *Fraxinus excelsior*. *Environmental Science and Pollution Research* 25(9): 8137–8147.

Couture, J.J., T.D. Meehan, K.F. Rubert-Nason, and R.L. Lindroth. 2017. Effects of elevated atmospheric carbon dioxide and tropospheric ozone on phytochemical composition of trembling aspen (*Populus tremuloides*) and paper birch (*Betula papyrifera*). *Journal of Chemical Ecology* 43: 26–38. DOI: 10.1007/s108 86-016-0798-4

Davison, A.W., and J.D. Barnes. 1998. Effects of ozone on wild plants. *New Phytologist* 139: 135–151.

Dawnay, L., and G. Mills. 2009. Relative effects of elevated background ozone concentrations and peak episodes on senescence and above-ground growth in four populations of *Anthoxanthum odoratum* L. *Environmental Pollution* 157: 503–510.

De Marco, A., P. Sicard, S. Fares, J.P. Tuovinen, A. Anav, and E. Paoletti. 2016. Assessing the role of soil water limitation in determining the Phytotoxic Ozone Dose (PODY) thresholds. *Atmospheric Environment* 147: 88–97.

Emberson, L.D., H. Pleijel, E.A. Ainsworth, M. Van den Berg, W. Ren, and S. Osborne. 2018. Ozone effects on crops and consideration in crop models. *European Journal of Agronomy* 100: 19–34.

Feder, W.A. 1981. Bio-assaying for ozone with pollen systems. *Environmental Health Perspectives* 37: 117–123.

Feder, W.A., and R. Shrier. 1990. Combination of UV-B and ozone reduces pollen-tube growth more than either stress alone. *Environmental and Experimental Botany* 30: 451–454.

Feng, Z., and K. Kobayashi. 2009. Assessing the impacts of current and future concentrations of surface ozone on crop yield with meta-analysis. *Atmospheric Environment* 43: 1510–1519.

Feng, Z., K. Kobayashi, and E.A. Ainsworth. 2008. Impact of elevated ozone concentration on growth, physiology, and yield of wheat (*Triticum aestivum* L.): a meta-analysis. *Global Change Biology* 14: 2696–2708.

Ferretti, M., G. Bacaro, G. Brunialti, M. Confalonieri, F. Cristolofini, A. Cristofori, E. Cristofori. L. Frati, A. Finco, G. Gerosa, S. Maccherini and E. Gottardini. 2018. Scarce evidence of ozone effect on recent health and productivity of Alpine forests – A case study in Trentino, N. Italy. *Environmental Science and Pollution Research* 25(9): 8217–8232.

Fiscus, E.L., F.L. Booker, and K.O. Burkey. 2005. Crop responses to ozone: uptake, modes of action, carbon assimilation and partitioning. *Plant, Cell and Environment* 28: 997–1011.

Foyer, C.H., H.L. Delgado, J.F. Dat, and I.M. Scott. 1997. Hydrogen peroxide and glutathione-associated mechanisms of acclimatory stress tolerance and signalling. *Physiologia Plantarum* 100: 241–254.

Francini, A., C. Nali, V. Picchi, and G. Lorenzini. 2007. Metabolic changes in white clover clones exposed to ozone. *Environmental and Experimental Botany* 60: 11–19.

Fuhrer, J. 2009. Ozone risk for crops and pastures in present and future climates. *The Science of Nature* 96: 173–194.

Fuhrer, J., M. Val Martin, G. Mills, C.L. Heald, H. Harmens, F. Hayes, and M.R. Ashmore. 2016. Current and future ozone risks to global terrestrial biodiversity and ecosystem processes. *Ecology and Evolution* 6: 8785–8799.

Fumagalli, I., B.S. Gimeno, D. Velissariou, T.L. De, and G. Mills. 2001. Evidence of zone-induced adverse effects on crops in the Mediterranean region. *Atmospheric Environment* 35: 2583–2587.

Gao, F., V. Calatayud, F. Garcia-Breijo, J. Reig-Arminana, and Z. Feng. 2016. Effects of elevated ozone on physiological, anatomical and ultrastructural characteristics of four common urban tree species in China. *Ecological Indicators* 67: 367–379. DOI: 10.1016/j.ecolind.2016.03.012

Ghosh, A., M. Agrawal, and S.B. Agrawal. 2020. Effect of water deficit stress on an Indian wheat cultivar (*Triticum aestivum* L. HD 2967) under ambient and elevated level of ozone. *Science of the Total Environment* 714: 136837.

Głodowska, M., and M. Wozniak. 2019. Changes in soil microbial activity and community composition as a result of selected agricultural practices. *Agricultural Sciences* 10: 330–351.

Glick, R.E., C.D. Schlagnhaufer, R.N. Arteca, and E.J. Pell. 1995. Ozone-induced ethylene emission accelerates the loss of ribulose-1,5-bisphosphate carboxylase/oxygenase and nuclear-encoded mRNAs in senescing potato leaves. *Plant Physiology* 109: 891–898.

Guidi, L., C. Nali, G. Lorenzini, F. Filippi, and G.F. Soldatini. 2001. Effect of chronic ozone fumigation on the photosynthetic process of poplar clones showing different sensitivity. *Environmental Pollution* 113: 245–254.

Gupta, S.K., M. Sharma, V.K. Maurya, F. Deeba, and V. Pandey. 2021. Effects of ethylenediurea (EDU) on apoplast and chloroplast proteome in two wheat varieties under high ambient ozone: an approach to investigate EDU's mode of action. *Protoplasma* (2021). https://doi.org/10.1007/s00709-021-01617-1

Han, Y.J., G. Amir, M. Inga, F. Nadja, B. Winston, and U. Christian. 2020. Effect of different durations of moderate ozone exposure on secondary metabolites of *Brassica campestris* L. ssp. chinensis. *Journal of Horticultural Science and Biotechnology* 96(1): 110–120.

Harmens, H., F. Hayes, K. Sharps, A. Radbourne, and G. Mills. 2019. Can reduced irrigation mitigate ozone impacts on an ozone-sensitive African wheat variety? *MDPI-Plants* 8: 220.

Harrison, B.H., and W.A. Feder. 1974. Ultrastructural changes in pollen exposed to ozone. *Phytopathology* 64: 257–258.

Heath, R.L. 2008. Modification of the biochemical pathways of plants induced by ozone: what are the varied routes to change? *Environmental Pollution* 155: 453–463.

Himanen, S.J., A. Nissinen, S. Auriola, G.M. Auriola, C.N. Stewart, and J.K. Holopainen. 2008. Constitutive and herbivore-inducible glucosinolate concentrations in oilseed rape (*Brassica napus*) leaves are not affected by Bt Cry1Ac insertion but change under elevated atmospheric CO_2 and O_3. *Planta* 227: 427–437.

Holopainen J.K., V. Virjamo, R.P. Ghimire, J.D. Blande, R. Julkunen-Tiitto, and M. Kivimäenpää. 2018. Climate change effects on secondary compounds of forest trees in the Northern Hemisphere. *Frontiers in Plant Science* 9: 1445. DOI: 10.3389/fpls.2018. 01445

Hormaza, J.I., K. Pinney, and V.S. Polito. 1996. Correlation in the tolerance to ozone between sporophytes and male gametophytes of several fruit and nut tree species (Rosaceae). *Sexual Plant Reproduction* 9: 44–48.

Husen, A. 2021a. The Harsh Environment and Resilient Plants: An Overview. In: Husen, A. (ed.). Harsh Environment and Plant Resilience. Springer, Cham. https://doi.org/10.1007/978-3-030-65912-7_1

Husen, A. 2021b. Plant Performance Under Environmental Stress (Hormones, Biostimulants and Sustainable Plant Growth Management). Springer, Cham. https://doi.org/10.1007/978-3-030-78521-5

Husen, A. 1997. Impact of air pollution on the growth and development of *Datura innoxia* Mill. M.Sc. dissertation, Jamia Hamdard, New Delhi, India.

Husen, A., and M. Iqbal. 2004. Growth performance of *Datura innoxia* Mill. under the stress of coal-smoke pollution. *Annals of Forestry* 12: 182–190.

Husen, A., S.T. Ali, and I.M. Mahmooduzzafar. 1999. Structural, functional and biochemical responses of *Datura innoxia* Mill. to coal-smoke pollution. *Proceedings of Academy of Environmental Biology* 8: 61–72.

Iqbal, M., J. Jura-Morawiec, and W.M. Włoch. 2010a. Foliar characteristics, cambial activity and wood formation in *Azadirachta indica* A. Juss. as affected by coal-smoke pollution. *Flora-Morphology, Distribution, Functional Ecology of Plants* 205: 61–71.

Iqbal M., Mahmooduzzafar, F. Nighat, and P.R. Khan. 2010b. Photosynthetic, metabolic and growth responses of *Triumfetta rhomboidea* to coal-smoke pollution at different stages of plant ontogeny. *Journal of Plant Interactions* 5: 11–19.

Jolivet, Y., M. Bagard, M. Cabané, M.N. Vaultier, A. Gandin, D. Afif, and D. Le Thiec. 2016. Deciphering the ozone-induced changes in cellular processes: A prerequisite for ozone risk assessment at the tree and forest levels. *Annals of Forest Science* 73: 923–943.

Jordan, D.N., T.H. Green, A.H. Chappelka, B.G. Lockaby, R.S. Meldahl, and D.H. Gjerstad. 1991. Response of total tannins and phenolics in loblolly pine foliage exposed to ozone and acid rain. *Journal of Chemical Ecology* 17: 505–513.

Joshi, N., A. Chauhan, and P.C. Joshi. 2009. Impact of industrial air pollutants on some biochemical parameters and yield in wheat and mustard plants. *The Environmentalist* 29(4): 398–404.

Kadono, T., Y. Yamaguchi, T. Furuichi, M. Hirono, J.P. Garrec, and T. Kawano. 2006. Ozone-induced cell death mediated with oxidative and calcium signaling pathways in tobacco Bel-W3 and Bel-B cell suspension cultures. *Plant Signaling Behavior* 1(6): 312–322.

Kangasjärvi, J., P. Jaspers, and H. Kollist. 2005. Signalling and cell death in ozone-exposed plants. *Plant, Cell and Environment* 28: 1021–1036.

Kanoun, M., M.J. Goulas, and J. Biolley. 2001. Effect of a chronic and moderate ozone pollution on the phenolic pattern of bean leaves (*Phaseolus vulgaris* L. cv Nerina): relations with visible injury and biomass production. *Biochemical Systematics and Ecology* 29: 443–457.

Karnosky, D.F., K.E. Percy, A.H. Chappelka, C. Simpson, and J. Pikkarainen (eds). 2003. Air Pollution, Global Change and Forests in the New Millennium. Elsevier, Oxford.

Kerstiens, G., and K.J. Lendzian. 1989. Interactions between ozone and plant cuticles. 1. Ozone deposition and permeability. *New Phytologist* 112: 13–19.

Kim, J.J., R. Fan, L.K. Allison, and T.L. Andrew. 2020. On-site identification of ozone damage in fruiting plants using vapor-deposited conducting polymer tattoos. *Science Advances* 6(36): eabc3296.

Kleanthous, S., M. Vrekoussis, N. Mihalopoulos, P. Kalabokas, and J. Lelieveld. 2014. On the temporal and spatial variation of ozone in Cyprus. *Science of Total Environment* 476: 677–687.

Krause, G.H.M., W.D. Riley, and W.A. Feder. 1975. Effects of ozone on petunia and tomato pollen tube elongation in vivo. *Proceedings of the American Phyto-pathological Society* 2: 100.

Krupa, S.V., and W.J. Manning. 1988. Atmospheric ozone: formation and effects on vegetation. *Environmental Pollution* 50: 101–137.

Laloi, C., K. Apel, and A. Danon. 2004. Reactive oxygen signalling: the latest news. *Current Opinion in Plant Biology* 7: 323–328.

Lavola, A., R. Julkunen-Tiitto, and E. Paakkonen. 1994. Does ozone stress change the primary or secondary metabolites of birch (*Betula pendula Roth.*)? *New Phytologist* 126: 637–642.

Lindroth, R.L. 2010. Impacts of elevated atmospheric CO_2 and O_3 on forests: phytochemistry, trophic interactions, and ecosystem dynamics. *Journal of Chemical Ecology* 36: 2–21.

Lindroth, R.L., B.J. Kopper, W.F.J. Parsons, J.G. Bockheim, D.F. Karnosky, G.R. Hendrey, K.S. Pregitzer, J.G. Isebrands, and J. Sobe. 2001. Consequences of elevated carbon dioxide and ozone for foliar chemical composition and dynamics in trembling aspen (*Populus tremuloides*) and paper birch (*Betula papyrifera*). *Environmental Pollution* 115: 395–404.

Li, P., Z. Feng, V. Catalayud, X.Yuan, Y. Xu, and E. Paoletti 2017. A meta-analysis on growth, physiological, and biochemical responses of woody species to ground-level ozone highlights the role of plant functional types. *Plant Cell and Environment* 40: 2369–2380. DOI: 10.1111/pce.13043

Li, Z., J. Yang, B. Shang, Y. Xu, J. J. Couture, X. Yuan, K. Kobayashi, and Z. Feng. 2020. Water stress rather than N addition mitigates impacts of elevated O_3 on foliar chemical profiles in poplar saplings. *Science of Total Environment* 707: 135935.

Matyssek, R., D.F. Karnosky, G. Wieser, K. Percy, E. Oksanen, T.E.E. Grams, and H. Pretzsch. 2010. Advances in understanding ozone impact on forest trees: messages from novel phytotron and free air fumigation studies. *Environmental Pollution* 158: 1990–2006.

Morgan, P.B., E.A. Ainsworth, and S.P. Long. 2003. How does elevated ozone impact soybean? A meta-analysis of photosynthesis, growth and yield. *Plant, Cell and Environment* 26: 1317–1328.

Mukherjee, A., S.B. Agrawal, and M. Agrawal. 2020. Responses of tropical tree species to urban air pollutants: ROS/RNS formation and scavenging. *Science of the Total Environment* 710: 136363.

Mumford, R.A., H. Lipke, D.A. Laufer, and W.A. Feder. 1972. Ozone-induced changes in corn pollen. *Environmental Science and Technology* 6: 427–430.

Munné-Bosch, S., and L. Alegre. 2004. Die and let live: leaf senescence contributes to plant survival under drought stress. *Functional Plant Biology* 31: 203–216.

NASA. 2018. NASA Ozone watch. https://ozonewatch.gsfc.nasa.gov/facts/SH.html (as accessed on April 17, 2020).

Noctor, G., and C.H. Foyer. 1998. Ascorbate and glutathione: Keeping active oxygen under control. *Annual Review of Plant Physiology and Plant Molecular Biology* 49: 249–279.

Novak, K., J.M. Skelly, M. Schaub, N. Kräuchi, C. Hug, W. Landolt, and P. Bleuler. 2003. Ozone air pollution and foliar injury development on native plants of Switzerland. *Environmental Pollution* 125: 41–52.

Overmyer, K., M. Brosche, and J. Kangasjärvi. 2003. Reactive oxygen species and hormonal control of cell death. *Trends in Plant Science* 8: 335–342.

Overmyer, K., M. Brosch, R. Pellinen, T. Kuittinen, H. Tuominen, R. Ahlfors, M. Keinanen, M. Saarma, D. Scheel, and J. Kangasjärvi. 2005. Ozone-induced programmed cell death in the *Arabidopsis* radical-induced cell death1 mutant. *Plant Physiology* 137: 1092–1104.

Pellegrini, E., A. Francini, G. Lorenzini, and C. Nali. 2015. Eco physiological and antioxidant traits of *Salvia officinalis* under ozone stress. *Environmental Science and Pollution Research* 22: 13083–13093.

Pellegrini, E., M.G. Carucci, A. Campanella, G. Lorenzini, and C. Nali. 2011. Ozone stress in *Melissa officinalis* plants assessed by photosynthetic function. *Environmental and Experimental Botany* 73: 94–101.

Persson, K., H. Danielsson, G. Sellden, and H. Pleijel. 2003. The effects of tropospheric ozone and elevated carbon dioxide on potato (*Solanum tuberosum* L. cv. Bintje) growth and yield. *Science of Total Environment* 310: 191–201.

Piechowiak, T., P. Antos, P. Kosowski, K. Skrobacz, R. Józefczyk, and M. Balawejder. 2019. Impact of ozonation process on the microbiological and antioxidant status of raspberry (*Rubus ideaeus* L.) fruit during storage at room temperature. *Agricultural and Food Science* 28: 35–44.

Pinto, D.M., J.D. Blande, R. Nykanen, W.X. Dong, A.M. Nerg, and J.K. Holopainen. 2007a. Ozone degrades common herbivore-induced plant volatiles: does this affect herbivore prey location by predators and parasitoids? *Journal of Chemical Ecology* 33 : 683– 694.

Pinto, D.M., A.M. Nerg, and J.K. Holopainen. 2007b. The role of ozone-reactive compounds, terpenes, and green leaf volatiles (GLVs), in the orientation of *Cotesia plutellae*. *Journal of Chemical Ecology* 33: 2218–2228.

Płażek, A., M. Rapacz, and A. Skoczowski. 2000. Effects of ozone fumigation on photosynthesis and membrane permeability in leaves of spring barley, meadow fescue and winter rape. *Photosynthetica* 38: 409– 413.

Polle, A. 2001. Dissection the superoxide dismutase-ascorbate-glutathione pathway by metabolic modeling: Computer analysis as a step towards flux analysis. *Plant Physiology* 126: 445–462.

Radman, R., C. Bucke, and T. Keshavarz. 2004. Elicitor effects on reactive oxygen species in liquid cultures of *Penicillium chrysogenum*. *Biotechnology Letters* 26: 147–152.

Ranieri, A., D. Giuntini, F. Ferraro, C. Nali, B. Baldan, G. Lorenzini, and G.F. Soldatini. 2001. Chronic ozone fumigation induces alterations in thylakoid functionality and composition in two poplar clones. *Plant Physiology and Biochem*istry 39: 999–1008.

Rich, S., P.E. Waggoner, and H. Tomlinson. 1970. Ozone uptake by bean leaves. *Science* 169: 79–80.

Riikonen, J., S. Kontunen-Soppela, V. Ossipov, A. Tervahauta, M., Tuomainen, E. Oksanen, E. Vapaavuori, J. Heinonen, and M. Kivimäenpää. 2012. Needle metabolome, freezing tolerance and gas exchange in Norway spruce seedlings exposed to elevated temperature and ozone concentration. *Tree Physiology* 32: 1102–1112. DOI: 10.1093/treephys/ tps072

Royal Society. 2008. Ground-level ozone in the 21st century: future trends, impacts and policy implications. Science Policy Report 15/08. http://royalsociety.org/displaypagedoc. asp?id=31506.

Runeckles, V.C., and B.I. Chevone. 1992. Crop Responses to Ozone. In: Lefohn, A.S. (ed.). Surface Level Ozone Exposures and Their Effects on Vegetation. CRC Press, pp. 189–270.

Sager, E.P., T.C. Hutchinson, and T.R. Croley. 2005. Foliar phenolics in sugar maple (*Acer saccharum*) as a potential indicator of tropospheric ozone pollution. *Environmental Monitoring and Assess*ment 105: 419– 430.

Saleem, A., J. Loponen, K. Pihlaja, and E. Oksanen. 2001. Effects of long-term open-field ozone exposure on leaf phenolics of European silver birch (*Betula pendula* Roth.) *Journal of Chemical Ecology* 27: 1049–1062.

Sandermann, H. 2008. Ecotoxicology of ozone: bioactivation of extracellular ascorbate. *Biochemical Biophysical Research Communi*cations 366: 271–274.

Sarkar, A., and S.B. Agrawal. 2012. Evaluating the response of two high yielding Indian rice cultivars against ambient and elevated levels of ozone by using open top chambers. *Journal of Environmental Management* 95: S19–S24.

Saxe, H. 2002. Physiological responses of trees to ozone-interactions and mechanisms. *Current Topics in Plant Biology* 3: 27–55.

Schauberger, B., S. Rolinski, S. Schaphoff, and C. Müller. 2019. Global historical soybean and wheat yield loss estimates from ozone pollution considering water and temperature as modifying effects. *Agriculture and Forest Meteorology* 265: 1–15.

Shadkami, F., R.J. Helleur, and R.M. Cox. 2007. Profiling secondary metabolites of needles of ozone-fumigated white pine (*Pinus strobus*) clones by thermally assisted hydrolysis/ methylation GC/MS. *Journal of Chemical Ecology* 33: 1467–1476.

Shakeri, A., A. Sahebkar, and B. Javadi. 2016. *Melissa officinalis* L. – a review of its traditional uses, phytochemistry and pharmacology. *Journal of Ethnopharmacology* 188: 204–228.

Sharma, A., S.K. Sharma, and T.K. Mandal. 2016. Influence of ozone precursors and particulate matter on the variation of surface ozone at an urban site of Delhi, India. *Sustainable Environmental Research* 26: 76–83.

Sharma A, A. Bachheti, P. Sharma, R.K. Bachheti, A. Husen. 2020. Phytochemistry, pharmacological activities, nanoparticle fabrication, commercial products and waste utilization of *Carica papaya* L.: a comprehensive review. *Current Research in Biotechnology* 2: 145–160

Sicard, P., A. De Marco, L. Dalstein-Richier, F. Tagliaferro, C. Renou, and E. Paoletti. 2016. An epidemiological assessment of stomatal ozone flux-based critical levels for visible ozone injury in Southern European forests. *Science of the Total Environment* 541: 729–741.

Singh S., and A. Husen. 2020. Behavior of Agricultural Crops in Relation to Nanomaterials under Adverse Environmental Conditions. In: Husen, A., and Jawaid, M. (eds). Nanomaterials for Agriculture and Forestry Applications. Elsevier, Cambridge, MA, pp. 219–256 https://doi.org/10.1016/B978-0-12-817852-2.00009-3

Sitch, S., P.M. Cox, W.J. Collins, and C. Huntingford. 2007. Indirect radiative forcing of climate change through ozone effects on the land-carbon sink. *Nature* 448: 791–794.

Sun, L., H. Su, Y. Zhu, and M. Xu. 2012. Involvement of abscisic acid in ozone-induced puerarin production of *Pueraria thomsnii* Benth. suspension cell cultures. *Plant Cell Rep*orts 31: 179–185.

Tamaoki, M. 2008.The role of phytohormone signaling in ozone-induced cell death in plants. *Plant Signaling & Behavior* 3: 166–174.

Unsworth, M.H., and D.P. Ormrod. 1982. Effects of Gaseous Air Pollutants in Agriculture and Horticulture. Butterworth-Heinemann. doi.org/10.1016/C2013-0-06285-8

US National Park Service. 2020. Ozone Effects on Plants. www.nps.gov/subjects/air/nature-ozone.htm (accessed July 12, 2021)

Valkama, E., J. Koricheva, and E. Oksanen. 2007. Effects of elevated O_3, alone and in combination with elevated CO_2, on tree leaf chemistry and insect herbivore performance: a meta-analysis. *Global Change Biology* 13: 184–201.

Vainonen, J.P., and J. Kangasjärvi. 2015. Plant signalling in acute ozone exposure. *Plant Cell and Environment* 38: 240–252. DOI: 10.1111/pce.12273

Vuorinen, T., A.M. Nerg, M.A. Ibrahim, G.V.P. Reddy, and J.K. Holopainen. 2004. Emission of *Plutella xylostella*-induced compounds from cabbages grown at elevated CO_2 and orientation behaviour of the natural enemies. *Plant Physiology* 135: 1984–1992.

Waldner, P., M. Schaub, E. Graf-Pannatier, M. Schmitt, A. Thimonier, and L. Walthert, 2007. Atmospheric deposition and ozone levels in Swiss forests: are critical values exceeded? *Environmental Monitoring and Assessment* 128: 5–17.

Wilkinson, S., G. Mills, R. Illidge, and W.J. Davies. 2012. How is ozone pollution reducing our food supply? *Journal of Experimental Botany* 63(2): 527–536.

Xu, D., M. Shi, B. Jia, Z. Yan, L. Gao, W. Guan, Q. Wang, and J. Zuo. 2019. Effect of ozone on the activity of antioxidant and chlorophyll-degrading enzymes during postharvest storage of coriander (*Coriandrum sativum* L.). *Journal of Food Processing and Preservation* 43: e14020.

Xu, M., B. Yang, J. Dong, D. Lu, H. Jin, L. Sun, Y. Zhu, and X. Xu. 2011. Enhancing hypericin production of *Hypericum perforatum* cell suspension culture by ozone exposure. *Biotechnology Progress* 27: 1101–1106.

Yadav, D.S., R. Rai, A.K. Mishra, N. Chaudhary, A. Mukherjee, S.B. Agrawal, and M. Agrawal. 2019. ROS production and its detoxification in early and late sown cultivars of wheat under future O_3 concentration. *Science of the Total Environment* 659: 200–210.

Zandra, C., A. Borgogni, and C. Marucchini. 2006. Quantification of Jasmonic acid by SPME in tomato plants stressed by ozone. *Journal of Agricultural and Food Chemistry* 54: 9317–9321.

Zhang, X., X. Zhang, T. Wang, and C. Li. 2021. Metabolic response of soybean leaves induced by short-term exposure of ozone. *Ecotoxicology and Environmental Safety* 213: 112033

3 Impact of UV Radiation on the Growth and Pharmaceutical Properties of Medicinal Plants

Deepti, Archana (Joshi) Bachheti, Kiran Chauhan,
Rakesh Kumar Bachheti, Azamal Husen

CONTENTS

3.1 INTRODUCTION

With the help of microbial proof, the origin of life is supposed to have started about a few billion years ago. The solar spectrum supported the origin of life as this is the main driving force for the exclusive ecosystem of Earth; it is embraced with the electromagnetic spectrum, which includes radio waves to gamma waves ranging from different wavelengths. The electromagnetic waves supported the formation of the early atmosphere and eventually led to the origin of life on Earth. The region of our concern is 'Ultraviolet (UV) radiation', which lies at the end of the short wavelength region, which is known for the more energetic regions. The infrared, visible, and ultraviolet rays are of the utmost importance for life on Earth. Being part of the solar spectrum, infrared light and visible light are responsible for the increment of temperature (up to the survival range) on Earth. About 8 to 9 per cent of the radiation

TABLE 3.1
Distribution of Solar Irradiance Energy

Spectral Regions	Wavelength (nm)	% of Total Energy
Infrared	≥ 700	49.4
Visible	400–700	42.3
UV-A	320–400	6.3
UV-B	290–320	1.5
UV-C	≤ 290	0.5

is ultraviolet radiation in total solar radiation which is the component of the non-ionizing region of the electromagnetic spectrum. Typically, UV radiation is divided into three wavelength ranges, namely (a) UV-A (320–400 nm), (b) UV-B (280–320 nm), and (c) UV-C (200–280 nm).

UV-A radiation is the least harmful of total UV radiation, while UV-B radiation is dangerous for plants and animals, and at the ecosystem level too, whereas UV-C radiation is immensely injurious for all organisms (Table 3.1) (Rai and Agarwal, 2017).

According to antediluvian society, the sun was the only source for warmth, visibility, health, and vitality, and this belief about the sun is also engrossed in cultural tradition and mythology, but the younger generations did not believe it which gave birth to the quest for new concepts, beliefs, and understanding. At the beginning of the nineteenth century, it was concluded by newer knowledge and principles that sunlight is not a single spur of only one wavelength, but rather a bundle of spurs of varied wavelengths. Before 1920, during these exploratory events, UV radiation was discovered. In 1801 Ritter detected invisible rays over the violet end of the spectrum and named them deoxidizing rays; later they were termed chemical rays. The notation of the spectral extent of UV radiation was given by Becquerel and Draper in 1842 when they observed that the wavelength between 240 and 400 nm induces a photochemical reaction (Rai and Agarwal, 2017). According to Kumari and Agrawal (2010), environmental factors are the prime focus of concern, considering the high demand for aromatic and medicinal plants in the global market. Pharmacists and researchers took an interest in UV-B and its impact on medicinal plants due to exhortation of volatile production and secondary metabolites such as flavonoids and phenolics which show the appreciating role of UV-B in the yield and quality of herbal products. This may be because of their significant participation in intensifying the number of phenylpropanoids and natural antioxidants.

Medicinal plants are the source of secondary metabolites which have great significance in the pharmaceutical industries and other disciplines. The production of secondary metabolites in medicinal plants is greatly influenced by both abiotic and biotic factors. Ultraviolet (UV) radiation is one of the abiotic factors which affect secondary metabolites concentration and hence the pharmaceutical properties of medicinal plants. Of the three types of UV radiation, UV-C is fully absorbed and scattered by ozone whereas all of UV-A and some of UV-B radiation reach the Earth's environment. UV radiation affects the rate of photosynthesis by reducing leaf area and total canopy leaf area, altering loss of water by transpiration, and leaf thickness in medicinal as well as crop plants. The concentration of secondary metabolites such as terpenoids, alkaloids, and phenolics is usually accelerated by exposure to UV-B. However, it has been noticed that the enhancement in secondary metabolites is mainly affected by the duration of radiation, and the dose of radiation at the various developmental stages of the plant tissue. Overall, the aim of this chapter is to give an overview of the impact of UV radiation on medicinal plant growth, development, secondary metabolites concentration, and reproductive processes.

3.2   EFFECTS OF UV RADIATION AND LIFE OF PLANTS

The first life on Earth was formed under water before stratospheric ozone formation as life on land was not possible. With the help of UV-C and other solar radiation, stratospheric ozone was formed, and after that UV-C was fully absorbed and scattered whereas all the UV-A radiation and some UV-B radiation reached the Earth's environment. UV radiation transformed the Earth's environment throughout the geological periods, hence it is concluded that UV radiation has regulatory properties. According to Turunen and Latola (2005), the plants present at higher elevations are less sensitive to very high levels of UV-B irradiance in comparison to plants present at lower elevations, and this is explained by the fact that terrestrial plants coevolved under various solar UV-B levels (Cockell and Horneck, 2001; Rozema et al., 2002). Terrestrial plants adjust to UV-B radiation at different levels and get varied UV-B doses as their UV-B environment fluctuates moderately in both space and time. Hence it is assumed that, on increment in doses of solar UV-B, terrestrial plants react in different ways (Rozema, 2000). The detrimental impact of UV-B radiation on terrestrial plants is validated by extensively significant biochemical machinery like photosystem II (PSII) and deoxyribonucleic acid (DNA) (Singh et al., 2008). Due to ozone layer depletion, the amount of UV-B radiation reaching the earth's surface increases (Madronich et al.,1998) which leads to changes in the metabolism or growth of plants. Lesser growth, biomass accumulation, carbon assimilation reduction, and photosynthetic pigment damage are some common consequences of UV-B radiation exposure on plants, which may affect their productivity. Caldwell et al. (2003) have reported that some plants are very susceptible to UV-B radiation and not able to tolerate these stress conditions, whereas some plants can tolerate the stress. The plants that fail to handle the stress attain some other defence mechanisms to survive in such environmental stress, such as more flavonoid production, antioxidant formation induction, activation of reactive species to quench free radicals, and increments in leaf thickness. Various physiological responses of plants are outcomes of increments of UV-B radiation in the environment (Tevini and Teramura, 1989). The decline in biomass (Tevini et al.,1981; Sullivan and Teramura,1988), increments in flavonoid levels, changes in the ultra-structure of leaves and stomatal conductance (Tevini et al., 1981,1991b; Beggs and Wellman, 1985), epidermal distortion (Tevini and Steinmuller, 1987), and abatement in pollen germination percentage (Flint and Cladwell, 1984) are among the effects noticed by various researchers after supplemental UV-B radiation.

3.3 UV RADIATION UPTAKE BY PLANTS

Plants respond to UV radiation just like all other living organisms on Earth. All types of UV radiation are known to harm the various plant processes. Such damage can be of two types, one in which the physiological process gets disturbed and the other in which the DNA (which can cause mutations that can be heritable) of the plant gets damaged (Caldwell et al., 1989). Photoreceptors in living beings work as responders towards light and ameliorate growth, as they stimulate many biological processes by absorbing the specific wavelength of radiation. There are many photoreceptors in plants and each photoreceptor has its specific absorption spectrum, some of them being especially known for absorption of ultraviolet radiation. UV resistance 8 (UVR8), phytochrome A (Shinomura et al., 1996), phototropins (Christie and Briggs, 2001; Vandenbussche et al., 2014), and cryptochromes (Ahmad et al., 2002) are accredited for UV-B-specific sensors in UV responses (Heijde and Ulm, 2012). Although there are no photoreceptors reported for UV-C till now, it is said by some researchers that some known photoreceptors such as cryptochromes and UVR8 were triggered by UV-C absorbance, hence it may be possible that they are playing a role in responses to UV-C in plants. The action spectrum of photoreceptors and absorption spectrum are so much correlated because of protein contexture in chromophores. So, it is well known that at any wavelength of the absorption range of chromophore the photoreceptors will be activated. The phototropins show activity at lower than 350 nm wavelengths and are credited as blue-light photoreceptors (Vandenbussche et al., 2014).

3.4 IMPACT OF UV RADIATION ON GROWTH, DEVELOPMENT, AND REPRODUCTIVE PROCESSES

Cell division, elongation, directional growth, and branching are the most important factors behind the morphological or vegetative growth of plants. All these processes are light-dependent. The plant responds and grows according to the intensity, time period, and direction of light (Thelier et al., 2015). The morphological responses are also species-dependent and UV radiation exposure (Figure 3.1) can alter the plant growth (Kakani et al., 2003), as in some species it can increase the dry or fresh biomass and leaf area (Sakalauskaite et al., 2013), while in the case of some other species it decreases shoot dry mass, shoot length and foliar area (Kuhlmann and Müller, 2009a, 2009b; Liu et al., 2013; Singh et al., 2011; Torre et al., 2012; Wargent et al., 2006; Zhang et al., 2014). Although the impact of UV-A radiation has not been explored like that of UV-B radiation, it has been shown that UV-A radiation can decrease the leaf area and height of plants even in the absence of UV-B (Krizek et al., 1997). According to Barnes et al. (1990) and Torre et al. (2012), exposure to UV-B radiation can increase the number of stems. The seed yield of the plant is certified by floral induction, flower determination, and fruit production which is influenced by UV radiation exposure as it can increase the size of the flower (Petropoulou et al., 2001; Sampson and Cane, 1999). In the past also various studies have shown that UV radiation has affected the growth and development of plants. In an experiment, it was concluded that the plants from different genotypes exposed to UV-B radiation had

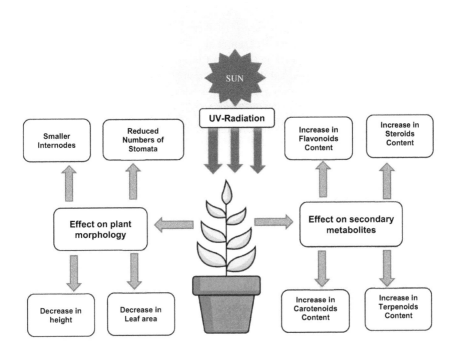

FIGURE 3.1 Effect of UV radiations on medicinal plant growth and developmental processes.

shorter hypocotyl while there was no impact on their cotyledon area and dry weight (Ballare et al., 1991). Gaba and Black (1983) reported that UV radiation also helps in regulating phototropism in various plants.

3.5 IMPACT OF UV RADIATION ON PHYSIOLOGICAL PROCESSES

Carbohydrate is the main by-product of water and carbon dioxide (CO_2) in the process of photosynthesis, which occurs in the presence of sunlight. The major components of the photosynthetic apparatus are the photosystems (PS) PSI and PSII, out of which the prime focus of UV-B radiation is PSII (Fiscus and Booker, 1995). It is composed of protein and pigment, which helps in the transportation of electrons from the splitting of water to plastoquinone (Barber et al., 1997; Mattoo et al., 1999). The core of PSII is formed by the proteins present in it, which are known as D1 and D2. These proteins are very sensitive to UV radiation, mainly UV-B, that can deteriorate both proteins and lead to impairment of PSII, which is evaluated by the drop in variable chlorophyll fluorescence (Fv/Fm) and oxygen (Rai and Agrawal, 2017). The rate of photosynthesis can be affected by the sensitivity of all components within the PSII, i.e., from manganese-binding sites to plastoquinone acceptor sites towards UV-B radiation. The factors which affect the rate of photosynthesis are (i) reduction in the individual leaf area, (ii) total canopy leaf area, (iii) alteration in loss of water by transpiration, (iv) the uptake rate of CO_2 for photosynthesis, (v) stomatal closure, (vi) modifications in anatomy, and (vii) leaf thickness (Vass et al., 1996; Yao and Liu, 2006). The destruction that occurs due to UV-B radiation first affects the acceptor side of PSII, then the donor side (Van Rensen et al., 2007). It was supposed that the stomatal closure could be generated by the loss of turgor pressure with ion leakage from the guard cells and it was noticed that even at low UV-B levels the stomatal closure occurred quickly, while the stomatal opening is slower at high UV-B levels (Zlatko et al., 2012). The uptake of CO_2 gets reduced, all because of the harmful impacts of UV-B radiation on leaf anatomy (Mansfield and Freer-Smith, 1984). Similarly, CO_2 fixation also gets affected because of a decrease in concentration and activity of ribulose-1,5-bisphosphate carboxylase-oxygenase (RuBisCo). The photoinhibition is caused by a decrease in fluorescence-measuring parameter (Fv/Fm) ratio in UV-B stress, hence the Fv/Fm can be considered as a potent physiological indicator of a plant's responses to stress (Husen, 2009, 2010, 2013). The decline in Fm (maximum fluorescence) in Fv/Fm and increment in initial fluorescence in the plants that are exposed to UV-B radiation lead to inactivation of the PSII reaction centre which is light-dependent (Long and Humphries, 1994). The reduction in the ratio of Fv/Fm was also found to be linked to increased production of superoxide radicals, and under UV-B stress it was found aligned with higher malondialdehyde (MDA) content. The higher magnitude of UV-B radiation causes collateral harm to the membrane in terms of MDA content (Jin and Tao, 2000).

3.6   IMPACT OF UV RADIATION ON BIOCHEMICAL PROCESSES

It has been reported that the biochemical and physiological properties of plants are severely affected by UV-B radiation (Miller et al., 1994; Rao et al., 1996; Ambhast and Agrawal, 2003). Photosynthetic apparatus or chromophores absorb the radiation which enters the plants. Approximately 90 per cent of UV-B radiation is taken up by leaves (Robberecht and Caldwell, 1980; Cen and Bornman, 1993; Gonzalez et al., 1996). Proteins, lipids, quinones, nucleic acids, and flavonoids (water-soluble phenolic pigments) are some cell components that are present in leaves and absorb UV-B radiation, which helps in protecting the plants as reactive oxygen species can be produced due to UV-B stress (Jordan, 1996). Nicotinamide adenine dinucleotide phosphate (NADPH) oxidase activated by UV-B becomes the cause of the production of reactive oxygen species (Rao et al., 1996). High molecular weight enzymes, such as catalase, superoxide dismutase (SOD), peroxidase (POD), and ascorbate peroxidase, and low molecular weight antioxidants like flavonoids, carotenoids, ascorbic acid, and phenols are an integral part of the active defence system which acts as a blueprint to metabolize and acclimatize reactive oxygen species (Cervilla et al., 2007). Polyamines play a major role in ruling the structure and function of photosynthetic apparatus and help in reducing cellular damage of proteins, lipids, and DNA of the plant (Sung et al., 2008; Agrawal et al., 2009). Also, for protection against UV-B stress, polyamines work very well with the help of signals provided by stress messengers, which provide safety against oxidative stress (Kramer et al., 1991; Deutsch et al., 2005). UV-B radiation stimulates the response of the regulatory key enzyme phenylalanine ammonia-lyase which is an important enzyme of the flavonoid biosynthesis pathway and is transcriptionally induced by UV treatment (Ravindran et al., 2010).

3.7 DEFENCE MECHANISMS UNDER UV RADIATION STRESS

The most usual type of defence mechanism adopted by plants for protection from increased levels of UV-B radiation is an accumulation of waxy material on leaves which makes leaf surfaces smooth and lustrous. It results in maximum reflectivity from plants. When plants become more vulnerable to UV-B, they can also show symptoms such as the reddening and bronzing of leaves because of the production of polyphenolic compounds (such as flavonoids) (Rai and Agarwal, 2017). High levels of UV radiation adversely affect photosynthesis and decrease crop yield and growth. Studies have shown that plants have their own defence systems to protect themselves from UV radiation through some natural plant chemicals such as phenolics. These compounds act as a natural sunscreen for the plant as it absorbs the unwanted UV radiation, leaving the radiation that is used in photosynthesis or plant development. The plants that are found at high altitudes possess protection all the time, but other plants produce this protection only when needed, i.e., when they are exposed to increased levels of UV-B radiation (Sharma et al., 2017). The compound that works behind this protection is a protein that can recognize the attack of UV radiation on the plant and is known as UVR8, i.e., UV resistance locus 8. The plants that lack this protein are not able to protect themselves from the sun and get severely damaged in the summer sunlight. Still, the fundamental mechanism of UVR8 which helps in controlling the plant's response to UV-B is one of the top research areas for researchers. Study till now reveals that the UVR8 absorbs UVB which leads to some sort of changes in the plant and helps the UVR8 protein to stack in the plant cell's nucleus. In protecting the plant from UV-B radiation and damage caused by it, this is an important action (Yin et al., 2016).

3.8 IMPACT OF UV RADIATION STRESS ON SECONDARY METABOLITES

Like other plants, medicinal plants are also exposed to all kinds of UV radiation (UV-A, UV-B, and UV-C) present in sunlight, as they use sunlight for photosynthesis. Of these, UV-C gets trapped by the ozone layer and UV-A and UV-B reach the Earth's surface. Plants can sense and respond to UV radiation. It is known that UV radiation can cause harm to the plant by damaging its physiological activities and genes but, in some studies, it has been found that UV radiation stress leads to an elevation in concentrations of secondary metabolites (Table 3.2). The effect of the UV-B study on secondary metabolites production is mainly focused on UV-B screening of phenolic compounds which are a kind of phytochemical that exhibits several human health benefits such as antioxidative and anticarcinogenic characteristics (Holst and Williamson, 2004; Schreiner et al., 2009). The oxidative stress is reduced by phenolics by balancing the redox status (Kumari et al., 2009b). Many studies have suggested the

TABLE 3.2
Effect of UV Radiations on Plant Secondary Metabolite Concentration

Medicinal plant (Family)	Plant part	Medicinal uses	Major class of secondary metabolite	Name of metabolite	Environmental condition	Concentration variation	Key references
Capsicum annuum (Solanaceae)	fruit	Used to treat osteoarthritis, shingles, rheumatoid arthritis, post-herpetic neuralgia, trigeminal neuralgia, diabetic neuropathy, fibromyalgia, and back pain	Cynaroside	Falvone	UV radiation stress	Increases	Ellenberger et al. (2020)
Catharanthus roseus (Apocynaceae)	flower	Used to relieve muscle pain, depression of the central nervous system, wasp stings, to heal wounds, stomach ache and diabetes	Indole alkaloid	Catharanthine and vindoline	UV radiation stress	Increases	Zhong, et. al. (2019)
Crocus sativus (Iridaceae)	Leaves Stigma	Used for asthma, cough, whooping cough (pertussis), and to loosen phlegm (as an expectorant) and baldness	Flavonoids Total phenolics and total flavonoids	Anthocyanins -	UV-B	Increases	Rikabad et al. (2019)
Cuminum cyminum (Apiaceae)	Seeds	For chronic diarrhoea and dyspepsia, acute gastritis, diabetes, and cancer	Terpenoids, phenols, flavonoids, and alkaloids	Anthocyanins, beta-carotene, and lycopene	UV-B	Increases	Ghasemi et al. (2018)

Species (Family)	Part	Uses			UV	Effect	Reference
Withania somnifera (Solanaceae)	Leaf and root	Used as an aphrodisiac, liver tonic, anti-inflammatory agent, and to treat asthma, ulcers, insomnia, and senile dementia, anxiety, cognitive and neurological disorders, inflammation, and Parkinson's disease	Phenols and alkaloids	Anthocyanins, lignin, alkaloids, saponins and phytosterols	UV-B	Increases	Takshak and Agrawal (2014)
Chrysanthemum (Asteraceae)	Flower	Used to treat chest pain (angina), high BP, type 2 diabetes, fever, cold, headache, dizziness, and swelling	Phenols	Phenolic acids	UV-B	Increases	Ma et al. (2016)
Flourensia cernua (Asteraceae)	Leaf	Used to treat respiratory infections, such as tuberculosis	Phenols	Phenolic acids	UV-B	No effect	Estell et al. (2016)
Nasturtium officinale (Brassicaceae)	Leaf	Used as antiscorbutic, depurative, diuretic, expectorant, purgative, hypoglycaemic, odontalgic, stimulant and stomachic	Phenols, others	Phenolic acids, Glucosinolate	UV-B	Increases	Estell et al. (2016) Reifenrath and Müller (2007)
Astragalus compactus (Fabaceae)	Leaf	Used as anti-inflammatory, immunostimulant, antioxidative, anticancer, antidiabetic, cardioprotective, hepatoprotective, and antiviral	Phenols	Phenolic acids	UV-B	Increases	Naghiloo et al. (2012)

(continued)

TABLE 3.2 (Continued)
Effect of UV Radiations on Plant Secondary Metabolite Concentration

Medicinal plant (Family)	Plant part	Medicinal uses	Major class of secondary metabolite	Name of metabolite	Environmental condition	Concentration variation	Key references
Asparagus officinalis (Asparagaceae)	Whole plant	Used for rheumatism, hormone imbalances in women, dryness in the lungs and throat, constipation, nerve pain (neuritis), AIDS, cancer, and diseases caused by parasites	Phenols	Flavonol quercetin-4'-O monoglucoside	UV-B	Increases	Eichholz et al. (2012)
Rosmarinus officinalis (Lamiaceae)	Whole plant	Used for treatment of inflammatory diseases including allergies, asthma and atherosclerosis and anti-inflammatory properties	–	Rosmarinic acid	UV-B	Increases	Luis et al. (2007)
Arnica montana (Asteraceae)	Flowering head	Used for bruises, sprains, muscle aches, wound healing, superficial phlebitis, joint pain, inflammation from insect bites, and swelling from broken bones	Phenols	Phenolic acids	UV-B	Increases	Spitaler et al. (2006)
Psychotria brachyceras (Rubiaceae)	Whole plant	Used for treating cough, ulcer, bronchitis, stomach ache and infections of female reproductive system	Monoterpene-indole alkaloid	Brachycerine	UV-B	Increases	Gregianini et al. (2003)

direct dependency of the amount of phenolic compound and antioxidant properties in blueberries (Eichholz et al., 2011), and sweet flag (Kumari et al., 2009b). Some research also provides a result that these secondary metabolites also act as a UV protector by absorbing light between 270 to 290 nm. *Rosmarinus officinalis*, which contains the bioactive compound rosemary acid, shows anti-inflammatory properties which have several health benefits on overcoming inflammatory diseases, including atheroscler-osis, allergies, and asthma. The overall yield of these substances can be increased by exposing these plants to UV-B treatment (Luis et al., 2007). For instance, cynaroside (falvone) concentrations get increased in *Capsicum annum* (Solanaceae) under UV radiation stress (Ellenberger et al., 2020). Zhong et al. (2019) also reported that in *Catharanthus roseus* (Apocynaceae), the phytochemicals catharanthine and vindoline (indole alkaloid) get increased in UV-B radiation stress. According to Higashio et al. (2007), the concentration of plants' secondary metabolism can be affected by UV-B treatment which is an application of abiotic stress. The content of ascorbic acid, flavonoids, and tocopherols in aromatic medical plants can be increased by ultraviolet supplements (Higashio et al., 2007; Kumari and Agrawal, 2010).

In a study, research was conducted on apple fruit via UV-B irradiation with visible light which enhances the phenolic activity. These treatments also affect the chlorogenic acid, quercetin glycosides, anthocyanins, and ascorbic acid in apple fruit (Hagen et al., 2007). Another scientific study also reported that there was an increment in the content of flavonoids in leaves, total phenolics, and total flavonoids in the stigma of *Crocus sativus* when the plant was kept under UV-B radiation stress (Rikabad et al., 2019). Also, Ghasemi et al. (2018) reported that, under UV-B stress, there was an enhancement in total phenolic, lycopene, carotenoids, and ascorbic acid (anthocyanins, beta-carotene, and lycopene) in seeds of *Cuminum cyminum*. Similarly, phenolic alkaloids, anthocyanins, lignin, tannins, saponins, and phytosterols in leaves and roots of *Withania somnifera* get increased under UV-B stress (Takshak and Agarwal, 2014). Reifenrath and Müller (2007) in their study reported that under the stress treatment of UV radiation the concentration of glucosinolate in leaves of *Nasturtium officinale* also rises.

The phenolic compounds' accumulation generally gets triggered by UV-B radiations. For example, the concentration of phenolic acids in flowers of *Chrysanthemum* (Ma et al., 2016), in the flowering head of *Arnica montana* (Spitaler et al., 2006), in leaves of *Nasturtium officinale* (Estell et al., 2016), and in leaves of *Astragalus compactus* (Naghiloo et al., 2012) gets increased in UV-B stress, whereas there was no effect on the concentration of phenolic acids in leaves of *Flourensia cernua* (Asteraceae) under the same stress (Estell et al., 2016). One another study by Eichholz et al. (2012) on *Asparagus officinalis* (Asparagaceae) proved the accumulation of phenols (flavonol, quercetin-4-O-monoglucoside) because of UV-B radiation stress. In 2007, Luis et al. found that in UV radiation exposure the yield of rosmarinic acid increases in *Rosmarinus officinalis* while in *Psychotria brachyceras* the production of brachycerine (monoterpene-indole alkaloid) gets doubled in UV exposure Gregianini et al., 2003). It has been suggested that, for the development of natural drug manipulation, it is necessary to understand the plant's responses to this potent abiotic elicitor.

Phenolics are positively associated with various health-promoting characteristics, such as antioxidative or anticarcinogenic. These chemicals invade the redox status which helps in amending the oxidative stress (Kumari et al., 2009b). Phenolics also work as UV protectors as their absorption range is between 270 and 290 nm. Eichholz et al. (2011) in blueberries and Kumari et al. (2009b) in the sweet flag have reported the enhancement of phenolic compounds and antioxidant activity under UV-B stress. Flavonoids are the major UV-B-regulated phenylpropanoid derivatives, rapidly induced by UV-B exposure. Flavonoids stabilize and protect the lipid phase of the thylakoid membrane, and are quenchers of the excited triplet state of chlorophyll and singlet oxygen generated under excessive stress (Agrawal and Rathore, 2007). Psoralens are absorbed under the ultraviolet (200–320 nm) band, which is naturally occurring furocoumarins, as furocoumarins and coumarins are induced by UV radiations. This photosensitizing property of furocoumarins and coumarins is responsible for their medical importance in the treatment of many dermatological diseases, such as leprosy and psoriasis, leukoderma, vitiligo, and melanocytes formation, i.e., re-pigmentation.

In Abelmoschus esculentus (a good source of minerals, vitamins, antioxidants, and fibre), the absorbance of psoralens was affected by variation in the UV-B environment (Kumari et al., 2009a) and in Pastinaca sativa the absorbance of coumarins and furocoumarins increases with increment in UV-B radiations (Zangerl and Berenbaum, 1987). It has been reported that the UV-B radiation influences the diterpenes carnosol and carnosic acid in *Rosmarinus officinalis* (Luis et al., 2007). The main bioactive compound of *Glycyrrhiza uralensis* (glycyrrhizin) can inhibit carcinogenicity, possess anti-tumour activity with the help of the activation of hepatotoxic metabolites, and shows keen activity in controlling replication of human immunodeficiency virus (HIV-1) and severe acute respiratory syndrome (SARS) (Chan et al., 2003). With inadequate UV treatment, the triterpenoids saponin and glycyrrhizic acid increase while they decrease at high UV doses (Afreen et al., 2005). In *Catharanthus roseus*, the production of dimeric terpenoid and indole alkaloid were influenced by UV radiation (Hirata et al., 1993; Ouwerkerk et al., 1999). Ramani and Chelliah (2007) also reported that UV-B exposure of *Catharanthus roseus* cell suspension cultures can increase the production of catharanthine by affecting signalling events. Schreiner et al. (2009) studied the effect of moderate and short-term UV-B radiation on *Tropaeolum majus* and reported that the concentration of aromatic glucosinolate glucotropaeolin was approximately six times that of control plants and the production of glucosinolates was invoked. When *Arabidopsis* leaves were irradiated by sharp UV-B (1.55 Wm-2) for about one hour, over 12 hours, the glucosinolate content was decreased and the expression of glucosinolate biosynthetic genes gets inhibited (Wang et al., 2011). Exposure of UV-B radiation on *Hypericum perforatum* leads to enhancing the concentration of flavonoids and tannins in its leaves (Germ et al., 2010).

3.9   CONCLUSION

Medicinal plants play an important role in human life and their medicinal value depends on their growth and development. The growth processes (cell division, elongation, directional growth, branching) and photosynthesis of medicinal plants are affected by UV radiation. Lesser growth, biomass accumulation, carbon assimilation reduction, and photosynthetic pigment damage are among the common consequences of UV-B radiation exposure on medicinal plants, which may affect their productivity. The medicinal value of plants mainly depends on the concentration of secondary metabolites which can be affected by UV radiation. Usually, the concentration of secondary metabolites accelerated due to the exposure of UV-B. However, it has been noticed that the enhancement in secondary metabolites is mainly affected by the duration of radiation, and the dose of radiation at the various developmental stages of the plant tissue. As UV-B radiation can increase the production of secondary metabolites (terpenoids, alkaloids, and phenolics), it can, in a controlled manner, be used as a tool in the production of medical plant-based secondary metabolic products for use in different herbal products. However, at the same time, such radiation is known to harm plants' growth and developmental processes; therefore, the overall risk factors need to be carefully investigated. Additionally, more investigation is needed to understand the potential interactive effects between UV radiation and other climatic factors and thereafter knowledge obtained can be applied to medicinal plant cultivation.

REFERENCES

Afreen, F., S.M.A. Zobayed, and T. Kozai. 2005. Spectral quality and UV-B stress stimulate glycyrrhizin concentration of *Glycyrrhiza uralensis* in hydroponic and pot system. *Plant Physiology and Biochemistry* 43: 1074–1081.

Agrawal, S.B., and D. Rathore. 2007. Changes in oxidative stress defense system in wheat (*Triticum aestivum* L.) and mung bean (*Vigna radiata* L.) cultivars grown with and without mineral nutrients and irradiated by supplemental ultraviolet-B. *Environmental and Experimental Botany* 59: 21–33.

Agrawal, S.B., S. Singh, and M. Agrawal. 2009. UV-B induced changes in gene expression and antioxidants in plants. *Advances in Botanical Research* 52: 47–86.

Ahmad, M., N. Grancher, M. Heil, R.C. Black, B. Giovani, and P. Galland. 2002. Action spectrum for cryptochrome-dependent hypocotyl growth inhibition in *Arabidopsis*. *Plant Physiology* 129: 774–785.

Ambasht, N.K., and M. Agrawal. 2003. Interactive effects of ozone and ultraviolet-B singly and in combination on physiological and biochemical characteristics of soybean plants. *Journal of Plant Biology* 30: 37–45.

Ballare, C.L., P.W. Barnes, and R.E. Kendrick. 1991. Photo morphogenic effects of UV-B radiation on hypocotyl elongation in wild type and stable phytochrome deficient mutant of cucumber. *Physiologia Plantarum* 83: 652–658.

Barber, J., J. Nield, E.P. Morris, D. Zheleva, and B. Hankamer. 1997. The structure, function and dynamics of photosystem 2. *Physiologia Plantarum.* 100: 817–827.

Barnes, P.W., S.D. Flint, and M.M. Caldwell. 1990. Morphological responses of crop and weed species of different growth forms to ultraviolet-B radiation. *American Journal of Botany* 77: 1354–1360.

Beggs, C.J., E. Wellman, and A. Stolzer-Jehle. 1985. Isoflavonoid formation as an indicator of UV stress in beans (*Phaseolus vulgaris* L.) leaves. *Plant Physiology* 79: 630–635.

Caldwell, M.M., A.H. Teramura, and M. Tevini. 1989. The changing solar ultraviolet climate and the ecological consequences for higher plants. *Trends in Ecology and Evolution* 4: 363–367.

Caldwell, M.M., C.L. Ballare, J.F. Bornman, S.D. Flint, L.O. Bjorn, A.H. Teramura, G. Kulandaivelu, and M. Tevini. 2003. Terrestrial ecosystems, increased solar ultraviolet radiation and interactions with other climate change factors. *Photochemical and Photobiological Sciences* 2: 29–38.

Cen, Y.P., and J.F. Bornman. 1993. The effect of exposure to enhanced UV-B radiation on the penetration of monochromatic and polychromatic UV-B radiation in leaves of *Brassica napus*. *Physiologia Plantarum* 87: 249–255.

Cervilla, L.M., B. Blasco, J.J. Rìos, L. Romero, and J.M. Ruiz. 2007. Oxidative stress and antioxidants in tomato (*Solanum lycopersicum*) plants subjected to boron toxicity. *Annals of Botany* 100: 747–756.

Chan, H.T., C. Chan, and J.W. Ho. 2003. Inhibition of glycyrrhizic acid on aflatoxin B1-induced cytotoxicity in hepatoma cells. *Toxicology* 188: 211–217.

Christie, J.M., and W.R. Briggs. 2001. Blue light sensing in higher plants. *Journal of Biological Chemistry* 276: 11457–11460.

Cockell, C.S., and G. Horneck. 2001. The history of the UV radiation climate of the earth – theoretical and space-based observations. *Photochemistry and Photobiology* 73: 447–451.

Deutsch, A.W.V., C.D. Mitchell, C.E. Williams, K.K. Dutt, N.A. Silvestrov, B.J. Klement, I.K. Abukhalaf, and D.A.V. Deutsch. 2005. Polyamines protect against radiation-induced oxidative stress. *Gravitional and Space Biology Bulletin* 18: 109–110.

Eichholz, I., S.H. Keil, A. Keller, D. Ulrich, L.W. Kroh, and S. Rohn. 2011. UV-B-induced changes of volatile metabolites and phenolic compounds in blueberries (*Vaccinium corymbosum* L.). *Food Chemistry* 126: 60–64.

Eichholz, I., S. Rohn, A. Gamm, N. Beesk, W.B. Herppich, L.W. Kroh, C. Ulrichs, and S.H. Keil. 2012. UV-B-mediated flavonoid synthesis in white asparagus (*Asparagus officinalis* L.). *Food Research International* 48: 196–201.

Ellenberger, J., N. Siefen, P. Krefting, J.B.S. Lutum, D. Pfarr, M. Remmel, L. Schroder, and S.R. Schmittgen. 2020. Effect of UV radiation and salt stress on the accumulation of economically relevant secondary metabolites in bell pepper plants. *Agronomy* 10: 142.

Estell, R.E., E.L. Fredrickson, and D.K. James. 2016. Effect of light intensity and wavelength on concentration of plant secondary metabolites in the leaves of *Flourensia cernua*. *Biochemical Systematics and Ecology* 65: 108–114.

Fiscus, E.L., and F.L. Booker. 1995. Is increased UV-B a threat to crop photosynthesis and productivity? *Photosynthesis Research* 43: 81–92.

Flint, S.D., and M.M. Caldwell. 1984. Partial inhibition of in vitro pollen germination by simulated solar ultraviolet-B radiation. *Ecology* 65: 792–795.

Gaba, V., and M. Black. 1983. The control of cell growth by light. In *Encyclopedia of Plant Physiology*, New Series (H. Mohr and W. Shropshire Jr, eds). Springer, 358–400.

Germ, M., V. Stibilj, S. Gaberščik, and I. Kreft. 2010. Flavonoid, tannin and hypericin concentrations in the leaves of St John's wort (*Hypericum perforatum* L.) are affected by UV-B radiation levels. *Food Chemistry* 122: 471–474.

Ghasemi. S., H.H. Kumleh and M. Kordrostami. 2018. Changes in the expression of some genes involved in the biosynthesis of secondary metabolites in *Cuminum cyminum* L. under UV stress. Springer.

Gonzalez, R. 1996. Responses to ultraviolet-B radiation (280–315 nm) of pea (*Pisum sativum*) lines differing in leaf surface wax. *Physiologia Plantarum* 98: 852–860.

Gregianini, T.S., V.C. de Silveira, D.D. Porto, V.A. Kerber, A.T. Henriques, and A.G. Fett-Neto. 2003. The alkaloid brachycerine is induced by ultraviolet radiation and is a singlet oxygen quencher. *Photochemistry and Photobiology* 78: 470–474.

Hagen, S.F., G.I.A. Borge, G.B. Bengtsson, W. Bilger, A. Berge, K. Haffner, and K.A. Solhaug. 2007. Phenolic contents and other health and sensory related properties of apple fruit

(*Malus domestica* Borkh, cv Aroma): effect of postharvest UV-B irradiation. *Postharvest Biology and Technology* 45: 1–10.

Heijde, M., and R. Ulm. 2012. UV-B photoreceptor-mediated signalling in plants. *Trends in Plant Science* 17: 230–237.

Higashio, H., H. Hirokane, F. Sato, S. Tokuda, and A. Uragami. 2007. Enhancement of functional compounds in *Allium* vegetables with UV radiation. *Acta Horticulturae* 744: 357–361.

Hirata, K., M. Asada, E. Yatani, K. Miyamoto, and Y. Miura. 1993. Effects of near-ultraviolet light on alkaloid production in *Catharanthus roseus* plants. *Planta Medica* 59: 46–50.

Holst, B., and G. Williamson. 2004. A critical review of the bioavailability of glucosinolates and related compounds. *Natural Product Report* 21: 425–447.

Husen, A. 2013. Growth characteristics, biomass and chlorophyll fluorescence variation of Garhwal Himalaya's fodder and fuel wood tree species at the nursery stage. *Open Journal of Forestry* 3: 12–16.

Husen, A. 2010. Growth characteristics, physiological and metabolic responses of teak (*Tectona grandis* Linn. f.) clones differing in rejuvenation capacity subjected to drought stress. *Silvae Genetica* 59: 124–136.

Husen, A. 2009. Growth, chlorophyll fluorescence and biochemical markers in clonal ramets of shisham (*Dalbergia sissoo* Roxb.) at nursery stage. *New Forests* 38: 117–129.

Jin, Y.H., and D.L. Tao. 2000. PS II photoinhibition and O2 production. *Acta Botanica Sinica* 42: 10–14.

Jordan, B.R. 1996. The effects of ultraviolet-B radiation on plants: a molecular perspective. *Advances in Botanical Research* 22: 98–138.

Kakani, V.G., K.R. Reddy, D. Zhao, and K. Sailaja. 2003. Field crop responses to ultraviolet-B radiation: a review. *Agriculture and Forest Meteorology* 120: 191–218.

Kramer, G.F., H.A. Norman, D.T. Krizek, and R.M. Mirecki. 1991. Influence of UV-B radiation on polyamines, lipid peroxidation and membrane lipids in cucumber. *Phytochemistry* 30: 2101–2108.

Krizek, D.T., R.M. Mirecki, and S.J. Britz. 1997. Inhibitory effects of ambient levels of solar UV-A and UV-B radiation on growth of cucumber. *Physiologia Plantarum* 100: 886–893.

Kuhlmann, F., and C. Müller. 2009a. Development-dependent effects of UV radiation exposure on broccoli plants and interactions with herbivorous insects. *Environmental and Experimental Botany* 66: 61–68.

Kuhlmann, F., and C. Müller. 2009b. Independent responses to ultraviolet radiation and herbivore attack in broccoli. *Journal of Experimental Botany* 60: 3467–3475.

Kumari, R., S. Singh, and S.B. Agrawal. 2009a. Combined effects of psoralens and ultraviolet-B on growth, pigmentation and biochemical parameters of *Abelmoschus esculentus* L. *Ecotoxicology and Environmental Safety* 72: 1129–1136.

Kumari, R., S.B. Agrawal, S. Singh, and N.K. Dubey. 2009b. Supplemental ultraviolet-B induced changes in essential oil composition and total phenolics of *Acorus calamus* L. (Sweet flag). *Ecotoxicology and Environmental Safety* 72: 2013–2019.

Kumari, R., and S.B. Agrawal. 2010. Comparative analysis of essential oil composition and oil containing glands in *Ocimum sanctum* L. (Holy basil) under ambient and supplemental level of UV-B through Gas chromatography–mass spectrometry (GC-MS) and Scanning electron microscopy (SEM). *Acta Physiologiae Plantarum* 33: 1093–1110.

Liu, B., X. Liu, Y. Li, and S.J. Herbert. 2013. Effects of enhanced UV-B radiation on seed growth characteristics and yield components in soybean. *Field Crops Research* 154: 158–163.

Long, S.P., and S. Humphries. 1994. Photoinhibition of photosynthesis in nature. *Annual Review of Plant Physiology and Plant Molecular Biology* 45: 633–662.

Luis, J.C., R.M. Perez, and F.V. Gonzalez. 2007. UV-B radiation effects on foliar concentrations of rosmarinic and carnosic acids in rosemary plants. *Food Chemistry* 101: 1211–1215.

Ma, C.H., J.Z. Chu, X.F. Shi, C.Q. Liu, and X.Q. Yao. 2016. Effects of enhanced UV-B radiation on the nutritional and active ingredient contents during the floral development of medicinal *Chrysanthemum*. *Journal of Photochemistry and Photobiology B* 158: 228–234.

Madronich, S., R.L. McKenzie, L.O. Bjorn, and M.M. Caldwell. 1998. Changes in biologically active ultraviolet radiation reaching the Earth's surface. *Journal of Photochemistry and Photobiology B: Biology* 46: 5–19.

Mansfield, T.A., and P.H. Freer-Smith. 1984. The role of stomata in resistance mechanism. In *Gaseous Air Pollutants and Plant Metabolism* (M.J. Koziol and F.R. Whatley, eds). Butterworths, 131–146.

Mattoo, A.K., M.T. Giardi, A. Raskind, and M. Edelman. 1999. Dynamic metabolism of photosystem II reaction center proteins and pigments. *Physiologia Plantarum* 107: 454–461.

Miller, J.E., F.L. Booker, E.L. Fiscus, A.S. Heagle, W.A. Pursley, S.F. Vozzo, and W.W. Heck. 1994. Ultraviolet-B radiation and ozone effects on growth, yield, and photosynthesis of soybean. *Journal of Environmental Quality* 23: 83–91.

Naghiloo, S., A. Movafeghi, A. Delazar, H. Nazemiyeh, S. Asnaashari, and M.R. Dadpour. 2012. Ontogenic variation of volatiles and antioxidant activity in leaves of *Astragalus compactus* Lam. (Fabaceae). *EXCLI Journal* 11: 436–443.

Ouwerkerk, P.B.F., D. Hallard, R. Verpoorte, and J. Memelink. 1999. Identification of UV-B light-responsive regions in the promoter of the tryptophan decarboxylase gene from *Catharanthus roseus*. *Plant Molecular Biology* 4: 491–503.

Petropoulou, Y., O. Georgiou, G.K. Psaras, and Y. Manetas. 2001. Improved flower advertisement, pollinator rewards and seed yield by enhanced UV-B radiation in the Mediterranean annual Malcolmia maritima. *New Phytologist* 152: 85–90.

Rai, K., and S.B. Agrawal. 2017. Effects of UV-B radiation on morphological, physiological and biochemical aspects of plants: an overview. *International Journal of Science and Research* 61: 87–113.

Ramani, S., and J. Chelliah. 2007. UV-B-induced signalling events leading to enhanced-production of catharanthine in *Catharanthus roseus* cell suspension cultures. *BMC Plant Biology* 7: 61.

Rao, M.V., G. Paliyath, and D.P. Ormrod. 1996. Ultraviolet-B- and ozone induced biochemical changes in antioxidant enzymes of *Arabidopsis thaliana*. *Plant Physiology* 110: 125–136.

Ravindran, K.C., A. Indrajith, V. Balakrishnan, K. Venkatesan, and G. Kulandaivelu. 2010. Determination of defense mechanism in *Phaseolus trilobus* Ait. seedlings treated under UV-B radiation. *African Crop Science Journal* 16: 111–118.

Reifenrath, K., and C. Müller. 2007. Species-specific and leaf-age dependent effects of ultraviolet radiation on two Brassicaceae. *Phytochemistry.* 68: 875–885.

Rikabad, M.M., L. Pourakbar, S.S. Moghaddam, and J. Popović-Djordjević. 2019. Agro-biological, chemical and antioxidant properties of saffron (*Crocus sativus* L.) exposed to TiO_2 nanoparticles and ultraviolet-B stress. *Industrial Crops and Products* 137: 137–143.

Robberecht, R., and M.M. Caldwell. 1980. Leaf ultraviolet optical properties along a latitudinal gradient in the Arctic-alpine life zone. *Ecology* 61: 612–619.

Rozema, J. 2000. Effects of solar UV-B radiation on terrestrial biota. In *Causes and Environmental Implications of Increased UV-B Radiation* (Hester, R.E., and Harrison, R.M. eds). *Environmental Science and Technology* 14: 86–105.

Rozema, J., and L.O. Bjorn. 2002. Evolution of UV-B absorbing compounds in aquatic and terrestrial plants. *Journal of Photochemistry and Photobiology B: Biology* 66: 1–2.

Sakalauskaite, J., P. Viskelis, E. Dambrauskiene, S. Sakalauskiene, G. Samuoliene, A. Brazaityte, P. Duchovskis, and D. Urbonaviciene. 2013. The effects of different UV-B radiation intensities on morphological and biochemical characteristics in *Ocimum basilicum* L. *Journal of the Science of Food and Agriculture* 93: 1266–1271.

Sampson, B.J., and J.H. Cane. 1999. Impact of enhanced ultraviolet-B radiation on flower, pollen, and nectar production. *American Journal of Botany* 86: 108–114.

Schreiner, M., A. Krumbein, I. Mewis, C. Ulrichs, and S. Huyskens-Keil. 2009. Short-term and moderate UV-B radiation effects on secondary plant metabolism in different organs of nasturtium (*Tropaeolum majus* L). *Innovative Food Science and Emerging Technologies* 10: 93–96.

Sharma, S., S. Chatterjee, S. Kataria, J. Joshi, S. Datta, M. Vairale, and V. Veer. 2017. A Review on Responses of Plants to UV-B Radiation Related Stress In *UV-B Radiation: From Environmental Stressor to Regulator of Plant Growth* (V. Singh, S. Singh, S.M. Prasad, and P. Parihar, eds). Wiley-Blackwell, 75–97.

Shinomura, T., A. Nagatani, H. Hanzawa, M. Kubota, M. Watanabe, and M. Furuya. 1996. Action spectra for phytochrome A- and B-specific photoinduction of seed germination in *Arabidopsis thaliana*. *Proceedings of National Academy of Sciences of United States of America* 93: 8129–8133.

Singh, S.K. 2008. Developing screening tools for abiotic stresses using cowpea (*Vigna unguiculata* L.) Walp.) as a model crop. PhD dissertation, Mississippi State University, MS.

Singh, S., R. Kumari, M. Agrawal, and S.B. Agrawal. 2011. Modification in growth, biomass and yield of radish under supplemental UV-B at different NPK levels. *Ecotoxicology and Environmental Safety* 74: 897–903.

Spitaler, R., P.D. Schlorhaufer, E.P. Ellmerer, I. Merfort, S. Bortenschlager, H. Stuppner, and C. Zidorn. 2006. Altitudinal variation of secondary metabolite profiles in flowering heads of *Arnica montana* cv. ARBO. *Phytochemistry* 67: 409–417.

Sullivan, J.H., and A.H. Teramura. 1988. Effects of UV-B irradiation on seedling growth in the Pinaceae. *American Journal of Botany* 75: 225–230.

Sung, J.K., S.Y. Lee, J.H. Park, S.M. Lee, Y.H. Lee, D.H. Choil, T.W. Kim, and B.H. Song. 2008. Changes in carbohydrate, phenolics and polyamines of pepper plants under elevated UV-B radiation. *Korean Journal of Soil Science and Fertilizer* 41: 38–43.

Takshak, S., and S.B. Agrawal. 2014. Secondary metabolites and phenylpropanoid pathway enzymes as influenced under supplemental ultraviolet-B radiation in *Withania somnifera* Dunal, an indigenous medicinal plant. *Journal of Photochemistry and Photobiology B* 140: 332–343.

Tevini, M., J. Braun, and G. Fieser. 1991a. The protective function of the epidermal layer of rye seedlings against ultraviolet-B radiation. *Photochemistry and Photobiology* 53:329–333.

Tevini, M., U. Mark, G. Fieser, and M. Saile. 1991b. Effects of enhanced solar UV-B radiation on growth and function of selected crop plant seedlings. In *Photobiology* (E. Riklis, ed.). Plenum Press, 635–649.

Tevini, M., W. Iwanzik, and U. Thoma. 1981. Some effects of enhanced ultraviolet irradiation on the growth and composition of plants: barley, maize, kidney beans, radishes. *Planta* 153: 388–394.

Tevini, M., and D. Steinmuller. 1987. Influence of light, UV-B radiation, and herbicides on wax biosynthesis of cucumber seedlings. *Journal of Plant Physiology* 131: 111–121.

Tevini, M., and A.H. Teramura. 1989. UV-B effects on terrestrial plants. *Photochemistry and Photobiology* 50: 479–487.

Thélier, L.H., L. Crespel, J.L. Gourrierec, P. Morel, S. Sakr, and N. Leduc. 2015. Light signaling and plant responses to blue and UV radiations – Perspectives for applications in horticulture. *Environmental and Experimental Botany* 121: 22–38.

Torre, S., A.G. Roro, S. Bengtsson, L. Mortensen, K.A. Solhaug, H.R. Gislerød, and J.E. Olsen. 2012. Control of plant morphology by UV-B and UV-B-temperature interactions. *Acta Horticulturae* 956: 207–214.

Turunen, M., and K. Latola. 2005. UV-B radiation and acclimation in timberline plants. *Environmental pollution* 137: 390–403.

Vandenbussche, F., K. Tilbrook, A.C. Fierro, K. Marchal, D. Poelman, and D. Van-Der Straeten. 2014. Photoreceptor-mediated bending towards UV-B in *Arabidopsis. Molecular Plant* 7: 1041–1052.

Van Rensen, J.J.S., W.J. Vredenberg, and G.C. Rodrigues. 2007. Time sequence of the damage to the acceptor and donor sides of photosystem II by UV-B radiation as evaluated by chlorophyll a fluorescence. *Photosynthesis Research* 94: 291–297.

Vass, I., L. Sass, C. Spetea, A. Bakou, D.F. Ghanotakis, and V. Petrouleas. 1996. UV-B induced inhibition of photosystem II electron transport studied by EPR and chlorophyll fluorescence. Impairment of donor and acceptor side components. *Biochemistry* 35: 8964–8973.

Wang, Y., W.J. Xu, X.F. Yan, and Y. Wang. 2011. Glucosinolate content and related gene expression in response to enhanced UV-B radiation in *Arabidopsis. African Journal of Biotechnology* 10: 6481–6491.

Wargent, J.J., A. Taylor, and N.D. Paul. 2006. UV supplementation for growth and disease control. *Acta Horticulturae* 711: 333–338.

Yao, X., and Q. Liu. 2006. Changes in morphological, photosynthetic and physiological responses of Mono Maple seedlings to enhanced UV-B and to nitrogen addition. *Plant Growth Regulation* 50: 165–177.

Yin, R., M.Y. Skvortsova, S. Loubéry, and R. Ulm. 2016. COP1 is required for UV-B induced nuclear accumulation of UVR8 photoreceptor. *Proceedings of National Academy of Sciences of United States of America* 113.

Zangerl, A.R., and M.R. Berenbaum. 1987. Furanocoumarins in wild parsnip: effects of photosynthetically active radiation, ultraviolet light and nutrition. *Ecology* 68: 516–520.

Zhang, L., L.H. Allen Jr., M.M. Vaughan, B.A. Hauser, and K.J. Boote. 2014. Solar ultraviolet radiation exclusion increases soybean internode lengths and plant height. *Agricultural and Forest Meteorology* 184: 170–178.

Zhong, Z., S. Liu, W. Zhu, Y. Ou, H. Yamaguchi, K. Hitachi, K. Tsuchida, J. Tian, and S. Komatsu. 2019. Phosphoproteomics reveals the biosynthesis of secondary metabolites in *Catharanthus roseus* under ultraviolet-B radiation. *Journal of Proteome Research* 18: 3328–3341.

Zlatko S.Z., J.C.L. Fernando, and M. Kaimakanova. 2012. Plant physiological responses to UV-B radiation. *Emirates Journal of Food and Agriculture* 24: 481–501.

4 Impact of Sulphur Dioxide Deposition on Medicinal Plants' Growth and Production of Active Constituents

Shakeelur Rahman, Azamal Husen

CONTENTS

4.1 INTRODUCTION

Industries, thermal power plants, and automobiles are the sources of atmospheric pollution, creating various pollutants by reacting with temperature, light and humidity (Husen 1997, 2021). The chemicals present in the air travel far away from their origin point and affect the ecosystems and broad regional locations (Figure 4.1). Sulphur dioxide (SO$_2$) and nitrogen oxides (NO$_x$) are the major chemical precursors leading to acidic conditions in the atmosphere by reacting with oxygen, water, and sunlight. The wet (acidic rain, fog, snow) and dry (acidic gas, particles) parts of acid deposition in the atmosphere fall onto plants. Acid deposition is an example of the accumulation of acids and oxides from the atmosphere in the form of gas, rain, snow, or particulates that begin mainly from human activities (Abbasi et al. 2013). As a result, sulphuric acid (H$_2$SO$_4$) and nitric acids (HNO$_3$) are deposited on vegetation. About 60–70 per cent of the acid deposition found throughout the world is SO$_2$. Only 10 per cent of sulphur in the atmosphere is from natural sources, the remaining 90 per cent being anthropogenic (Ramadan 2004).

DOI: 10.1201/9781003178866-4

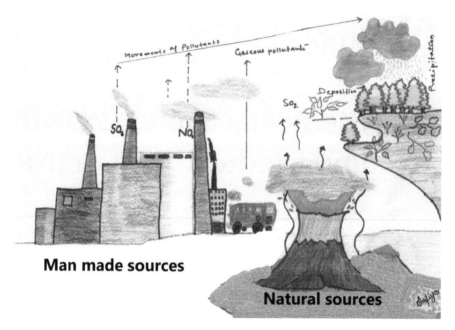

FIGURE 4.1 Sources of sulphur dioxide/nitrogen oxide pollutants and their movements towards vegetation.

Various investigations have been conducted to observe the effect of air pollution on different medicinal plants. These are mainly focused on plant growth, development, foliar morphology, biochemical changes, and various enzymatic activities (Iqbal et al. 1986, 2010a, 2010b; Ghouse et al. 1989; Husen 1997, 2021; Husen et al. 1999; Husen and Iqbal 2004). The effects of sulphuric acid have been observed on plants and many of them are very complex, including visible symptoms of injury such as chlorosis and necrosis, and invisible impacts like reduced photosynthesis, loss of nutrients from leaves, water imbalance, change in enzymatic activities, modification in pollen viability, physiology and structure (Bellani et al. 1997; Husen and Iqbal 2004). Medicinal plants are severely affected by sulphuric acid that comes into the leaf tissue through the cuticle and generates morphological and anatomical changes in plants (Amthor 1984). The major visible changes are changes in colour and leaf margin deformation as well as the emergence of brown, spotted necrotic lesions (Evans et al. 1997). It has been observed that acid rain usually leads to sluggish plant growth by developing abnormalities in plant metabolism, such as chlorophyll content, photosynthesis, production of reactive oxygen in species and nitrogen metabolism. Contrarily, exceptional cases of growth promotion have also been observed (Zhang et al. 2020). Sulphuric acid can develop alterations of the cuticle thickness (Cape 1986), obstruction of stomatal cells, cellular deformation, cell collapse in the mesophyll, loss of trichomes in the epidermis, and creation of scar tissue (Da Silva et al. 2005). The hydrophilic condition generated by leaf surface changes, mainly of the cuticle, could produce a hydrophilic situation that increases the water permeability, the leaf becomes more sensitive to water loss or to the entry of acidic substances into the cells (Baker and Hunt 1986). Moreover, it is probable that as water drops evaporate, acid concentration increases, ultimately becoming stronger acid that causes lesions in the epidermal cells (Evans and Curry 1979). When the acidic substances come into contact with the chloroplasts of the palisade chlorenchyma, the photosynthetic pigments undergo changes in their structure that lead to a decrease in the concentration of photosynthetic pigments and reduced efficiency in catching electrons (Hu et al. 2014). Consequently, the leaf chlorophyll content is a significant indicator of direct leaf damage and is related to plant growth and productivity that can be considerably reduced by sulphuric acid (Du et al. 2017).

In general, most of the scientific literature has been paying attention to plant species of agricultural importance. There has been much less focus on the impact of sulphuric acid or acid rain on important medicinal plant species to enhance the quality of herbal medicines. There is a need for time to evaluate the impact of SO_2 on the morphology, anatomy, physiology, and phytochemicals of medicinally important plant species. The plant species provide not only essential active constituents but also good indicators of adverse environmental conditions that can cause hazards to human and veterinary health.

4.2 SO_2 UPTAKE

Sulphur is one of the important elements for the growth and reproduction of plants. Sulphate anions are absorbed by plant roots from the soil while leaves suck up SO_2 from the atmosphere (Figure 4.2). SO_2 comes into the stomata of leaves by following the same diffusion mechanism as carbon dioxide. The atmospheric SO_2 is an alternative source when soil sulphur is insufficient. It has been observed that uptake of excess sulphur from the soil as well as the atmosphere can result in plant injury. It also interrupts the physiological or metabolic process that impacts growth and yield reduction. Increasing levels of atmospheric SO_2 can lead to increasing leaf sulphate contents (Dwivedi et al. 2008). In some plant species, exposure to SO_2 causes stomatal closure that guards the leaf against more entry of the pollutant but also limits photosynthesis. Agrawal and Singh (2000) observed the effects of SO_2 discharge of coal-fired power plants on the nutrient status of *Cassia siamea*. It was found there was an increase in the foliar sulphate content after the emerging of new leaves during the summer because of translocation of sulphur from woody plant parts. The dissolved SO_2 in plant cells forms bisulphate and sulphite ions, the toxic sulphite at low concentrations being metabolized by chloroplasts to non-toxic sulphate. Therefore, SO_2 as an air pollutant provides sulphur nutrients for the plant. The high concentration of these polluting gases in urban areas cannot be detoxified fast enough to avoid plant injury. In addition, many biotic and abiotic factors cause damage when plants are exposed to SO_2. Biotic factors include genetic make-up, plant nutrients, growth stages, pests, and diseases. Abiotic factors are temperature, light, moisture availability, relative humidity, and other pollutants. Water stress can also affect reduced stomatal opening that causes reduced SO_2 uptake. SO_2 uptake increases when air movement decreases the diffusive resistance of the leaf boundary. SO_2 dissolves in the aqueous phase of the cell wall to form sulphite or bisulphate, then undergoes an

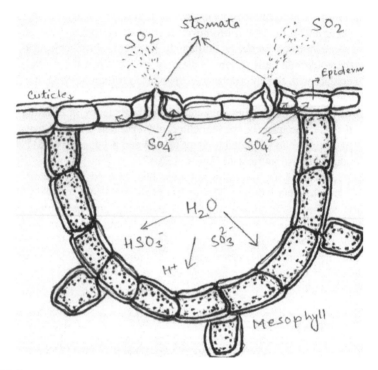

FIGURE 4.2   Uptake of sulphur dioxide in plant leaves from the atmosphere.

enzymatic change to sulphate. The sulphate is further carried into leaf cells where it is assimilated into organic molecules, including amino acids such as methionine and cysteine, then amalgamated into proteins (Rennenberg and Herschbach 1996). Qifu and Murray (1991) recommended that a reduced response to water stress when plants are exposed to SO_2 might be a result of increased stomatal resistance that may have reduced SO_2 uptake.

4.3   IMPACTS OF SO_2 ON GROWTH AND BIOMASS PRODUCTION

According to Pandey (2005), high levels of SO_2 can have adverse effects on plant growth such as reduced leaf area, but the effects of air pollution on leaf area are species-dependent and are based on a plant's ability to protect itself or adapt to its surrounding environment (Wuytack et al. 2011). The impact of SO_2 on the seedlings of *Pinus resinosa* in the cotyledon stage resulted in reduced chlorophyll content and dry weight (Constantinidou et al. 1976). The growth of seedlings of the SO_2 tolerant pin oak (*Quercus palustris*) and SO_2 sensitive white birch (*Betula papyrifera*) were compared. It was observed that the height of the birch seedlings was also considerably superior in the high SO_2 plots. The increase in the growth of birch was due to the supplementary sulphur supply. However, pin oak height was significantly less and dry weight was reduced in the high SO_2 compared to the low SO_2 treatment (Roberts

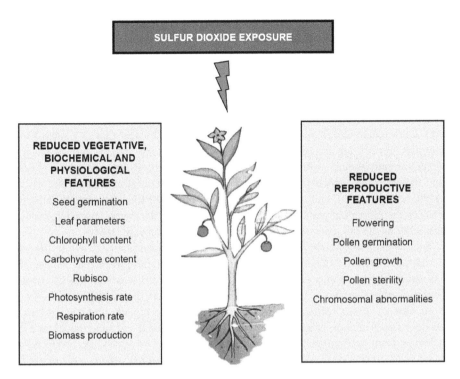

FIGURE 4.3 Impact of sulphur dioxide on medicinal plants.

1975). Loss of dry matter production was mainly an effect of loss of green leaf area in SO$_2$ exposed plants (Kropff 1990). Overall, the impact of SO$_2$ on medicinal as well as other plants is presented in Figure 4.3.

The ideal pH range for plant growth is a pH value between 5 and 8; out of these ranges in soils, plants face difficulties in germinating or growing. Plants cannot grow if pH is less than 3.7 (Larssen et al. 2006). Therefore, acid rain with pH<3 is measured as harmful for the germination and growth of various plant species. Morphological changes were observed in plants exposed to low pH acid rain, including leaf necrotic spot, chlorosis, and dehydration, suppression of leaf production, withering of leaves, leaf curling, leaf abscission, and even death of plants (Silva et al. 2006). All the parameters of plant growth, like plant height, leaf area, and fresh weight, get reduced significantly at all acidity levels with respect to the controlled one; the maximum decrease has been observed at pH 2.0 level (Evans et al. 1997; Liu et al. 2010). The relative growth rate and harvest index of plants became lowest at pH 2.0 and pH 3.0 (Iglesias et al. 1994).

Acid deposition impacts the proper growth and health of plants (Wyrwicha and Sklodowska 2006). Sulphuric acid has a negative impact on the growth and yield performance because of a decrease in photosynthesis which causes chlorosis, necrosis, and leaf abscission. The plant growth and yield reduce in acid rain due to foliar injury (Balasubramanian et al. 2007). Sulphuric acid rain decreases the essential nutrients of soil formed by magnesium (Mg), potassium (K), nitrogen (N), and calcium (Ca), etc. The non-toxic form of manganese (Mn), aluminium (Al), lead (Pb), mercury (Hg), and cadmium (Cd) in the soil become toxic in the presence of sulphuric acid rain that causes damage or death to plant parts (Curtis and Childs 2010; Liu et al. 2011). Red spruce trees become less cold tolerant and exhibit winter injury and may even die when calcium is decreased from the needles (Lazarus et al. 2006). Sulphuric acid rain alters the chemistry of the leaf surface; pollen germination, fertilization, and seed development and fruit formation are diminished. The acid rain kills useful soil microbes that liberate nutrients from decaying organic matter into the soil; as a result, plants lose nutrients. Even though some plants can stay alive in the aftermath of acid rain, they become weak and incapable of surviving in natural calamities like strong winds, heavy rainfall, and drought. It has been observed that the genetic variation between the individual plant's sensitivity to SO_2 may be the cause for the augmented variability as the SO_2 concentration is increased (Coleman et al. 1990).

The lowest plant height of *Aloe vera* was reported from plants grown in the soil of rich acid sulphate (Chowdhury et al. 2018). The highest fresh leaf gel weight was obtained from plants grown in calcareous soil, which was identical to the gel weight of plants grown in acid soil. Zaman et al. (2015) recorded similar findings in the case of *Stevia rebaudiana* as the tallest plant and highest numbers of leaves were counted from non-calcareous soil and the shortest plant and an identical number of leaves from acid sulphate soil. An identical number of leaves are produced in saline soils. Improved presentation of non-calcareous and calcareous soils might be due to their moderate pH, good soil texture, less water holding capacity, and higher nutrient contents compared to other soils. These findings were in good agreement with the results reported by Khanom et al. (2008) and Zaman et al. (2015) for *Stevia rebaudiana*. The plant grown in acid sulphate soil has the lowest fresh weight, which was identical to the fresh weight of the plant grown in peat soil. Better performance of acid soil might be due to having pH less than 7, strongly acid in reaction with the moderate status of organic matter and low moisture-holding capacity.

Long-term exposure to SO_2 of *Eucalyptus rudis* increased the height, average leaf area, and average dry weight of leaves. The increased leaf area and dry weight were attributed to the increased size of leaves, as there was no effect on the total number of leaves (Clarke and Murray 1990). Seed treatment of *Cassia fistula* in sulphuric acid for two minutes and then soaking in hot water at 100°C for six minutes was the best method for breaking the seed dormancy of *Cassia fistula* that resulted in an increased germination percentage to 96 per cent and gave a high quality of golden shower seedlings. It was also observed that the plant of the treated seed was tallest; roots were longest and the number of leaves was increased (Amira and Abbas 2013). Seeds and seedlings of *Cinnamomum camphora* when exposed to sulphuric acid resulted in inhibited seed germination, decrease in chlorophyll contents, foliar damage, and retardation of seedling growth (Fan and Li 1999). Further, Table 4.1 summarizes the effect of SO_2 exposure on some selected medicinal plants.

TABLE 4.1
Impact of SO$_2$ Exposure on Some Selected Medicinal Plants

Medicinal plant (Family)	Photographs	Common name	Part used	Medicinal uses	Impact of SO$_2$ exposure	References
Abies alba (Pinaceae)		Silver fir	Bud, bark, needle, resin	Leucorrhoea, joint pain, muscular pain, cold, and flu	Lower SO$_2$ concentration decreased pollen germination while total inhibition of germination at the highest SO$_2$ level.	Keller and Beda 1984
Aloe vera (Asphodelaceae)		Aloe vera	Leaf	Cosmetics, constipation, burns, and wounds	Lowest plant height grown in the soil of rich acid sulphate.	Chowdhury et al. 2018

(continued)

TABLE 4.1 (Continued)
Impact of SO$_2$ Exposure on Some Selected Medicinal Plants

Medicinal plant (Family)	Photographs	Common name	Part used	Medicinal uses	Impact of SO$_2$ exposure	References
Azadirachta indica (Meliaceae)		Neem, Indian lilac	Leaf, flower, fruit, bark	Cancer, hepatoprotective, skin diseases, and intestinal worms	Increased in oxidation of proteins, superoxide dismutase and peroxidase activities.	Rao and Dubey 1990
Betula papyrifera (Betulaceae)		White birch	Wood, bark	Skin problems, joint pain, burns, and dysentery	Superior seedlings in the high SO$_2$ exposure.	Roberts 1975

Cassia fistula (Fabaceae)		Golden shower	Bark, fruit	Dysentery, leprosy, jaundice, syphilis, and heart disease	Seed treatment in sulphuric acid increased germination percentage, high quality of seedlings, tallest plant, longest roots and increased number of leaves.	Amira and Abbas 2013
Cassia siamea (Fabaceae)		Siamese cassia	Leaf, bark	Constipation, diabetes, insomnia, asthma, and typhoid	Increase in nutrient status of the plant.	Agrawal and Singh 2000

(continued)

TABLE 4.1 (Continued)
Impact of SO_2 Exposure on Some Selected Medicinal Plants

Medicinal plant (Family)	Photographs	Common name	Part used	Medicinal uses	Impact of SO_2 exposure	References
Cinnamomum camphora (Lauraceae)		Camphor tree	Leaf, stem, bark	Cough, cold, skin diseases, joint, and muscle pain	Exposure to sulphuric acid inhibited seed germination, decrease in chlorophyll contents, foliar damage and retardation of seedling growth.	Fan and Li 1999
Eucalyptus rudis (Myrtaceae)		Flooded gum	Leaf	Antioxidant, cold, dry skin, joint pain, and gum bleeding	Long-term exposure to SO_2 increased the height, average leaf area, and average dry weight of leaves.	Clarke and Murray 1990

Ficus carica (Moraceae)		Common fig	Bark, fruit, leaf, root, latex	Laxative, paralysis, liver diseases, leprosy, aphrodisiac, piles, and leucoderma	Significant effects on the genotoxicities and contents phenolics and flavonoids.	Asmaa et al. 2018
Geranium carolinianum (Geraniaceae)		Geranium	Whole plant parts	Hepatitis B, and diarrhoeal infection	Damaging effect on pollen germination and growth.	Du Bay and Murdy 1983

(continued)

TABLE 4.1 (Continued)
Impact of SO$_2$ Exposure on Some Selected Medicinal Plants

Medicinal plant (Family)	Photographs	Common name	Part used	Medicinal uses	Impact of SO$_2$ exposure	References
Mangifera indica (Anacardiaceae)		Mango	Leaf, bark, flower, fruit, root	Anticancer, antidiabetic, antioxidant, antibacterial, antifungal, anthelmintic, gastroprotective, and immunomodulator	Increased in oxidation of proteins, superoxide dismutase and peroxidase activities.	Rao and Dubey 1990
Morus alba (Moreae)		White mulberry	Fruit, root, leaf	Elephantiasis, insomnia, premature ageing, influenza, eye infections, and nosebleeding	Photosynthesis processes are sensitive to SO4^{2-}	Kaiser et al., 1986

Petunia hybrida (Solanaceae)		Petunia	Leaf	Antibacterial	Exposure to SO_2 resulted in significant reductions in both pollen germination and tube length.	Linskens et al. 1985
Pinus banksiana (Pinaceae)		Jack pine	Needle, bark, resin	Kidney, bladder complaints, rheumatism, and respiratory diseases	Sugar decreases with increased SO_2 concentration.	Malhotra and Sarkar 1979

(*continued*)

TABLE 4.1 (Continued)
Impact of SO$_2$ Exposure on Some Selected Medicinal Plants

Medicinal plant (Family)	Photographs	Common name	Part used	Medicinal uses	Impact of SO$_2$ exposure	References
Pinus resinosa (Pinaceae)		Red pine	Needle, bark, resin	Coughs, colds, and influenza,	Reduced chlorophyll content and dry weight of the seedlings in cotyledon stage.	Constantinidou et al. 1976
Pongamia pinnata (Fabaceae)		Pongameoiltree	Flower, seed, bark, leaf, stem, root	Piles, ulcer, bronchitis, leprosy, arthritis, eczema, and gonorrhoea	Hazy shapes of stomata.	Kondo et al. 1980

Quercus palustris (Fagaceae)		Pin oak	Bark	Toothache, diarrhoea, and dysentery	Significant linear decrease in stomatal conductance, photosynthetic activity, water use efficiency and transpiration rate.	Kropff 1990; Lorenzini et al. 1995
Rumex obtusifolius (Polygonaceae)		Broad-leafed dock	Leaf, root	Blisters, burns, blood purifier, jaundice, and whooping cough	Small changes in peroxidase to next SO_2 exposure.	Horsman and Wellburn 1977

(continued)

TABLE 4.1 (Continued)
Impact of SO$_2$ Exposure on Some Selected Medicinal Plants

Medicinal plant (Family)	Photographs	Common name	Part used	Medicinal uses	Impact of SO$_2$ exposure	References
Solanum nigrum (Solanaceae)		Black nightshade	Leaf, fruit, root	Jaundice, asthma, whooping cough, and ulcer	Impact of SO$_2$ on pollen chromosomes.	Agrawal et al. 1995
Stevia rebaudiana (Asteraceae)		Candy leaf	Leaf	Strengthening heart, liver disease, and gastric hyperacidity	Shortest plant and identical number of leaves from acid sulphate soil.	Zaman et al. 2015

 *Syzygium cumini* (Myrtaceae)	Black plum	Seed, fruit, bark	Diabetes, dysentery, and mouth ulcers	Increased in oxidation of proteins, superoxide dismutase and peroxidase activities.	Rao and Dubey 1990
 *Tagetes erecta* (Asteraceae)	Marigold	Leaf, flower	Constipation, coughs, eczema, earaches, skin diseases, and conjunctivitis	SO_2 is able to induce apoptosis and reduce the viability of guard cells.	Unsworth and Black 1981

(continued)

TABLE 4.1 (Continued)
Impact of SO$_2$ Exposure on Some Selected Medicinal Plants

Medicinal plant (Family)	Photographs	Common name	Part used	Medicinal uses	Impact of SO$_2$ exposure	References
Terminalia arjuna (Combretaceae)		Arjuna	Bark	Heart diseases, diarrhoea, bronchitis, and urinary tract infections	Photosynthesis processes are sensitive to SO4^{2-}	Kaiser et al. 1986
Zizyphus mauritiana (Rhamnaceae)		Indian jujube	Fruit, seed	Muscular strength, anticancer, expectorant, palpitations, and insomnia	Increase in oxidation of proteins, superoxide dismutase and peroxidase activities.	Rao and Dubey 1990

4.4 IMPACTS OF SO_2 ON REPRODUCTIVE FEATURES

It was evaluated that tree reproduction could be affected in the field where SO_2 is higher (Keller and Beda 1984). Several experiments demonstrated that SO_2 has a damaging effect on pollen germination and growth. It was found in the whole plant studied with geranium (*Geranium carolinianum*) that, at the maximum relative humidity, SO_2 dissolved into the moist surface of the stigma and pollen grains, and affected the capability of the pollen to germinate (DuBay and Murdy1983). The impacts of lower SO_2 concentrations on the pollen of silver fir (*Abies alba*) resulted in the germination of pollen decreasing while total inhibition of germination was observed at the highest SO_2 level. SO_2 exposure to petunia (*Petunia hybrida*) resulted in significant reductions from the controls in both pollen germination and tube length because of reductions in pH of the medium (Linskens et al. 1985).

Agrawal et al. (1995) examined a nightshade (*Solanum nigrum*) complex, which demonstrated three natural cytotypes in diploid (*S. americanum*), tetraploid (*S. villosum*), and hexaploid (*S. nigrum*)] to evaluate the impact of SO_2 on pollen chromosomes. It was observed that, when pollen mother cells (PMC) were scrutinized, those meiotic chromosomal abnormalities were highest in diploid plants and least in hexaploid plants. Further, the length of exposure increased the abnormalities for all plant species. The chromosomal abnormalities impact the sterility of the pollen. It was concluded that the abnormalities might have resulted from free-radical splitting of phosphodiester linkages of DNA or from bisulphite combining with cytosine or uracil which may result in alteration of DNA or RNA functions. In the case of barrel medic (*Medicago truncatula*), the impact of SO_2 resulted in a reduction in flowering as SO_2 concentration increased (Murray and Wilson 1991).

4.5 VISIBLE INJURY

Symptoms of SO_2 injury can be explained as either acute or chronic. Short-term exposure to high SO_2 concentrations results in acute injury because of cellular death of plant parts. Chronic injury is the impact of long-term exposure to sub-lethal concentrations that cause alterations in cellular metabolism and may or may not exhibit visible injury (Legge et al. 1998). The visible injuries of SO_2 are plasmolysis, chlorosis, membrane damage, metabolic inhibition, necrosis, and death (Alseher et al. 1987). It has been observed that medicinal herbs are more sensitive to direct injury by sulphuric acid than trees (Heck et al. 1986). The surface of the leaf wax layer, chlorophyll, and other active constituents of the cells are destroyed because of sulphuric acid.

Necrotic streaks appear on parallel-veined plants between the veins near the leaf tip and extending toward the base with an increase in severity of injury. Conifers' needle tips show injury in the form of browning that extends toward the base with an increase in severity of injury. High levels of SO_2 exposure result in a dark lining on the brown necrotic part of the needle. Broadleaf plants in response to acute SO_2 exposure suffer marginal injury (Legge and Krupa 2002). It was observed that new leaves are more sensitive than older leaves to SO_2 (Horsman and Wellburn 1977). Most of the plants' leaf susceptible region impacted by SO_2 became flaccid and subsequently dried in most plant leaves (Taylor 1973). Brown leaf spots are chronic injury expressed from leaf margins, causing a decline in leaf area and plant yield (Kropff 1991). Acute injury symptoms emerge at the tip of leaves, causing necrotic and chlorotic streaks with irregular reddish pigmentation in monocot plants (Rai et al. 2011).

The leaf tissue injured by SO_2 exposed by microscopic assessment that the mesophyll cells affected with the chloroplasts become plasmolysed and the spongy cells are frequently more affected than the palisade. It has also been seen that the epidermis cells are also plasmolysed under severe conditions. The genotypes of timothy (*Phleum pratense*) from the location of a gas plant were examined for SO_2 sensitivity. The plants were measured resistant when they expressed no signs of visible injury and no significant decreases in shoot or root dry weights compared to control plants (Clapperton and Reid 1994). Hazy shapes of stomata found in *Pongamia pinnata* exposed to drain pollution resulted from lowering of pH in the cytoplasm of guard cells and thus change in the turgor relations of the stomata complex (Kondo et al. 1980) due to physiological injury within the leaf (Ashenden and Mansfield 1978). Rai and Mishra (2013) added that the plants growing along the roadsides have modified leaf surface characteristics, including stomata and epidermal cells, due to the stress of vehicular pollution.

4.6   IMPACTS ON METABOLIC ACTIVITIES

Several studies have been carried out on the impacts of SO_2 on metabolic processes in plants that can affect photosynthesis. The following observations have been made of one or more of the processes directly related to photosynthesis, including photochemical efficiency, chlorophyll content, stomatal conductance, carbon dioxide assimilation, carbohydrate metabolism, and dark respiration.

Exposure of plants to SO_2 resulted in the interruption of normal metabolic activity. Photosynthesis and respiration can be altered or enzymatic activity can be increased or decreased. The concentration of SO_2 is responsible for the scale of metabolic change. There is a possibility that metabolic disorders can happen without visible signs of injury. In green plants, chlorophyll is one of the main vital parts of energy production and its quantity is significantly affected by the environmental situation. Reduction in chlorophyll causes a decrease in plant productivity and then causes plants to display poor strength (Speeding and Thomas 1973). Bell and Mudd (1976) suggested that the resistance of plants to SO_2 might be related to the synthesis of deprivation of chlorophyll. However, plants balancing their chlorophyll under polluted circumstances are said to be tolerant (Singh and Verma 2007). The important tool to evaluate the effects of SO_2 is the measurement of chlorophyll as it has an important role in plant metabolism. Plant growth is directly affected if there is any decrease in chlorophyll content (Agbaire and Esiefarienhre 2009). Generally, the impact of increased SO_2 on plant growth is indicated by the net photosynthetic rate (Woo et al. 2007).

The gaseous SO_2 shattered chlorophyll-a more quickly than chlorophyll-b of *Spinacia oleracea* leaves. The corresponding increase in pheophytin-a was not accompanied by the loss of chlorophyll-a. Because of free-radical oxidation, SO_2 mostly destroys chlorophyll, and chlorophyll-a breakdown was repressed by superoxide dismutase. The superoxide dismutase activity is inhibited by SO_2 in the fumigated cells (Shimazaki et al. 1980). Sometimes plant growth may be promoted by a small dosage of SO_2, accepting the fact that sulphur is an essential constituent of biomolecules such as hormones, proteins, and chlorophyll. SO_2 helps in the proficient fixation of nitrogen by legumes (Li and Yi 2012).

Common effects of SO_2 observed are reduction in chlorophyll content, photosynthetic rate, biomass production, and respiration rate (Padhi et al. 2013). Plants are not competent in assimilating the SO_2 at the same rate as it is being absorbed when the exposure to SO_2 is at toxic levels (Stratigakos and Ormrod 1985) and this results in acute injury (Thomas 1961). The gaseous form of SO_2 received through stomata is metabolized by plants; as a result, the key effect of high SO_2 exposure is the modification of stomata and damaging of guard cells (Bytnerowicz et al. 2007). A study conducted by Kozioł and Whatley (2016) reported that a high concentration of SO_2 aborted the stomata. Once it has gone through the intercellular spaces, SO_2 dissolves in water on the cellular layers, resulting in substantial cellular K^+ loss. The epicuticular wax layer around the stomata is erased by sulphur ion that results in stomatal injury. SO_2 is able to induce apoptosis and reduce the viability of guard cells in *Tagetes erecta* in a controlled dose (Unsworth and Black 1981).

Plant metabolism is affected by SO_2 by either accepting or donating electrons that impact the cellular electron transport system in plants (Osmond and Avadhani 1970). Lipid peroxidation in plants is guided by exposure to a higher concentration of SO_2 (Li and Yi, 2012). The chlorophyll content of plants works as a biomarker for air pollution levels (Darrall and Jager 1984). Chlorophyll is converted into phaeophytin by SO_2 with the release of Mg^{2+} by reducing the pH and altering the spectral characters (Rao and Leblanc 1965). Some of the enzymes become inactivated because of breaking of the disulfide bridges (Cecil and Wake 1962) or, by inducing some conformational changes, some hydrolytic enzymes may even be activated (Malhotra and Hocking 1976). SO_2 also affects the photosynthetic CO_2 fixation; as a result, RuBisCO is inhibited by the higher concentration of SO_2 by competing with CO_2 or bicarbonate for the binding sites in RuBP carboxylase (Zeigler 1972). Different findings show that SO_2 also had an impact on respiration (Gheorghe and Ion 2011). Glutathione reductase (GR) enzymes engaged in the antioxidant defence system are shown to be expressed at elevated levels when exposed to SO_2. Lorenzini et al. (1995) reported the gas-exchange reaction of two-year-old oak seedlings (*Quercus pubescens*) exposed to SO_2 showed a significant linear decrease in stomatal conductance, photosynthetic activity, water use efficiency, and transpiration rate. However, total sulphur content and foliar starch increased linearly with increasing SO_2 concentration. It was observed that the genetic differences between the ecotypes of *Geranium carolinianum* affected physiological expression of different biochemical threshold levels of response to SO_2 (Taylor et al. 1986).

The effects of SO_2 exposure on enzymatic activities of broad-leafed dock (*Rumex obtusifolius*) was studied in plant samples collected from areas of high and low concentration of SO_2. None of the plants was reported with visible injury and fresh weights of leaf were unaffected. It was also observed that ribulose-diphosphate carboxylase (RuDPC) was reduced in the low concentration area. However, the plants from the high concentration area had significantly higher levels of peroxidase than the low concentration plants and observed comparatively small changes in peroxidase to next SO_2 exposure. It was summarized that the differences in enzyme activities between low and high concentration plants could be a result of an increase of sulphite tolerance in the high concentration plants and could be the reason for the continuation of RuDPC activity (Horsman and Wellburn 1977).

A study on the impact of SO_2 on antioxidant production and its role in protecting four tropical medicinally valuable tree species (*Zizyphus mauritiana, Syzygium cumini, Azadirachta indica, Mangifera indica*) reported that oxidation of proteins, superoxide dismutase activities, and peroxidase activities increased in all four species. The scale of the reaction varied with species and was correlated to the SO_2 concentration. It was noticed that the increased peroxidase and superoxide dismutase activities increase SO_2 tolerance under field conditions (Rao and Dubey 1990). Kaiser et al. (1986) reported that the photosynthesis process in *Terminalia arjuna* and *Morus alba* are sensitive to SO_4^{2-} because it acts as a competitive inhibitor of ribulose-1,5-biphosphate carboxylase and inhibits the photophosphorylation process.

The physiological activity of a plant is indicated by the concentration of soluble sugars that decide the sensitivity of plants toward SO_2 pollution. Reduction in soluble sugar content in plants at polluted sites can be endorsed to the increased rate of respiration and decreased CO_2 fixation because of chlorophyll decline (Wilkinson and Barnes 1973). The carbohydrate content is reduced in an SO^{2-} exposed plant, due to sulphite reaction with ketones and aldehydes of carbohydrates (Dugger and Ting 1970). SO_2 fumigation increased the content of the reducing sugars in *Pinus banksiana* due to a slowdown of polysaccharides rich in reducing sugars (Malhotra and Sarkar 1979).

4.7   IMPACTS ON ACTIVE CONSTITUENTS

Phytochemicals and biologically active compounds or constituents are formed naturally in plants (Table 4.2) which are useful for human health and nutrition (Hasler and Blumberg 1999). Phytochemicals protect plants against disease and injury as well as providing aroma, colour, and flavour to plants. Generally, phytochemicals or active plant constituents protect plant cells from pathogenic attack and environmental hazards, such as pollution, stress, drought, UV radiation, etc. A catalogue of more than 4,000 phytochemicals has been prepared and classified by chemical characteristics, protective function, and physical characteristics (Meagher and Thomson 1999). Phytochemicals are found in different parts of the plants including leaves, stems, roots, flowers, fruits, and seeds (Mamta et al. 2013). Phytochemicals have been classified as primary or secondary constituents, depending on their contributions to plant metabolism. Primary constituents are the amino acids, common sugars, proteins, purines, and pyrimidines of nucleic acids, chlorophylls, etc. Secondary constituents are the remaining plant chemicals, such as flavonoids, alkaloids, plant steroids, terpenes, lignans, curcumins, saponins, phenolics, glucosides, and flavonoids (Hahn 1998).

In recent decades, the use of herbal medicines and herb-based nutrition has been popularized to a large extent. More than 65–80 per cent of people in the world use herbal cosmetics and medicines as treatment options for many diseases and skincare (Awodele et al. 2013). High levels of toxic agents are found in herbal medicines when they are harvested from high pollution sites, such as metal mining sites, and

TABLE 4.2
Important Bioactive Constituents Available in Medicinal Plants

Classification	Main groups of compounds	Biological function
NSA (Non-starch polysaccharides)	Cellulose, hemicellulose, pectins, lignins, gums, and mucilages	Binding toxins, bile acids water holding capacity, and hindrance in nutrient absorption
Antibacterial & Antifungal	Alkaloids, terpenoids, and phenolics	Decrease the risk of fungal infection, and inhibitors of micro-organisms
Antioxidants	Flavonoids, polyphenolic compounds, ascorbic acid, carotenoids, and tocopherols	Obstacle in lipid peroxidation, and quench oxygen free radical
Anticancer	Flavonoids, carotenoids, polyphenols, and curcumin	Inhibited development of lung cancer, inhibitors of tumour, and anti-metastatic activity
Detoxifying Agents	Carotenoids, flavones, retinoids, reductive acids, tocopherols, phenols, indoles, aromatic isothiocyanates, coumarins, phytosterols, and cyanates	Inducers of drug binding of carcinogens, inhibitors of tumourogenesis, and inhibitors of procarcinogen activation
Other	Terpenoids, alkaloids, biogenic amines, and volatile flavour compounds	Cancer chemoprevention, neuropharmacological agents, and antioxidants

sites close to industrial areas, roadways, and oil refineries (Vasudevan et al. 2009). The exposure of human beings to high levels of toxic agents constitutes a significant health risk (Beyersmann and Hartwig 2008). Few studies on plants grown in polluted areas have been done to determine the toxic effects of pollutants (Menezes 2013). It was reported in a study that SO_2 pollution has significant effects on the genotoxicities of *Ficus carica* and *Schinusmolle*, and on their contents of total phenolics and flavonoids (Asmaa et al. 2018). An experiment was conducted on jack pine (*Pinus banksiana*) seedlings with gaseous SO_2 found in a transfer between the reducing and non-reducing sugars. Sugar decreases with increased SO_2 concentration. Increasing concentrations of gaseous SO_2 increased the amino acid content of the whole tissues. Gas-liquid chromatographic analyses of the amino acids indicated SO_2 impact in an increase in the content of valine, alanine, glycine, leucine, isoleucine, threonine, lysine, arginine, and aspartic acid tyrosine, and reduction in the content of glutamic acid and serine (Malhotra and Sarkar 1979). SO_2 exposure also decreases protein content and deactivates the enzymes involved in protein synthesis (Nandi et al. 1990; Agrawal and Deepak 2003; Hamid and Jawaid 2009).

4.8 CONCLUSION

Sulphur is one of the important elements needed by plants for proper growth and reproduction. Sulphate anions are absorbed by plant roots from the soil; plant leaves also absorb SO_2 from the atmosphere. Atmospheric SO_2 is used as an alternate source of sulphur when there is a sulphur deficiency in the soil. It has been observed that uptake of excess sulphur from the atmosphere or the soil, or both, can result in injury to plants or their parts. Overall, SO_2 exposure has been reported to alter important morpho-physiological (biomass, leaf traits, gas-exchange characteristics, etc), anatomical, biochemical/metabolic, and enzymatic activities of some medicinal plants. Further, the incidence of frequent acid rains has increased because of man-made activities and natural causes. Power plants, industries, and automobiles use coal, oil, and gas to generate power, which produces SO_2 and NOx that create acid rain. Acid rain is one of the sources of sulphuric acid that affects the growth, reproduction, and production of chemical constituents of plants. The visible damage to plant parts can be seen at pH 2.5 and anatomical and biochemical injuries are detected at pH 3.8. Plant growth and reproduction are mainly affected by necrosis, chlorosis, and disorders of metabolic or physiological processes. However, further study on the impact of SO_2 deposition on active chemical constituents of medicinal plants and their mechanisms in exhibiting certain biological activities is needed to understand the complete phytochemical report and the complex pharmacological effects. In addition, further study on the toxicity of phytochemicals of the plants obtained from polluted sites should be carried out to ensure their safety to be used as sources of drugs. There is also a need for time to reduce acid rain and other sources of sulphuric acid in the atmosphere to get safe plant produce and activate their constituents.

REFERENCES

Abbasi, T., P. Poornima, and T. Kannadasan. 2013. Acid rain: past, present, and future. *International Journal of Environmental Engineering* 5: 229–272.

Agbaire, P.O., and E.J. Esiefarienrhe. 2009. Air pollution tolerance indices of some plants around Otorogungas plant in Delta state, Niger. *Journal of Applied Sciences and Environmental Management* 13(1): 11–14.

Agrawal. M., and S.S. Deepak. 2003. Physiological and biochemical responses of two cultivars of wheat to elevated levels of CO_2 and SO_2, singly and in combination. *Environmental Pollution* 121: 189–197.

Agrawal, S., R. Singh, and A. Sahi.1995. Cytogenetic effect of sulphur dioxide on cytotypes of *Solanum nigrum* complex. *CytoBioScience* 83: 41–47.

Agrawal, M., and J. Singh.2000. Impact of coal power plant emission on the foliar elemental concentrations in plants in a low rainfall tropical region. *Environmental Monitoring and Assessment* 60: 261–282.

Alseher, R., M. Franz, and C.W. Geske. 1987. Sulphur dioxide and chloroplast metabolism. In Saunders J.A., L. Kosak-Channing, and E.E. Conn (eds). Phytochemical Effects of Environmental Compounds, 1–28. New York, Plenum Publishing Corporation.

Amira, S.S., and M.S. Abbas. 2013. Effects of sulfuric acid and hot water pre-treatments on seed germination and seedlings growth of *Cassia fistula* L. *American-Eurasian Journal of Agricultural and Environmental Sciences* 13(1): 7–15.

Amthor, J. 1984. Does acid rain influence plant growth? Some comments and observations. *Environmental Science and Pollution Research* 36: 1–6.

Ashenden, T.W., and T.A. Mansfield. 1978. Extreme pollution sensitivity of grasses when SO2and NO_2 are present in the atmosphere together. *Nature* 273: 142–143.

Asmaa, M.R., F.R. Naglaa, E.R. Abd, M. Donia, and A.M. Ganaie. 2018. Comparative studies on the effect of environmental pollution on secondary metabolite contents and genotoxicity of two plants in Asir area, Saudi Arabia. *Tropical Journal of Pharmaceutical Research* 17(8): 1599–1605.

Awodele, O., T.D. Popoola, and K.C. Amadi. 2013. Traditional medicinal plants in Nigeria remedies or risks. *Journal of Ethnopharmacology* 150: 614–618.

Baker, E.A., and G.M. Hunt. 1986. Erosion of waxes from leaf surfaces by simulated rain. *New Phytologist* 102: 161–173.

Balasubramanian, G., C. Udayasoorian, and P.C. Prabu. 2007. Effects of short term exposure of simulated acid rain on the growth of *Acacia nilotica. Journal of Tropical Forest Science* 19(4): 198–206.

Bell, J.N.B., and C.H. Mudd.1976. Sulphur dioxide resistance in plants: a case study of *Lolium perenne.* In Mansfield, T.A. (ed.). Effects of Air Pollution on Plants, 82–103. Cambridge: Cambridge University Press.

Bellani, L.M., C. Rinallo, S. Muccifora, and P. Gori. 1997. Effects of simulated acid rain on pollen physiology and ultrastructure in the apple. *Environmental Pollution* 95: 357–362.

Beyersmann, D., and A. Hartwig. 2008. Carcinogenic metal compounds: recent insight into molecular and cellular mechanisms. *Archives of Toxicology* 82: 493–512.

Bytnerowicz, A., K. Omasa, and E. Paoletti. 2007. Integrated effects of air pollution and climate change on forests: a northern hemisphere perspective. *Environmental Pollution* 147: 438–445.

Cape, J.N. 1986. Effects of air pollution on the chemistry of surface waxes of Scots Pine. *Water, Air, and Soil Pollution* 31: 393–399.

Cecil, R., and R.G. Wake. 1962. The reactions of inter-and intra-chain disulphide bonds in proteins with sulphite. *Biochemical Journal* 82(3): 401.

Chowdhury, T., M. Rahman, K. Nahar, M.Chowdhury, and M. Khan. 2018. Growth and yield performance of *Aloe vera* grown in different soil types of Bangladesh. *Journal of the Bangladesh Agricultural University* 16: 448–456.

Clapperton, M.J., and D.M. Reid. 1994. Effects of sulphur dioxide (SO2) on growth and flowering of SO$_2$ tolerant and non-tolerant genotypes of *Phleum pratense. Environmental Pollution* 86: 251–258.

Clarke, K., and F. Murray. 1990. Stimulatory effects of SO2 on growth of *Eucalyptus rudis* Endl. *New Phytologist* 115: 633–637.

Coleman, J.S., H.A. Mooney, and W.E. Winner. 1990. Anthropogenic stress and natural selection: variability in radish biomass accumulation increases with increasing SO2 dose. *Canadian Journal of Botany* 68: 102–106.

Constantinidou, H., T.T. Kozlowski, and K. Jensen. 1976. Effects of sulphur dioxide on *Pinus resinosa* seedlings in the cotyledon stage. *Journal of Environmental Quality* 5(2): 141–144.

Curtis, N., and C. Childs. 2010. Chemicals and soils. *Plant Nutrition* II: 2–3.

Da Silva, L., A. Alves, E. Da Silva, and M.E. Oliva. 2005. Effects of simulated acid rain on the growth of five Brazilian tree species and anatomy of the most sensitive species (*Joannesia princeps*). *Australian Journal of Botany* 53: 789–796.

Darrall, N.M., and H.J. Jager. 1984. Biochemical diagnostic tests for the effects of air pollution on plants. In Kozioł, M.J., and F. R. Whatley (eds). Gaseous Air Pollutants and Plant Metabolism, 333–350. London: Butterworth.

DuBay, D.T., and W.H. Murdy. 1983. Direct adverse effect of SO2 on seed set in *Geranium carolinianum* L.: a consequence of reduced pollen germination on the stigma. *Botanical Gazette* 144: 376–381.

Du, E., D. Dong, X. Zeng, Z. Sun, X. Jiang, and W. De Vries. 2017. Direct effect of acid rain on leaf chlorophyll content of terrestrial plants in China. *Science of the Total Environment* 605: 764–769.

Dugger, W.M., and I.P. Ting. 1970. Air pollution oxidants – their effects on metabolic processes in plants. *Annual Review of Plant Biology* 21: 215–234.

Dwivedi, A.K., B.D. Tripathi, and Shashi. 2008. Effect of ambient air sulphur dioxide on sulphate accumulation in plants. *Journal of Environmental Biology* 29 (3): 377–379.

Evans, L.S., N.F. Gmor, and F. Dacosta. 1997. Leaf surface and histological perturbations of leaves of *Phaseolus vulgaris* and *Helianthus annuus* after exposure to simulated acid rain. *American Journal of Botany* 4: 304–313.

Evans, L.S., and T.M. Curry. 1979. Differential responses of plant foliage to simulated acid rain. *American Journal of Botany* 66: 953–962.

Fan, H.B., and C.R. Li. 1999. Effects of simulated acid rain on seedling emergence and growth of five broad-leaved species. *Journal of Forestry Research* 10: 83–86.

Ghouse A.K.M., Mahmooduzzafar, M. Iqbal, P. Dastgiri.1989. Effect of coal-smoke pollution on the stem anatomy of *Cajanus cajan* L. Mill. *Indian Journal of Applied and Pure Biology* 4: 147–149.

Gheorghe, I.F., and B. Ion. 2011. The effects of air pollutants on vegetation and the role of vegetation in reducing atmospheric pollution. In The Impact of Air Pollution on Health, Economy, Environment and Agricultural Sources, 256–259. Shanghai: Intech Publisher.

Hahn, N.I. 1998. Are phytoestrogens nature's cure for what ails us? A look at the research. *Journal of the American Dietetic Association* 98: 974–976.

Hamid, N., and F. Jawaid. 2009. Effect of short-term exposure of two different concentrations of sulphur dioxide and nitrogen dioxide mixture on some biochemical parameters of soybean (*Glycine max* L. Merr.) *Pakistan Journal of Botany* 41: 2223–2228.

Hasler, C.M., and J.B. Blumberg. 1999. Symposium on Phytochemicals: biochemistry and Physiology. *Journal of Nutrition* 129: 756S–757S.

Heck, W.W., A.S. Heagle, and D.S. Shriner. 1986. Effects on vegetation: native, crops, forests. In A.C. Stern (ed.). Air Pollution Supplement to Air Pollutants, Their Transformation, Transport and Effects, 3rd Ed., Vol. VI: 248–333. New York: Academic Press.

Horsman, D.C., and A.R. Wellburn. 1977. Effect of SO2 polluted air upon enzyme activity in plants originating from areas with different annual mean atmospheric SO_2 concentrations. *Environmental Pollution* 13(1): 33–39.

Hu, H., L. Wang, C. Liao, C. Fan, Q. Zhou, and X. Huang. 2014. Combined effects of lead and acid rain on photosynthesis in soybean seedlings. *Biological Trace Element Research* 161: 136–142.

Husen, A. 2021. Morpho-anatomical, physiological, biochemical and molecular responses of plants to air pollution. In Husen, A. (ed). Harsh Environment and Plant Resilience, 203–234. Cham: Springer. https://doi.org/10.1007/978-3-030-65912-7_9

Husen, A., and M. Iqbal. 2004. Growth performance of *Datura innoxia* Mill. under the stress of coal-smoke pollution. *Annals of Forestry* 12 : 182–190.

Husen, A., S.T. Ali, Mahmooduzzafar, M. Iqbal. 1999. Structural, functional and biochemical responses of *Datura innoxia* Mill. to coal-smoke pollution. *Proceedings of Academy of Environmental Biology* l8: 61–72.

Husen, A. 1997. Impact of air pollution on the growth and development of *Datura innoxia* Mill. M.Sc. dissertation, Jamia Hamdard, New Delhi, India.

Iglesias, C.A., F. Calle, C. Hershey, G. Jaramillo, and E. Mesa. 1994. Sensitivity of cassava (*Manihot esculetna* Crantz) clones to environmental changes. *Field Crops Research* 36: 213–220.

Iqbal, M., Mahmooduzzafar, F. Nighat, and P.R. Khan. 2010a. Photosynthetic, metabolic and growth responses of *Triumfetta rhomboidea* to coal-smoke pollution at different stages of plant ontogeny. *Journal of Plant Interactions* 5: 11–19.

Iqbal, M., J. Jura-Morawiec, W. Włoch, and Mahmooduzzafar.2010b. Foliar characteristics, cambial activity and wood formation in *Azadirachta indica* A. Juss. as affected by coal-smoke pollution. *Flora: Morphology, Distribution, Functional Ecology of Plants* 205: 61–71.

Iqbal M, Z. Ahmad, I. Kabeer, Mahmooduzzafar, and Kaleemullah. 1986. Stem anatomy of *Datura inoxia* Mill. in relation to coal-smoke pollution. *Journal of Scientific Research* 8: 103–105.

Kaiser, W.M., M.G. Schroppel, and E. Wirth. 1986. Enzyme activities in an artificial stroma medium: an experimental model for studying effects of dehydration on photosynthesis. *Planta* 161: 292–299.

Keller, T., and H. Beda.1984. Effects of SO_2 on the germination of conifer pollen. *Environmental Pollution Series* A33: 237–243.

Khanom, S., M.A.H. Chowdhury, M.T. Islam, and B.K. Saha. 2008. Influence of organic and inorganic fertilizers on mineral nutrition of *Stevia* in different types of soil. *Journal of the Bangladesh Agricultural University* 6: 27–32.

Kondo, N., I. Maruta, and K. Sugahara. 1980. Research report from the National Institute for Environmental Studies, Yatabe, Japan: 127–136.

Kozioł, M.J., and F.R. Whatley. 2016. Gaseous air pollutants and plant metabolism. London: Butterworth-Heinemann.

Kropff, M.J. 1991. Long term effects of SO_2 on plants, metabolism and regulation of intracellular pH. *Plant Soil* 131: 235–245.

Kropff, M.J. 1990. The effects of long-term open-air fumigation with SO_2 on a field crop of broad bean (*Vicia faba* L.). Quantitative analysis of damage components. *New Phytologist* 115: 357–365.

Larssen, T., E. Lydersen, D. Tang, and Y. He. 2006. Acid rain in China. Environmental *Science and technology. American Chemical Society* 40(2): 418–425.

Lazarus, B.E., P.G. Schaberg, G.J. Hawley, and D.H. De Hayes. 2006. Landscape-scale spatial patterns of winter injury to red spruce foliage in a year of heavy region-wide injury. *Canadian Journal of Forest Research* 36(1): 142–152.

Legge, A.H., and S.V. Krupa. 2002. Effects of sulphur dioxide. *Air Pollution and Plant Life* 2: 135–162.

Legge, A.H., A. Jager, and S.V. Krupa. 1998. Sulphur dioxide. In R.B. Flagler (ed.). Recognition of Air Pollution Injury to Vegetation. A Pictorial Atlas. Second edition, 3–42. Pittsburgh, Pennsylvania: Air and Waste Management Association.

Linskens, H.F., Y. van Megen, P.L. Pfahler, and M. Wilcox. 1985. Sulfur dioxide effects on petunia pollen germination and seed set. *Bulletin of Environmental Contamination and Toxicology* 34: 691–695.

Li, L., and H. Yi. 2012. Plant physiology and biochemistry effect of sulfur dioxide on ROS production, gene expression and antioxidant enzyme activity in *Arabidopsis* plants. Plant Physiology and Biochemistry 58: 46–53.

Liu, X., D. Lei, J. Mo, E. Du, J. Shen, X. Lu, Y. Zhang, X. Zhou, C. He, and F. Zhang. 2011. Nitrogen deposition and its ecological impact in China: an Overview. *Environmental Pollution* 159(10): 2251–2264.

Liu, K.H., R.S. Mansell, and R.D. Rhue. 2010. Cation removal during application of acid solution into air dry soil columns. *Soil Science Society of America Journal* 4: 1747–1753.

Lorenzini, G., A. Panicucci, and C. Nali. 1995. A gas-exchange study of the differential response of *Quercus* species to long-term fumigations with a gradient of sulphur dioxide. *Water Air and Soil Pollution* 85: 1257–1262.

Malhotra, S.S., and S.K. Sarkar. 1979. Effects of sulphur dioxide on sugar and free amino-acid content of pine seedlings. *Physiologia Plantarum* 47: 223–228.

Malhotra, S.S., and D. Hocking. 1976. Biochemical and cytological effects of sulphur dioxide on plant metabolism. *New Phytologist* 76: 227–237.

Mamta, S., J. Saxena, R. Nema, D. Singh, and A. Gupta. 2013. Phytochemistry of medicinal plants. *Journal of Pharmacognosy and Phytochemistry* 1: 168–182.

Meagher, E., and C. Thomson. 1999. Vitamin and mineral therapy. In G. Morrison and L. Hark (eds). Medical Nutrition and Disease, 2nd ed, 33–58. Malden, Massachusetts: Blackwell Science.

Menezes, A.P.S., J. Da Silva, and J. Rollof. 2013. *Baccharis trimera* (Less) DC as genotoxicity indicator of exposure to coal and emissions from a thermal power plant. *Archives of Environmental Contamination and Toxicology* 65: 434–441.

Murray, F., and S. Wilson. 1991. The effects of SO_2 on the final growth of *Medicago truncatula*. *Environmental and Experimental Botany* 31: 319–325.

Nandi, P.K., M. Agrawal, S.B. Agrawal, and D.N. Rao. 1990. Physiological responses of *Vicia faba* plants to sulfur dioxide. *Ecotoxicology and Environmental Safety* 19: 64–71.

Osmond, C.B., and P.N. Avadhani. 1970. Inhibition of the β-carboxylation pathway of CO_2 fixation by bisulfite compounds. *Plant Physiology* 45(2): 228.

Padhi, S.K., M. Dash, and S.C. Swain. 2013. Effect of sulphur dioxide on growth, chlorophyll and sulphur contents of tomato (*Solanum lycopersicum* L.). *European Scientific Journal* 9(36).

Pandey, J. 2005. Evaluation of air pollution phytotoxicity downwind of a phosphate fertilizer factory in India. *Environmental Monitoring and Assessment* 100(1–3): 249–266.

Qifu, M., and F. Murray. 1991. Responses of potato plants to sulphur dioxide, water stress and their combination. *New Phytologist* 118: 101–109.

Rai, P., R.M. Mishra. 2013. Effect of urban air pollution on epidermal traits of road side tree species, *Pongamia pinnata* (L.) Merr. *Journal of Environmental Science, Toxicology and Food Technology* 2(6): 2319–2402.

Rai, R., M. Rajput, M. Agrawal, and S.B. Agrawal. 2011. Gaseous air pollutants: a review on current and future trends of emissions and impact on agriculture. *Journal of Scientific Research* 55: 77–102.

Ramadan, A.E.K. 2004. Acid deposition phenomena . *TESCE* 30(2): 1369–1389.

Rao, D.N., F. Leblanc. 1965. Effects of sulfur dioxide on the lichen alga, with special reference to chlorophyll. *Bryologist* 69: 69–75.

Rao, M.V., and P.S. Dubey. 1990. Explanations for the differential response of certain tropical tree species to SO_2 under field conditions. *Water, Air, and Soil Pollution* 51: 297–305.

Rennenberg, H., and C. Herschbach. 1996. Responses of plants to atmospheric sulphur. In M. Yunus and M. Iqbal (eds). Plant Response to Air Pollution, 285–293. Chichester: John Wiley & Sons.

Roberts, B.R. 1975. The influence of sulphur dioxide concentration on growth of potted white birch and pin oak seedlings in the field. *Journal of American Society for Horticultural Science* 100(6): 640–642.

Shimazaki, K., T. Sakaki, N. Kondo, and K. Sugahara.1980. Active oxygen participation in chlorophyll destruction and lipid peroxidation in SO_2 fumigated leaves of spinach. *Plant and Cell Physiology* 21: 1193–1204.

Silva, L.C., M.A. Oliva, A.A. Azevedo, and J.M. Araujo. 2006. Responses of resting plant species to pollution from an iron pelletization factory. *Water Air and Soil Pollution* 175(1–4): 241–256.

Singh, S.N., and A. Verma. 2007. Phytoremediation of air pollutants: a review. In: S.N. Singh, and R.D. Tripathi (eds). Environmental Bioremediation Technology, 293–314. Berlin/ Heidelberg: Springer.

Speeding, D.J., and W.J. Thomas.1973. Effect of sulphur dioxide on the metabolism of glycolic acid by barley (*Hardeum vulgare*) leaves. *Australian Journal of Biological Sciences* 6: 281–286.

Stratigakos, A., and D.P. Ormorod. 1985. Response of tomato to sulphur nutrition and SO_2. *Water, Air, and Soil Pollution* 247(1): 19–26.

Taylor, G.E.J., D.T. Tingey, and C.A. Gunderson.1986. Photosynthesis, carbon allocation and growth of sulfur dioxide ecotypes of *Geranium carolinianum* L. *Oecologia* 68: 350–357.

Taylor, O.C. 1973. Acute responses of plants to aerial pollutants. *Advances in Chemistry Series (US)* 122: 9–20.

Thomas, M.D. (1961). Effects of air pollution on plants. *Monograph Series World Health Organization* 46: 233–277.

Unsworth, M.H., and V.J. Black. 1981. Stomatal response to pollutants. In P.E. Jarvis, and T.A. Mansfield (eds). Stomatal Physiology, 187–203. Cambridge: Cambridge University Press.

Vasudevan, D.T., K.R. Dinesh, and S. Gopalakrishnan. 2009. Occurrence of high levels of cadmium, mercury and lead in medicinal plants of India. *Pharma Magazine* 5: 15–18.

Wilkinson, T.G., and R.L. Barnes. 1973. Effect of ozone on CO_2 fixation patterns in pine. *Canadian Journal of Botany* 9: 1573–1578.

Woo, S.Y., D.K. Lee, and Y.K. Lee. 2007. Net photosynthetic rate, ascorbate peroxidase and glutathione reductase activities of *Erythrina orientalis* in polluted and non-polluted areas. *Photosynthetica* 45(2): 293–295.

Wyrwicha, A., and M. Sklodowska. 2006. Influence of repeated acid rain treatment on antioxidative enzyme activities and on lipid peroxidation in cucumber leaves. *Environmental and Experimental Botany* 56: 198–204.

Wuytack, T., K. Wuyts, S. Van Dongen, L. Baeten, F. Kardel, K. Verheyen, and R. Samson. 2011. The effect of air pollution and other environmental stressors on leaf fluctuating asymmetry and specific leaf area of *Salix alba* L. *Environmental Pollution* 159: 2405–2411.

Zaman, M.M., M.A.H. Chowdhury, and T. Chowdhury. 2015. Growth parameters and leaf biomass yield of stevia (*Stevia rebaudiana*, Bertoni) as influenced by different soil types of Bangladesh. *Journal of the Bangladesh Agricultural University* 13(1): 31–37.

Zeigler, I. 1972. The effect of SO3⁻ on the activity of ribulose1,5 diphosphate carboxylase in isolated spinach chloroplasts. *Planta* 103: 155–163.

Zhang, C., X. Yi, X. Gao, M. Wang, C. Shao, Z. Lv, J. Chena, Z. Liu, and C. Shen. 2020. Physiological and biochemical responses of tea seedlings (*Camellia sinensis*) to simulated acid rain conditions. *Ecotoxicology and Environmental Safety* 192: 110–315.

5 Effect of Elevated CO$_2$ Conditions on Medicinal Plants
A Relatively Unexplored Aspect

Anuj Choudhary, Antul Kumar, Harmanjot Kaur, Mandeep Singh, Gurparsad Singh Suri, Gurleen Kaur, Sahil Mehta

CONTENTS

5.1 INTRODUCTION

In the last five decades, food security has seen a major threat from the influences of climate change on agriculture as well as the exponentially increasing world population. Global agricultural productivity has already been hit by climate change and, in a recent study, Ortiz-Bobea et al. (2021) found that climate change brought about by anthropogenic activities has reduced the world's agricultural total factor productivity since 1961 by about 21 per cent, with a greater impact on warm and dry regions, including Africa (-34 per cent), than for cooler regions such as Europe and Central Asia (-7.1 per cent).

Over the coming decades, climate change is expected to have an increased influence on the world's supply of medicinal plants. Further, their quantity and quality will also be impacted due to the complex effects of elevated atmospheric carbon dioxide, changing temperature, and rainfall patterns (Wang et al. 2018; Husen 2021).

DOI: 10.1201/9781003178866-5

Several investigations have reported that a massive increase in the level of the atmospheric CO_2 concentration will alter both plant biomass and chemistry in most plant species (Nogia et al. 2016). Although there is a variable response of individual plant species under elevated carbon dioxide conditions, more likely an enhancement of plant biomass as well as C:N ratio in tissues of medicinal plants is observed (Gifford et al. 2000).

Under elevated carbon dioxide levels, there is a positive effect on net primary production and photosynthetic rate as well as total plant biomass (Panchsheela et al. 2016). Increased photosynthetic rate subsequently enhances biomass and net primary productivity under elevated carbon dioxide levels (Thomson et al. 2017). All three factors – i.e., net primary productivity, photosynthesis, and biomass – are interconnected.

During plant development, there are a variety of secondary metabolites that get produced, which serve a cascade of cellular functions vital for physiological processes of the plant and are also used for many pharmaceutical purposes. The concentration, as well as type, of secondary metabolite production by plants is mostly dependent on the environmental factors affecting growth (Yang et al. 2018; Bachheti et al. 2021). The production of secondary metabolites and corresponding physiological adaptive responses are employed by various plant species against stress and defensive stimuli. A surge in studies enumerating the impact of abiotic environment conditions on *in vitro* and *in vivo* plant secondary metabolites has been witnessed in recent decades.

5.2   GLOBAL CO_2 IMPACT IN PAST DECADES

Emissions of CO_2 are the key regulator of global climate change. Over the past 170 years, anthropogenic activities have increased atmospheric CO_2 concentration by 48 per cent above the preindustrial level reported in 1850. The data shows a more drastic impact over a 20,000-year time interval, i.e., glacial maximum to 1850, or 185ppm to 280ppm. Meanwhile, the earth's temperature has significantly risen by about 2.12°F/1.18°C since the end of the nineteenth century. Consequently, two major ice sheets (Greenland and Antarctica) have declined in mass, and Antarctica has lost about 148 billion tonnes of ice, while Greenland has lost almost 279 billion tonnes per year for 26 years (1993 to 2019). Since the emergence of the Industrial Revolution, increased CO_2 levels have directly impacted the acidity of ocean water by 30 per cent. The ocean has absorbed approximately 30 per cent of total e[CO_2] in the past decade, i.e., 7.1 to 10.8 billion metric tonnes annually). Global CO_2 emission drastically reduced by 5.8 per cent in the last year, this being recorded as the highest ever decline apart from 2009 (Westerhold et al. 2020) (Figures 5.1–5.2).

5.3   EFFECT OF GASES ON PLANT GROWTH AND DEVELOPMENT

The increase of greenhouse gases in the atmosphere comes as a big environmental challenge for the growth and development of medicinal plants like *Digitalis lanata, Papaver setigerum,* and many other important plants (Table 5.1). Among these greenhouse gases, carbon dioxide is one of the major gases that have abruptly risen due to industrialization. Almost half of the anthropogenic emission of carbon dioxide

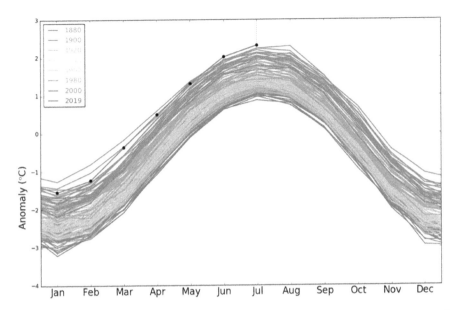

FIGURE 5.1 Diagrammatic representation of geographic information system (GIS) seasonal cycle from a period of 1880–2019 (to July only) based on MERRA2 reanalysis. The data have been adapted from the National Aeronautics and Space Administration (https://data.giss.nasa.gov/gistemp/graphs_v3/).

between 1750 and 2011 has occurred in the last four decades (IPCC 2014). Until now, the potential effects of a sudden rise of this gas on medicinal plants have remained undetermined. This sudden rise in carbon dioxide, along with various trace gases, is believed to be a prima facie component steering global climate change (IPCC 2017) (Figures 5.1–5.3).

Apart from climate change, there will be a direct impact on vegetation, and various studies have shown the positive response of plants to elevated levels of carbon dioxide (Taub et al. 2008). Being a major photosynthetic substrate, e[CO$_2$] invigorates photosynthesis which leads to increased uptake of carbon, and assimilation, thereby positively affecting plant growth. Generally, there is a difference in responses between C3 and C4 plants, as there is a greater growth response in C3 as compared to the C4 pathway (Wang, W.H. et al. 2012). It has been concluded that elevated carbon dioxide in the atmosphere enhanced plant growth with a C3 (33–40 per cent elevation) versus a C4 (10–15 per cent rise) via a photosynthetic pathway (Wang, C. et al. 2012). It has been observed in *Ananas comosus* that exposure to elevated carbon dioxide for four months enhances dry matter accumulation and total plant biomass (Zhu et al. 2002).

In addition to stimulating photosynthesis and plant biomass, there are also alterations in carbon partitioning under e[CO$_2$] conditions. In several cases, a large fraction of increased biomass produced following elevated CO$_2$ has been observed in the below-ground part – i.e., roots – often leading to an increased root-to-shoot ratio. Interestingly, plants direct their allocated photosynthates to various plant parts required to attain the most restraining resource; whenever carbon dioxide is

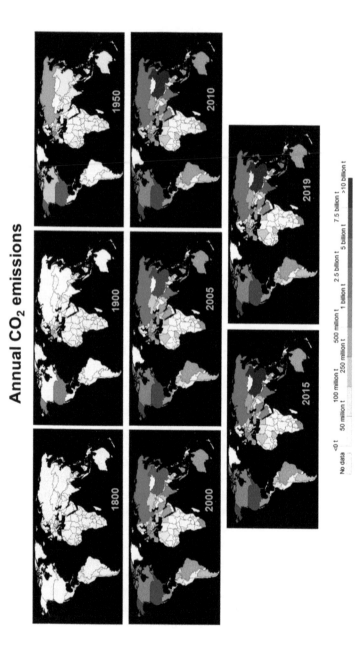

FIGURE 5.2 Timescale depiction of annual CO_2 emissions from the advent of the Industrialization era to the year 2019. In this timeline plot, the growth of global emissions prior to the Industrial Revolution was very low and stayed relatively slow until the mid-20th century. However, in 1950 the world emitted 6 billion tonnes of CO_2 which nearly quadrupled by 1990. Unfortunately, the emissions have continued to grow rapidly with emissions of over 36 billion tonnes each year. The data have been adapted from Our World in Data (https://ourworldindata.org/co2-emissions).

TABLE 5.1
Effect of e[CO$_2$] on Morphology and Physiology of Medicinal Plants

Plant name	Family	Morphological changes	Physiological changes	References
Populus tremuloides	Salicaceae	Increase in leaf structure by spongy mesophyll layer thickness and enhanced chloroplast cover index	Promotes inter-cellular air space in the mesophyll, starch accumulates in chloroplasts, reduces plastoglobuli number in chloroplasts	Oksanen et al. (2001)
Betula papyrifera	Betulaceae	Increased leaf biomass, plant height, root collar diameter, stem biomass	Thinning of leaves, cell walls, and increased volume for vacuoles	Oksanen et al. (2001)
Trigonella foenum-graecum	Fabaceae	Enhancement in plant shoot biomass	Increased C:N ratio, available nutrient content, higher net photosynthetic rate, lower stomatal conductance and stromal condensation in chloroplasts	Jain et al. (2007)
Aster tripolium	Asteraceae	Increased leaf area and photosynthesis	Formation of many mesophyll cell layers, thicker leaves due to structural changes in chloroplasts, and decline in water-use efficiency	Geissler et al. (2008)
Pteridium revolutum	Dennstaedtiaceae	Promotes root elongation and root branching	Increased C supply, more root exudates, and alters plant root symbionts	Zheng et al. (2008)
Lolium multiflorum	Poaceae	Increased total plant biomass and growth	Increased net assimilation rate and internal CO$_2$ concentration	Jia et al. (2011)
Lolium perenne	Poaceae	Increased tiller number, root and shoot dry weight, and total plant biomass	Enhanced photosynthesis and reduced stomatal resistance	Jia et al. (2011)

(continued)

TABLE 5.1 (Continued)
Effect of e[CO_2] on Morphology and Physiology of Medicinal Plants

Plant name	Family	Morphological changes	Physiological changes	References
Eruca sativa	Brassicaceae	Lower fresh weight in seedlings	Influences host-pathogen interaction, pathogenicity, and plants more prone to fungal diseases	Chitarra et al. (2015)
Ocimum basilicum	Lamiaceae	Promotes average plant biomass, enhances the rates of dark respiration and photosynthesis	Improved respiratory rate, enhanced sugar concentration, and causes alteration in tissue chemistry	Al Jauni et al. (2018)
Mentha piperita	Lamiaceae	Root and shoot biomass increased	Improvement in the concentration of antioxidant metabolites, including ascorbate, glutathione, and flavonoids	Al Jauni et al. (2018)

increased, the most limiting resource becomes nutrients or water. More rooting is observed under an e[CO$_2$] environment. For instance, in the medicinal plant *Solanum melongena,* it was observed that an extensive rooting system was developed under higher carbon dioxide conditions (Madegowda and Hatfield 2013).

Highly extensive robust root systems lead to increased carbohydrate storage, thereby enhancing a wider evaluation of the soil for absorption of water and nutrients to fulfil the needs of plant growth during periods of peak requirement – namely, boll development and filling. Not only an enhancement in rooting, but also roots colonization with mycorrhizae (a symbiotic relationship of fungi and roots of higher plants) and rhizobia is observed under elevated carbon dioxide which leads to nitrogen fixation and reduced dependency on organic fertilizers (Tiwari and Bhatia 2019; Tiwari et al. 2019; Tiwari et al. 2021a; Tiwari et al. 2021b). Similar results have been reported in *Trifolium repens* (an antirheumatic and antiscrophulatic plant) where enhanced mycorrhizal association and colonization with fungus, *Glomus mosseae,* was observed during higher carbon dioxide levels (Staddon et al. 2001). For plants to utilize a higher level of atmospheric carbon dioxide, they must have a means of harbouring the additional carbohydrates accumulated. It has been reported that plants with a tuberous rooting system tend to respond to carbon dioxide enrichment to a greater extent than plants with smaller and fibrous rooting systems (Prior et al. 2011).

Furthermore, the effects of e[CO$_2$] on carbon allocation and photosynthesis can also impact growth via alleviated plant–water relations. Elevated carbon dioxide slows down the transpiration rate through temporary closure of stomatal guard cells of a leaf which subsequently improves plant–water relations, coupled with increased photosynthesis (Xu, Z. et al. 2016).

Although much research is focused on the negative impacts of e[CO$_2$] on plants, medicinal plants have still received less attention than other forest and agronomic species. Even though most of the medicinal plant species will likely benefit from this rising carbon dioxide, research to support this contention is still lacking.

5.4   SOURCE AND ADVERSE EFFECTS OF E[CO$_2$]

This global rise in greenhouse gases is primarily attributable to the Industrial Revolution and population expansion, which have led to increased fossil fuel burning and changing land-use patterns. As reports have shown, there has been an abrupt rise in CO$_2$ level from 280 ppm, to above 419.13 ppm in the current period of May 2021 (Canadell et al. 2007) (Figure 5.3).

Based on recent NASA studies, it was concluded that e[CO$_2$] concentrations in the atmosphere can improve water-use efficiency in crops and appreciably mitigate yield losses due to climate change. Some of the studies have shown that higher CO$_2$ levels affect plant species in two important ways, the first being to improve crop quality as well as quantity traits by enhancing the photosynthesis rate, which subsequently spurs plant growth and development. Secondly, it brings down the amount of water that crops lose by transpiration. Therefore, both positive and negative impacts are observed on plant growth and their yield.

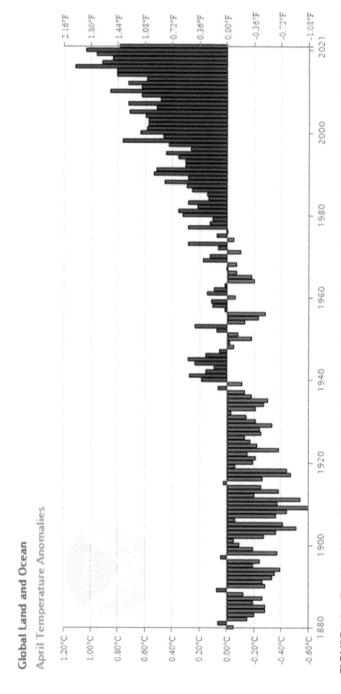

FIGURE 5.3 Graphical illustration showing global surface temperature variations over the period of 1880–2021 (to April). The April 2021 global surface temperature was 0.79°C (1.42°F) above the 20th-century average of 13.7°C (56.7°F). This was the smallest value for April since 2013 and was the ninth warmest April in the 142-year record. The data have been adapted from National Oceanic and Atmospheric Administration (www.ncdc.noaa.gov/sotc/global/202104).

Elevated CO$_2$ concentrations cause an increase in the photosynthetic activity of plants, which consequently results in positive effects like greater growth, above-ground biomass, and yield. Similar results have been seen in Himalayan Paris medicinal plants – i.e., *Paris polyphylla* – where elevated carbon dioxide conditions boost the photosynthetic activity and total plant growth rate (Qiang et al. 2020). On the other hand, a higher level of carbon dioxide also causes negative effects on plants, which may badly affect crop quality, such as a decrease in levels of protein and sugar levels as well as the nutritional quality of important crops. There is an approximately 15 per cent reduction in the protein content of sugarbeet (*Beta vulgaris*) under higher CO$_2$ conditions (Taub et al. 2008).

Elevated carbon dioxide boosts photosynthetic activity in plants, which consequently results in the significant production of carbohydrates along with biomass. This allocation of carbohydrates varies not only between different species of plant, but within species also. These biomolecules also act as signalling molecules along with good energy sources in plants. However, each plant species shows variable response patterns concerning changes in temperature and carbon dioxide concentration. Due to all these positive and negative effects, understanding plant responses to elevated carbon dioxide is very important. These plant responses to e[CO$_2$] will be of great concern in the upcoming years, as the level of carbon dioxide is expected to rise continuously over time.

5.5 EFFECT OF e(CO$_2$) LEVEL ON MEDICINAL PLANTS

5.5.1 MORPHOLOGICAL AND PHYSIOLOGICAL CHANGES

Elevated carbon dioxide has a direct effect on plant species because of photosynthetic gas exchange while the indirect effect is its contribution towards global warming due to potent greenhouse gas. There is a positive correlation between net primary production (NPP) and gross primary productivity (GPP) that have both increased under higher CO$_2$ levels as compared to normal conditions (Cheng et al. 2000).

This abrupt rise in carbon dioxide stimulates photosynthetic carbon assimilation rates of about 31 per cent across 40 plant species (Dusenge et al. 2019). In *Lolium perenne* elevated CO$_2$ results in an increased tiller number, root and shoot dry weight, and total plant biomass. It also affects physiological processes such as enhanced photosynthetic activity and reduced stomatal resistance (Jia et al. 2018). In the family Asteraceae, elevated carbon dioxide enhanced photosynthetic rate and leaf area, and reduced water-use efficiency in *Aster tripolium* (Geissler et al. 2009) (Table 5.1). Elevated carbon dioxide also promotes root elongation and root branching in *Pteridium revolutum* (Zheng, J. et al. 2008), and enhances plant shoot biomass in *Trigonella foenum-graecum* and *Ocimum basilicum* (Jain et al. 2007; Tursun et al. 2020). Under higher CO$_2$ levels, due to increment in carbon supply, more root exudates are accumulated inside plants (Zheng, Y. et al. 2008). According to various reports, carbon accumulation is more in roots than leaves and stem parts (Singh et al. 2018). Studies conducted by Zhu et al. (2002) showed that under elevated conditions there are more relative growth rates and net assimilation rates in *Ananas comosus*, a plant used in analgesic medicines. *Kalanchoe blossfeldiana* is an

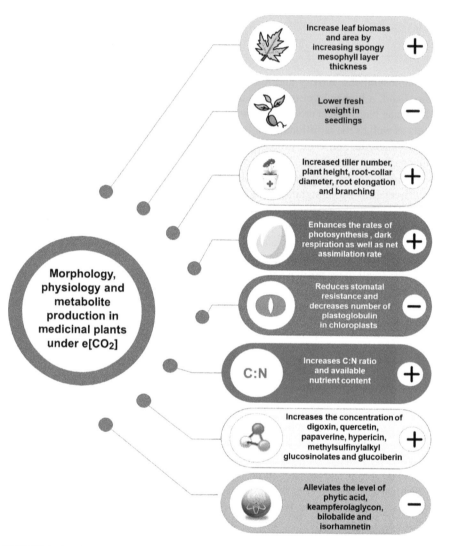

FIGURE 5.4 Adverse effects of enhanced CO_2 level on morphology, phenological and metabolite production in medicinal plants.

active immunosuppressive plant that responded positively to more leaves and nodes production under a CO_2-rich environment (Cho et al. 2020) (Figure 5.4).

In *Withania somnifera*, at 600 μmol mol^{-1} CO_2 and 800 μmol mol^{-1}, the dry weight of leaves was enhanced by up to 53.19 per cent and 90.66 per cent respectively. Also, the root dry weight and shoot dry weight were elevated by 14.24 per cent and 93.06 per cent at 800 μmol mol^{-1}, along with 32.59 per cent and 8.75 per cent at 600 μmol mol^{-1} CO_2 concentrations (Sharma et al. 2018). Rising atmospheric carbon dioxide also significantly increases the photosynthetic rate as well as the total dry biomass of

Xanthium strumarium (Ziska 2001). Similarly, in the case of *Cirsium arvense,* above-ground dry weight and photosynthesis rate were stimulated (Ziska 2002). In addition, some negative effects were also observed on the plant *Eruca sativa* in which there are more chances of host-pathogen interaction and pathogenicity and of the plant becoming more prone to fungal diseases (Chitarra et al. 2015).

5.5.2 PRODUCTION OF SECONDARY METABOLITES

Medicinal plants are enriched sources of secondary metabolites, such as alkaloids, flavonoids, glycosides, amines, and steroids, which have been enormously used in drugs and pharmaceuticals. These medicinal plants attain considerable metabolic plasticity to adapt to changing environmental conditions. However, this metabolic plasticity occurring due to plant secondary metabolites may affect the production of other major secondary metabolites which are typically the basis for their medicinal activity (Yang et al. 2018).

As an example, when the medicinal plant *Digitalis lanata* faced elevated carbon dioxide, there was about a 3.5-fold rise in the production of cardenolide glycoside, digoxin (Rahimtoola 2004). Along with this, during the enhancement of digoxin, the other three glycosides – i.e. digitoxigenin, digitoxin, and digoxin-mono-digitoxoside – showed a slowdown in their production level (Rahimtoola 2004). So, there is a differential response of elevated carbon dioxide on the production of secondary metabolites. In addition, the period is also one of the important factors in determining the metabolic flux concerning plant secondary metabolites. An example is *Hymenocallis littoralis*; in this plant bulbs possess antiviral properties. Under elevated carbon dioxide conditions, in the first year, there is an increase in the levels of three alkaloids, i.e. ,7-deoxynarciclasine, 7-deoxy-trans dihydronarciclasin, and pancratistatin, whereas there is a decrease in alkaloid levels in subsequent years (Idso et al. 2000).

In the case of *Papaver setigerum*, a shift of CO_2 from 300 μmol mol⁻¹ to 600 μmol mol⁻¹ enhances the alkaloids such as papaverine, morphine, noscapine, and codeine (Ziska et al. 2008). Furthermore, recent studies also showed that elevated carbon dioxide leads to a higher ratio between carbon and nutrients, due to the surplus production of non-structural carbohydrates. These non-structural carbohydrates are used for incorporation in carbon-based secondary metabolites.

Similarly, increased levels of phenolics (hypericin, pseudohypericin, and hyperforin) were observed in *Hypericum perforatum* under elevated carbon dioxide conditions (Tusevki et al. 2014). In the broccoli plant also, secondary metabolites like glucosinates show potential effects during the treatment of cancer and cardiovascular diseases (Ramakrishna et al. 2021). In this plant, it was predicted that higher levels of carbon dioxide enhanced the glucosinates (methylsulfinylalkyl glucosinolates, glucoraphanin, and glucoiberin) (Almuhayawi et al. 2020).

Due to anticancerous, antiviral, and diuretic properties, *Catharanthus roseus* is an important medicinal plant. In an experiment in which the *Catharanthus* plant was treated with higher carbon dioxide levels, there was an elevation in most secondary metabolites in the plant, including phenolics, alkaloids, flavonoids, and tannin (Yang et al. 2018). It was studied that, with e[CO_2], *Zingiber officinale* showed enhancement in flavonoid and phenolic content (Ghasemzadeh et al. 2018). In *Quercus ilicifolia*, elevated carbon dioxide increased flavonoid and phenolic contents (Kumar et al. 2017). It was reported that enhanced content of phenols and flavonoids subsequently incremented to phenylalanine (primary metabolite), which acts as a metabolic precursor in the formation of many secondary metabolites. Other studies carried out on *Labisia pumila* by Ibrahim et al. (2014) also showed elevated levels of flavonoids and phenols under higher carbon dioxide levels. In addition, in the medicinal plant *Pseudotsuga manziesii*, there was a sudden decline in monoterpenes under e[CO_2] conditions (Snow et al. 2003).

Another classical example is the traditional Chinese medicinal plant, *Ginkgo biloba*, which is being utilized in the treatment of Alzheimer's disease and vascular dementia. In this, elevated carbon dioxide increases the level of quercetin aglycon (by approximately15 per cent) and decreases bilobalide, kaempferol aglycon, and isorhamnetin (by approximately 10–15 per cent) (Huang et al. 2010). In *Paris polyphylla*, glycosides diosgenin and pennogenin levels rise under conditions of higher carbon dioxide levels (Qiang et al. 2020).

By reviewing the plant responses in different experiments, besides assessing the consequences concerning exposure period, seasonal variation, temperature, etc., it becomes crucial to focus on the entire secondary metabolome of medicinal plants. Whether these parameters work singly or in combination, it plays a major role in the alteration of the metabolic plasticity of medicinal plants. The research on medicinal plants concerning climate change is very sporadic and insignificant as compared to other commercial crops. To prevent the loss of bioactive components by metabolic alterations, appropriate conservatory practices are the need of the hour.

5.6 ADAPTIVE RESPONSES AND SURVIVAL STRATEGIES

Overall, it can be summarized that processes like photosynthesis, water-use efficiency, soil carbon, and biomass are elevated under higher carbon dioxide conditions which render physiological adaptation to the plant. Elevation in net NPP may help in overcoming impact by sequestering the e[CO_2]. Various studies indicate that, for climatic adjustments, plant species not only undergo phenology shifting but also make alterations at physiological as well as metabolic levels (Walther et al. 2002). Yet, there is a dearth of knowledge concerning the physiological responses and adaptations of medicinal plants under enhanced carbon dioxide levels in the atmosphere. It is crucial to understand the plants' adaptive behaviour, affecting biomass as well as yield in future conditions (Figure 5.5).

Studies have been carried out to observe the response to elevated CO_2 of various plants, such as *Catharanthus roseus*, *Vernonia herbacea*, *Thymus vulgaris*, *Mentha spicata*, *Digitalis lanata*, *Hizikia fusiforme*, *Papaver setigerum*, and *Ocimum basilicum*. *Catharanthus roseus* is an important medicinal plant, having the alkaloids vincristine and vinblastine in its sap which are an effective treatment for leukaemia and lymphoma (Table 5.2). Under elevated CO_2 conditions, HPLC analysis has been carried out for the screening of alkaloids in hairy roots and their iridoid precursors have been developed. It has been observed that, at the end of the sixth month, the alkaloids vincristine and vinblastine were present only in the

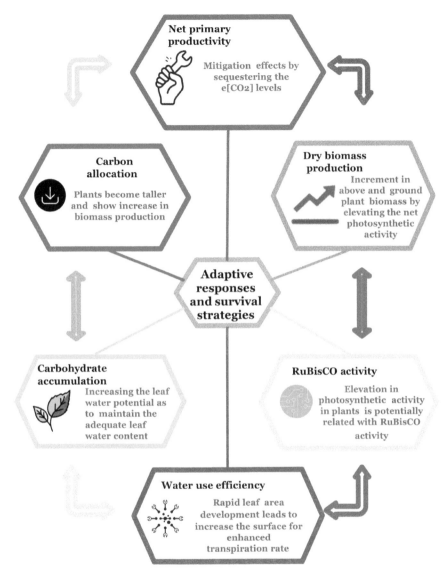

FIGURE 5.5 Adaptation strategies and survival response of medicinal plants under e[CO$_2$] levels.

control conditions (Saravanan and Karthi 2014). In the case of dry mass, it was reported that the higher carbon dioxide levels almost tripled the dry biomass production of *Digitalis purpurea* up to about 83 per cent from the average biomass (Stuhlfauth and Fock 1990).

Later, an investigation was done on *Hymenocallis littorali*, where 48 per cent and 56 per cent increments in above and ground plant biomass were observed (Idso et al. 2000). Similarly, Zobayed et al. (2003) reported an elevation in net photosynthetic

TABLE 5.2
Effect on the Production of Secondary Metabolites under e[CO$_2$] Levels in Medicinal Plants

Plant Species	Secondary metabolites		Roles	References
	Increase	Decrease		
Digitalis lanata	Digoxin	Digitoxin, digoxin-mono-digitoxoside and digitoxigenin	Use in cardiovascular diseases	Rahimtoola (2004); Mishra (2016)
Hymenocallis littoralis	7-deoxy-trans dihydronarciclasin, 7-deoxynarciclasine and pancratistatin	—	Treatment of ovary sarcoma and lymphocytic leukaemia	Idso et al. (2000)
Panax ginseng	Total phenolic and flavonoid contents	Saponin	Stress-related ailments and anticancerous and immunomodulation	Chang et al. (2003); Ali et al. (2005); Wang et al. (2007)
Pseudotsuga manziesii	Morphine, codeine, cocaine	Monoterpenes	Used as analgesia, sedation, euphoria, and for respiratory problems	Snow et al. (2003)
Hypericum perforatum	Hypericin, hyperforin, and pseudohypericin	—	Used in depression, anxiety, swelling, cancer, viral and bacterial diseases	Zobayed and Saxena (2004)
Quercus ilicifolia	Phenol and tannin content	—	Antioxidant activity	Stiling and Cornelissen (2007)
Brassica oleracea	Methylsulfinyl-alkyglucosinolates, glucoraphanin, and glucoiberin	—	Antibiotic, antioxidant, and cancer preventive activity	Schonhof et al. (2007)

Plant species	Compounds		Uses	References
Papaver setigerum	Papaverine, codeine, morphine, and noscapine	—	Antispasmodic drug, treatment of visceral spasm, vasospasm, and acute mesenteric ischaemia	Ziska et al. (2008)
Zingiber officinale	Total flavonoid and phenolic content	Phytic acid	Possesses anti-inflammatory, antioxidant, analgesic and antipyretic properties	Ghasemzadeh et al. (2010)
Ginkgo biloba	Quercetin and terpenoids	Keampferol aglycon, bilobalide and isorhammetin	Used in vascular dementia and Alzheimer's disease treatment	Huang et al. (2010); Weinmann et al. (2010)
Eleais guineensis	Phenylalanine, total flavonoids and phenolics	Phenylalanine lyase	Anti-microbial and anti-oxidant activity	Ibrahim and Jaafar (2012)
Catharanthus roseus	Total phenolics, ascorbic acid, and total alkaloids contents	—	Antitumoral, antimicrobial, analgesic, anticancerous, diuretic, and anti-viral properties	Saravanan and Karthi (2014); Ezuruike and Prieto (2014)
Labisia pumila	Enhances phenolics and flavonoids content	—	Used in gonorrhoea, dysentery, dysmenorrheal and flatulence	Ibrahim et al. (2014)

activity of spider lily plants of 124 per cent, and about 107 per cent increment in dry weight of the plant at 1,000 ppm CO_2 level.

Elevation in photosynthetic activity in plants is potentially related to rising inactivity of RuBisCO (ribulose 1,5-bisphosphate carboxylase enzyme) which can alter growth as well as production of secondary metabolites (Portis et al. 2007). Under higher carbon dioxide conditions, there is an increment in the rate at which carbon is subsumed into carbohydrates; this continues as RuBisCO becomes one of the limiting factors. The higher CO_2 levels result in rapid leaf area development, which further increases the surface for transpiration and an enhanced rate of transpiration (Betts et al. 2007). There is an abrupt increase in stomatal conductance which causes the rapid opening of stomatal guard cells under e[CO_2] in the atmosphere. There is also a significant improvement in the water-use efficiency of plants through increased turgor pressure which is important in the growth and development process.

The major mechanism for the observed response under increased carbon dioxide levels could potentially be that elevated CO_2 helps in the hyperaccumulation and bio-synthesis of carbohydrates, which subsequently raises the osmotic potential in leaf tissues. Ultimately, it helps to maintain adequate leaf water content for other phenomena associated with photosynthesis and growth traits. Along with this, the increasing climatic carbon dioxide level also affects the processes of photosynthesis, metabolism, growth, and development in various medicinal plants (Gamage et al. 2018).

Another adaptive physiological phenomenon has been studied in the plant *Withania somnifera* in which under higher carbon dioxide conditions there is an increasing trend of intrinsic carboxylation efficiency and intrinsic mesophyll efficiency as compared to controlled conditions (Sharma et al. 2018). Keidel et al. (2015) also reported an increase of 20 per cent of soil respiration in temperate grassland ecosystems. This rise in the rate of soil respiration is probably due to more availability of carbon for the activity of microbes due to elevated carbon dioxide (Adair et al. 2011). Net exchange of carbon dioxide also rises between atmosphere and biosphere, due to increased photosynthetic activity, water, and nitrogen-use efficiency of the plant (Baldocchi 2008).

In addition, there is an enhanced allocation of carbon to the leaves, stems, and roots of a plant. As a result of this, plants become taller and produce maximum biomass during higher carbon dioxide conditions. Singh et al. (2018) observed that night leaf respiration (NLR) was suppressed under elevated carbon dioxide conditions. It was reported that there is a rapid decline in leaf respiration when the partial pressure of atmospheric CO_2 increased (Ghannoum et al. 2000). The mechanism behind this respiration reduction is not fully elucidated yet, but it may be due to dark carbon dioxide fixation PEPc (phosphoenolpyruvate carboxylase), change in cytosolic pH, and alteration in enzyme membranes (Tan et al. 2020).

5.7 CONCLUDING REMARKS

In the recent past, carbon dioxide and temperature levels have been elevated, mainly due to climate change, piquing the interest of plant biologists to unravel plants' interaction with the changing environment. Therefore, a lacuna in understanding needs to be filled about what abrupt challenges plants might face and how they respond and acclimatize towards this climate change. Recent studies have concluded that medicinal plants could manage fluctuating climatic conditions, especially rising atmospheric CO_2 concentration. However, in such conditions of e[CO_2] levels, various physiological processes such as transpiration, water-use efficiency, photosynthetic rate, and stomatal conductance, may see an uprise which ultimately leads to better plant growth and increased root-shoot ratio, particularly the root biomass in most medicinal plants. Elevation in CO_2 levels results in changing patterns of secondary metabolite production, which could be critical for the economy of local communities that rely on medicinal plants for health and medical industries.

REFERENCES

Adair, E.C., P.B. Reich, J.J. Trost, and S.E. Hobbie. 2011. Elevated CO_2 stimulates grassland soil respiration by increasing carbon inputs rather than by enhancing soil moisture. *Global Change Biology* 17: 3546–3563.

Al Jaouni, S., A.M. Saleh, M.A.M. Wadaan, W.N. Hozzein, S. Selim, and H. Abd Elgawad. 2018. Elevated CO_2 induces a global metabolic change in basil (*Ocimum basilicum* L.) and peppermint (*Mentha piperita* L.) and improves their biological activity. *Journal of Plant Physiology* 224: 121–131.

Ali, M., E. Hahn, and K. Paek. 2005. CO_2-induced total phenolics in suspension cultures of *Panax ginseng* C. A. Mayer roots: role of antioxidants and enzymes. *Plant Physiology and Biochemistry* 43: 449–457.

Almuhayawi, M.S., H. AbdElgawad, S.K. Al Jaouni, S. Selim, A.H.A. Hassan, and G. Khamis. 2020. Elevated CO_2 improves glucosinolate metabolism and stimulates anticancer and anti-inflammatory properties of broccoli sprouts. *Food Chemistry* 30: 127102.

Bachheti, A., Deepti, R.K. Bachheti, and A. Husen. 2021. Medicinal plants and their pharmaceutical properties under adverse environmental conditions. In: A. Husen (ed.). Harsh Environment and Plant Resilience, pp. 457–502.

Baldocchi, D. 2008. 'Breathing' of the terrestrial biosphere: lessons learned from a global network of carbon dioxide flux measurement systems. *Australian Journal of Botany* 56: 1–26.

Betts, R.A., O. Boucher, M. Collins, P.M. Cox, P.D. Falloon, N. Gedney, M. Webb. 2007. Projected increase in continental runoff due to plant responses to increasing carbon dioxide. *Nature* 448: 1037–1041.

Canadell, J.G., C. Le Quéré, M.R. Raupach, C.B. Field, E.T. Buitenhuis, P. Ciais, and G. Marland. 2007. Contributions to accelerating atmospheric CO_2 growth from economic activity, carbon intensity, and efficiency of natural sinks. *Proceedings of the National Academy of Sciences* 104: 18866–18870.

Chang, Y., E. Seo, and C. Gyllenhaal. 2003. *Panax ginseng*: a role in cancer therapy? *Integrative Cancer Therapies* 2: 13–33.

Cheng, K.W., S.J. Kuo, M. Tang, and Y.P. Chen. 2000. Vapor-liquid equilibria at elevated pressures of binary mixtures of carbon dioxide with methyl salicylate, eugenol, and diethyl phthalate. *Journal of Supercritical Fluids* 18: 87–99.

Chitarra, W., I. Siciliano, I. Ferrocino, M.L. Gullino, and A. Garibaldi. 2015. Effect of elevated atmospheric CO_2 and temperature on the disease severity of rocket plants caused by *Fusarium* wilt under phytotron conditions. *PloS ONE* 10: e0140769.

Cho, A.R., H.J. Yang, E. Kim, and Y.J. Kim. 2020. Growth responses and flowering of kalanchoe cultivars under elevated CO_2 concentration and varying daylength. *Flower Research Journal* 28: 30–39.

Dusenge, M.E., A.G. Duarte, and D.A. Way. 2019. Plant carbon metabolism and climate change: elevated CO_2 and temperature impacts on photosynthesis, photorespiration and respiration. *New Phytologist* 221: 32–49.

Ezuruike, U., and J.M. Prieto. 2014. The use of plants in the traditional management of diabetes in Nigeria: pharmacological and toxicological considerations. *Journal of Ethnopharmacology* 155: 857–924.

Gamage, D., M. Thompson, M. Sutherland, N. Hirotsu, A. Makino, and S. Seneweera. 2018. New insights into the cellular mechanisms of plant growth at elevated atmospheric carbon dioxide concentrations. *Plant, Cell and Environment* 41: 1233–1246.

Geissler, N., S. Hussin, and H.W. Koyro. 2009. Elevated atmospheric CO_2 concentration ameliorates effects of NaCl salinity on photosynthesis and leaf structure of *Aster tripolium* L. *Journal of Experimental Botany* 60: 137–151.

Ghannoum, O., S.V. Caemmerer, L.H. Ziska, and J.P. Conroy. 2000. The growth response of C4 plants to rising atmospheric CO_2 partial pressure: a reassessment. *Plant, Cell and Environment* 23: 931–942.

Ghasemzadeh, A., H. Jaafar, and A. Rahmat. 2010. Elevated carbon dioxide increases contents of flavonoids and phenolic compounds, and antioxidant activities in Malaysian young ginger (*Zingiber officinale* Roscoe.) varieties. *Molecules* 15: 7922.

Gifford, R., Barrett, D., and J. Lutze. 2000. The effects of elevated $[CO_2]$ on the C:N and C:P mass ratios of plant tissues. *Plant and Soil* 5: 224.

Huang, W., X. He, and C. Liu. 2010. Effects of elevated carbon dioxide and ozone on foliar flavonoids of *Ginkgo biloba*. *Advanced Materials Research* 113: 165–169.

Husen, A. 2021. Morpho-anatomical, physiological, biochemical and molecular responses of plants to air pollution, In: A. Husen (ed.). Harsh Environment and Plant Resilience. Springer, Cham, pp. 203–234.

Ibrahim, M., and H. Jaafar. 2012. Impact of elevated carbon dioxide on primary, secondary metabolites and antioxidant responses of *Eleais guineensis* Jacq. (Oil Palm) seedlings. *Molecules* 17: 5195–5211.

Ibrahim, M., H. Jaafar, and E. Karimi. 2014. Allocation of secondary metabolites, photosynthetic capacity, and antioxidant activity of Kacip Fatimah (*Labisia pumila* Benth) in response to CO_2 and light intensity. *Science World Journal* 2014: 360290.

Idso, S., B. Kimball, and G. Pettit. 2000. Effects of atmospheric CO_2 enrichment on the growth and development of *Hymenocallis littoralis* (amaryllidaceae) and the concentrations of several antineoplastic and antiviral constituents of its bulbs. *American Journal of Botany* 87: 769–773.

IPCC. 2017. IPCC 46th Session of the IPCC – Decisions Adopted by the Panel, 6–10 September 2017, Montreal, Canada.

IPCC. 2014. Summary for policymakers. In: Climate Change 2014: Impacts, Adaptation, and Vulnerability. Part A: Global and Sectoral Aspects. Contribution of Working Group II to the Fifth Assessment Report of the Intergovernmental Panel on Climate Change.

Jain, V., M. Pal, A. Raj, and S. Khetarpal. 2007. Photosynthesis and nutrient composition of spinach and fenugreek grown under elevated carbon dioxide concentration. *Biologia plantarum* 51: 559–562.

Jia, Y., S.R. Tang, X.H. Ju, L.N. Shu, S.X. Tu, R.W. Feng, and L. Giusti. 2011. Effects of elevated CO(2) levels on root morphological traits and Cd uptakes of two Lolium species under Cd stress. *Journal of Zhejiang University. Science B* 12: 313–325.

Jia, Y., Y. Lu, D. Elsworth, Y. Fang, and J. Tang. 2018. Surface characteristics and permeability enhancement of shale fractures due to water and supercritical carbon dioxide fracturing. *Journal of Petroleum Science and Engineering* 165: 284–297.

Keidel, L., C. Kammann, L. Grünhage, G. Moser, and C. Müller. 2015. Positive feedback of elevated CO_2 on soil respiration in late autumn and winter. *Biogeosciences* 12: 1257–1269.

Kumar, V., T. Khare, S. Arya, V. Shriram, and S.H. Wani. 2017. Effects of toxic gases, ozone, carbon dioxide, and wastes on plant secondary metabolism. In: *Medicinal Plants and Environmental Challenges*. Springer, Cham, pp. 81–96.

Madegowda, M., and J. Hatfield. 2013. Dynamics of plant root growth under increased atmospheric carbon dioxide. *Agronomy Journal* 105: 657.

Mishra, T. 2016. Climate change and production of secondary metabolites in medicinal plants: a review. *International Journal of Herbal Medicine* 4 : 27–30.

Nogia, P., G.K. Sidhu, R. Mehrotra, S. Mehrotra. 2016. Capturing atmospheric carbon: biological and non-biological methods. *International Journal of Low-Carbon Technologies* 11: 266–274.

Oksanen, E., J. Sober, and D.F. Karnosky. 2001. Impacts of elevated CO$_2$ and/or O$_3$ on leaf ultrastructure of aspen (*Populus tremuloides*) and birch (*Betula papyrifera*) in the aspen FACE experiment. *Environmental Pollution* 115: 437–446.

Ortiz-Bobea, A., T.R. Ault, and C.M. Carrillo. 2021. Anthropogenic climate change has slowed global agricultural productivity growth. *Nature Climate Change* 11: 306–312.

Panchsheela, N., G.K. Sidhu, R. Mehrotra, and S. Mehrotra. 2016. Capturing atmospheric carbon: biological and nonbiological methods. *International Journal of Low-Carbon Technologies* 11: 266–274.

Portis, A.R., and M.A. Parry. 2007. Discoveries in Rubisco (Ribulose 1,5-bisphosphate carboxylase/oxygenase): a historical perspective. *Photosynthesis Research* 94: 121–143.

Prior, S.A., G.B. Runion, S.C. Marble, H.H. Rogers, C.H. Gilliam, and H.A. Torbert. 2011. A review of elevated atmospheric CO$_2$ effects on plant growth and water relations: implications for horticulture. *Horticulture Science* 46: 158–162.

Qiang, Q., Y. Gao, Y. Buzhu, M. Wang, W. Ni, S. Li, T. Zhang, W. Li, and L. Lin. 2020. Elevated CO$_2$ enhances growth and differentially affects saponin content in *Paris polyphylla* var. yunnanensis. *Industrial Crops and Products* 147: 112124.

Rahimtoola, S. 2004. Digitalis therapy for patients in clinical heart failure. *Circulation* 109: 2942–2946.

Ramakrishna, W., A. Kumari, N. Rahman, and P Mandave. 2021. Anticancer activities of plant secondary metabolites: rice callus suspension culture as a new paradigm. *Rice Science* 28: 13–30.

Saravanan, S., and S. Karthi. 2014. Effect of elevated CO$_2$ on growth and biochemical changes in *Catharanthus roseus* – an valuable medicinal herb. *World Journal of Pharmacology and Pharmaceutical Science* 3: 411–422.

Schonhof, I., H. Klaring, and A. Krumbein. 2007. Interaction between atmospheric CO$_2$ and glucosinolates in Broccoli. *Journal of Chemical Ecology* 33 : 105–114.

Sharma, R., H. Singh, M. Kaushik, R. Nautiyal, and O. Singh. 2018. Adaptive physiological response, carbon partitioning, and biomass production of *Withania somnifera* (L.) Dunal grown under elevated CO$_2$ regimes. *Biotech* 3: 8.

Singh, H., R. Sharma, R. Savita, A. Singh, M.P. Kumar, A.Verma, M.W. Ansari, and S.K. Sharma. 2018. Adaptive physiological response of *Parthenium hysterophorus* to elevated atmospheric CO2 concentration. *Indian Forester* 144: 1–14.

Snow, M., R. Bard, and D. Olszyk. 2003. Monoterpenes levels in needles of Douglas fir exposed to elevated CO$_2$ and temperature. *Physiologiae Plantarum* 117: 352–358.

Staddon, P., A. Fitter, and J. Graves. 2001. Effect of elevated atmospheric CO$_2$ on mycorrhizal colonization, external mycorrhizal hyphal production and phosphorus inflow in *Plantago lanceolata* and *Trifolium repens* in association with the arbuscular mycorrhizal fungus *Glomus mosseae*. *Global Change Biology* 5: 347–358.

Stiling, P., and T. Cornelissen. 2007. How does elevated carbon dioxide (CO$_2$) affect plant–herbivore interactions? a field experiment and meta-analysis of CO$_2$- mediated changes on plant chemistry and herbivore performance. *Global Change Biology* 13: 1823–1842.

Stuhlfauth, T., and H.P. Fock. 1990. Effects of whole season CO_2 enrichment on the cultivation of a medicinal plant, *Digitalis lanata*. *Journal of Agronomy and Crop Science* 164: 168–173.

Tan, X.J., and C.M. Cheung. 2020. A multiphase flux balance model reveals flexibility of central carbon metabolism in guard cells of C3 plants. *The Plant Journal* 104: 1648–1656.

Taub, D.R., B. Miller, and H. Allen. 2008. Effects of elevated CO_2 on the protein concentration of food crops: a meta-analysis. *Global Change Biology* 14: 565–575.

Thompson, M., D. Gamage, N. Hirotsu, A. Martin, and S. Seneweera. 2017. Effects of elevated carbon dioxide on photosynthesis and carbon partitioning: a perspective on root sugar sensing and hormonal crosstalk. *Frontiers in Physiology* 8: 578.

Tiwari, M. and S. Bhatia. 2019. Expression profiling of miRNAs indicates crosstalk between phytohormonal response and rhizobial infection in chickpea. *Journal of Plant Biochemistry and Biotechnology* 29: 380–394.

Tiwari, M., M. Yadav, B. Singh, V. Pandey, K. Nawaz, and S. Bhatia. 2019. Global analysis of Two Component System (TCS) member of chickpea and other legume genomes implicates its role in enhanced nodulation. *bioRxiv* 667741.

Tiwari, M., V. Pandey, B. Singh, and S. Bhatia. 2021a. Dynamics of miRNA mediated regulation of legume symbiosis. *Plant, Cell and Environment* 44: 1279–1291.

Tiwari, M., B. Singh, M. Yadav, V. Pandey, and S. Bhatia. 2021b. High throughput identification of miRNAs reveal novel interacting targets regulating chickpea-rhizobia symbiosis. *Environmental and Experimental Botany* 186: 104469.

Tursun, A.O., and I. Telci. 2020. The effects of carbon dioxide and temperature on essential oil composition of purple basil (*Ocimum basilicum* L.). *Journal of Essential Oil Bearing Plants* 23: 255–265.

Tusevski, O., S.J. Petreska, and M. Stefova. 2014. Identification and quantification of phenolic compounds in *Hypericum perforatum* L. transgenic shoots. *Acta Physiologiae Plantarum* 36: 2555–2569.

Walther, G.R., E. Post, P. Convey, A. Menzel, C. Parmesan, T.J. Beebee, and F. Bairlein. 2002. Ecological responses to recent climate change. *Nature* 416: 389–395.

Wang, C., L. Guo, Y. Li, and Z. Wang. 2012. Systematic comparison of C3 and C4 plants based on metabolic network analysis. *BMC Systems Biology* 6: S9.

Wang, J., S.K. Vanga, R. Saxena, V. Orsat, and V. Raghavan. 2018. Effect of climate change on the yield of cereal crops: a review. *Climate* 6: 41.

Wang, W.H., J.F. Hull, J.T. Muckerman, E. Fujita, and Y. Himeda. 2012. Second-coordination-sphere and electronic effects enhance iridium (III)-catalyzed homogeneous hydrogenation of carbon dioxide in water near ambient temperature and pressure. *Energy and Environmental Science* 5: 7923–7926.

Wang, W., Y. Zhao, and E. Rayburn. 2007. In vitro anti-cancer activity and structure–activity relationships of natural products isolated from fruits of *Panax ginseng*. *Cancer Chemotherapy Pharmacology* 59: 589–601.

Westerhold, T., N. Marwan, A. Drury, D. Liebrand, et al. 2020. An astronomically dated record of Earth's climate and its predictability over the last 66 million years. *Science* 369: 1383–1387.

Weinmann, S., S. Roll, and C. Schwarzbach. 2010. Effects of *Ginkgo biloba* in dementia: systematic review and meta-analysis. *BMC Geriatrics* 10: 14.

Xu, Z., Y. Jiang, B. Jia, and G. Zhou. 2016. Elevated-CO_2 Response of stomata and its dependence on environmental factors. *Frontiers in Plant Science* 7: 657.

Yang, L., K.S. Wen, X. Ruan, Y.X. Zhao, F. Wei, and Q. Wang. 2018. Response of plant secondary metabolites to environmental factors. *Molecules (Basel, Switzerland)* 23: 762.

Zheng, J., H. Wang, Z. Li, S. Tang, and Z. Chen. 2008. Using elevated carbon dioxide to enhance copper accumulation in *Pteridium revolutum*, a copper-tolerant plant, under experimental conditions. *International Journal of Phytoremediation* 10: 159–170.

Zheng, Y., Z. Yang, and X. Chen. 2008. Effect of high oxygen atmospheres on fruit decay and quality in Chinese bayberries, strawberries and blueberries. *Food Control* 19: 470–474.

Zhu, J., D.P. Bartholomew, and G. Goldstein. 2002. Photosynthetic gas exchange and water relations during drought in 'Smooth Cayenne' pineapple (*Ananas comosus* (L.) Merr.) grown under ambient and elevated CO$_2$ and three day/night temperatures. In IV *International Pineapple Symposium* 666: 161–173.

Ziska, L.H. 2002. Influence of rising atmospheric CO2 since 1900 on early growth and photosynthetic response of a noxious invasive weed, Canada thistle (*Cirsium arvense*). Functional Plant Biology 29: 1387–1392.

Ziska, L.H. 2001. Changes in competitive ability between a C4 crop and a C3 weed with elevated carbon dioxide. *Weed Science* 49(5): 622–627.

Ziska, L., S. Panicker, and H. Wojno. 2008. Recent and projected increases in atmospheric carbon dioxide and the potential impacts on growth and alkaloid production in wild poppy (*Papaver setigerum* DC.). *Climate Change* 91: 395–403.

Zobayed, S.M.A., S.J. Murch, H.P.V. Rupasinghe, and P.K. Saxena. 2003. Elevated carbon supply altered hypericin and hyperforin contents of St. John's wort (*Hypericum perforatum*) grown in bioreactors. *Plant Cell Tissue Organ Culture* 75: 143–149.

Zobayed, S., and P. Saxena. 2004. Production of St. John's wort plants under controlled environment for maximizing biomass and secondary metabolites. *In Vitro Cellular and Developmental Biology Plant* 40: 108–114.

6 Water Pollution and Medicinal Plants
Insights into the Impact and Adaptation Responses

Antul Kumar, Anuj Choudhary, Harmanjot Kaur,
Ritesh Kumar, Radhika Sharma, Himani Gautam,
Sahil Mehta

CONTENTS

6.1   INTRODUCTION

Over the last century, pollution has become an aggravated threat due to rapid urbanization, anthropogenic activities, and industrialization which disturb the health of plants, animals, and especially humankind. Considering the human perspective, at present, industrialization is acting as a major cause of water scarcity and water pollution, especially for humans living in developing countries. For example, as per sources of World water, the water per human, as well as water quality, is decreasing exponentially in India (Figure 6.1). Humans and animals can tackle this alarming state because of their ability to move from one state to another. However, in this situation, the plants under cultivation, as well as forest regions, suffer most. Plants especially of economic importance cannot escape from the impacts of these stressful conditions due to their sessile nature. Pollution from toxic gases, agricultural waste, industrial runoff, sewage sludge, and many environmental perturbations are also degrading the health of medicinal plants.

As a result, they have developed different strategies to mitigate these conditions. These strategies include morphological, biochemical, and molecular changes. Some of the known plants' approaches in mediating abiotic stresses include evapotranspiration

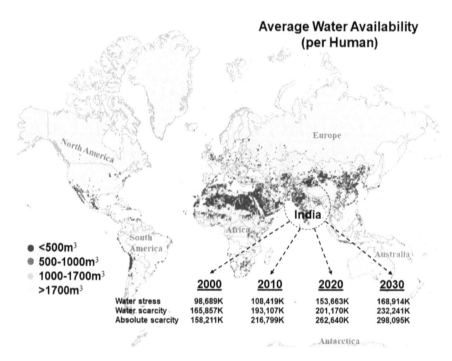

FIGURE 6.1 Diagrammatic representation of average water availability per human for the period of 2000–2030 (https://worldwater.io/).

of available water, high accumulation of different osmoprotectants, events, and their mechanisms in closings and openings of stomata as per water availability, induction of antioxidant machinery regulation control of osmoregulation, fine-tuning of transcriptional controls counter-balancing ionic concentration and post-transcriptional regulations of gene expressions (Shriram et al. 2016; Wani et al. 2017). Additionally, this enhanced pollution level elevates the biosynthesis of sugars, amino acids, saponins, glycosides, proteins, flavonoids, lignin, phenolics, terpenes, alkaloids, glucosinolates, waxes, and other secondary metabolites which are known to play an important role as a defence system (Kumar and Khare 2016).

6.2 TOWARDS MEDICINAL PLANTS UNDER WATER POLLUTION

From the Vedic period, medicinal plants have had cosmetic, nutritional, pharmaceutical, and other economic values. They exhibit a diverse range of commercial importance, represented by several families such as Apocynaceae, Lamiaceae, Rutaceae, Asphodelaceae, Hypericaceae, Solanaceae, Rubiaceae, Papaveraceae, Asteraceae, Piperaceae, Apiaceae, Araliaceae, Araceae, Ginkgoaceae, Rhamnaceae, and Zingiberaceae which serve as major sources of medicinally active compounds (for more detail see Sarma et al. 2011). Being utilized as a remedy for conventional disease cures from historic decades, medicinal plants have gained significant attention in every industrial aspect (Luo et al. 2021). In China, it was noted that approximately 60,000 COVID-19 patients (almost 85 per cent of total cases) were cured by the Chinese medicinal system (CMS) with impressive results in various infection stages, such as symptom management, poor deterioration, and mortality rate, quick response and recovery speed, and overall disease control on February 17, 2020 (Ministry of Science and Technology 2020).

Sewage water and sludge contain different inorganic and organic metal ions which create stress conditions for medicinal plants (Sarma et al. 2011). The effect of pollutants on the production of secondary metabolites has been reported and shows plasticity to maintain balance with environmental conditions. The upregulation of plant secondary metabolites solely depends upon the amount and time interval during exposed conditions of pollutants. The concentration of alkaloids like 7-deoxynarciclasine, pancratistatin, and 7-deoxy-trans-dihydronarciclasin increases with the exposed condition of greenhouse gases in the bulbs of *Hymenocallis littoralis* (Idso et al. 2000). For instance, in *Ginkgo biloba*, a high-altitude Chinese traditional herb for curing Alzheimer's disease, this resulted in fluctuated terpenoid content, 10 per cent decline in kaempferol aglycon, 15 per cent increase in bilobalide and isorhamnetin, and 15 per cent increase in quercetin aglycone (Huang et al. 2010). Enhancement in the production of artemisinin in *Artemisia annua* using hairy root cultures was reported in the application with silver (Ag) nanoparticles (900 mg L−1) for 20 days. Improvement in diosgenin content in fenugreek and steroidal saponin was observed under the influence of silver compounds (Zhang et al. 2013; Jasim et al. 2017).

The 3.9-fold increased silver nanoparticles can be correlated with the uplift amount of hydrogen peroxide, catalase activity, lipid peroxidation levels, and other signalling molecule production in *Arabidopsis*. The exposure of cadmium oxide was proved to be responsible for the increment in the content of isovitexin and ferulic acid. In *Prunella vulgaris*, callus tissue of such important antiviral properties containing herb grown in silver and naphthalene acetic acid (NAA), similar findings of high accumulation of flavonoids and phenolics along with the maximum growth in callus induction were reported (Fazal et al. 2016). Another medicinal plant, *Matricaria chamomilla*, was grown for ten days in different concentrations of copper solution (Kováčik et al. 2008). A total of 11 secondary metabolites were examined, including caffeic, salicylic acid, chlorogenic, o-coumaric acid, sinapic, protocatechuic, vanillic, p-hydroxybenzoic, gentisic, and syringic acid obtained from methanolic extract (Kováčik et al. 2008).

Ocimum tenuiflorum L. was cultivated in Hoagland solution (5 per cent) having different chromium concentrations (10–100 μM) to evaluate the amount of eugenol, a major chemical constituent. Significant enhancement in eugenol content of up to 100 μM in comparison to the control was observed.

An average 25 per cent increase was observed in eugenol content and two other secondary metabolites, hypophyllanthin and phyllanthin, from *Phyllanthus amarus* (Rai and Mehrotra, 2008). Enhancement in the production of hypophyllanthin was observed under high concentrations ranging from 20 to 100 μM without hyperaccumulation of chromium in leaves of *Phyllanthus amarus*. Cadmium (Cd) is a heavy metal that significantly increased the hypophyllanthin and phyllanthin levels in *Phyllanthus amarus* under different concentrations. The amount of both secondary metabolites increased at up to 50 ppm treatment but decreased significantly at 100 ppm (Rai et al. 2005). In another experiment, cadmium treatment was proved to enhance the biosynthesis of artemisinin, arteannuin B, and artemisinic acid, in the Himalayan herb *Artemisia annua* (Zhou, L. et al. 2016).

Nickel (Ni) is another toxic heavy metal that is abundant in industrial waste and contaminated runoff as well as pristine soils. On screening, the secondary metabolite profile of *Hypericum perforatum* L. (St. John's wort medicinal plant) cultivated in nickel-rich solution showed a 15–20 times decrease in the concentration of hypericin and pseudohypericin (Murch et al. 2002). In *digitalis lantana* (cardiovascular plant) it was shown to enhance the concentration of digoxin, a cardenolide glycoside, up to 3.5 times (Mishra 2016). Similarly, elevated digoxin amounts with three other glycosides, viz. digoxin-mono-digitoxoside, digitoxigenin, and digitoxin, were also reported under the effect of pollutants (Stuhlfauth and Fock 1990; Rahimtoola 2004) (Figure 6.2).

The content of secondary metabolites enhanced on prime response to heavy metal dissolved in growing water media (Stuhlfauth et al. 1987). Several successful results have been reported in recent years, including quercetin and tropane content being increased in *Pluchea lanceolata* (antispasmodic, sedative, diuretic, laxative, antipyretic, bronchitis, inflammations and piles) 200 μM Znso4 and 150μM (Kumar et al. 2004), *Mentha arvensis* yield of menthol (colic, rhinitis, antibacterial, cough, soreness, and antifebrile) content increased under Zn exposure (Preeti et al. 2007), sitosterol from *Costus specious* at 30mgL^{-1} of Pb (Kartosentono et al. 2002), eugenol from *O. tenuiflorum* (antibacterial, antiseptic, and antispasmodic) at 20 μM of Cr (Rai et al. 2004), *Hypericum perforatum* yield of hypericin (nerve pain, skin inflammations and wounds) at Cr(VI) 0.1 μM (Tirillini et al. 2006), hyoscyamine from *Datura stramonium* (bronchial asthma, peptic ulcers, and diarrhoea) at Cd 1 mM (Furze et al. 1991), berberine from *Thalictrum rugosum* (diarrhoea and dysentery) at 20–500 μM

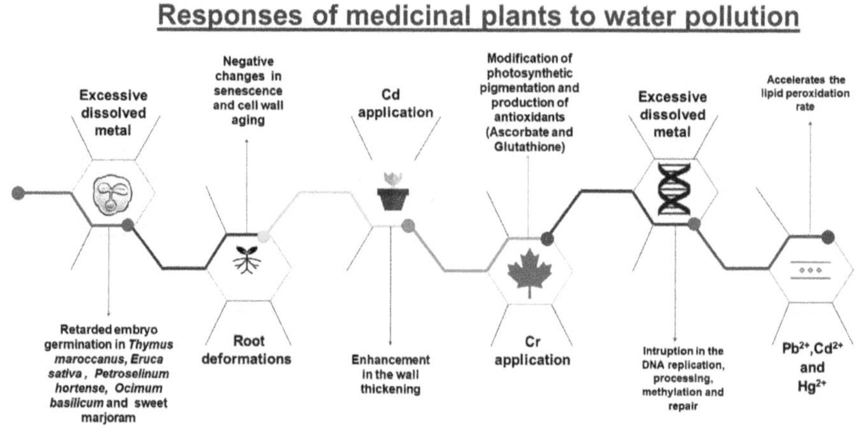

Responses of medicinal plants to water pollution

FIGURE 6.2   Types of water pollutants and their effect on various growth stages of medicinal plants.

Cu (Mahalanbis et al. 1991), flavonoids from *Ononis arvensis* (cardioprotective, antiproliferative, antioxidant, anticancerous, and anti-inflammatory) at 6.3 µmol Cr-Ni and Co (Tumova et al. 2001), herniarin from *Matricaria chamomilla* (spasmodic and anti-inflammatory) at 60–120 µM Cd, 50mg Kg^{-1} Zn, ajmalicine from *Catharathus roseus* (antidiabetic and anticancerous) at 0.05-0.4mM Cd (Zheng and Wu 2004) (Table 6.1).

6.3 RESPONSE OF MEDICINAL PLANTS TO VARIOUS POLLUTANTS

Excessive metal content in polluted water affects the early phases of germination by disturbing pH or drastically inhibits embryonic growth to a certain extent. Embryo germination was retarded in *Thymus maroccanus, Eruca sativa, Petroselinum Hortense, O. basilicum,* and sweet marjoram when exposed to contaminated water or soil (Belaqziz et al. 2009; Ali et al. 2007). The metal allocation within plant species alters with organ type; for example, in *Paspalum notatum*, the level of sugars and amino acids was noted as significantly high on the allocation of Ni, Cu, Al, and Fe ions. The alterations of metal distribution within roots could be tentatively due to the variation in architecture and metabolite concentration inside the root (Oves et al. 2016). In some cases, Fe and Al were mainly translocated to the cell wall with cellulose, hemicellulose, and pectin in both older and young roots. Moreover, certain older roots have indicated a large part of Al that was well stored intracellularly, indicating a poor compartmentalization of Al in the cytoplasm of the older root. However, the level of Cu was enhanced in intracellular components in older and younger roots, as available from nutrients. The above findings concluded that the *P. notatum* in roots suffered from negative changes in senescence and cell wall ageing (Oves et al. 2016) (Figure 6.2). Conclusively, application of Cd enhanced the wall thickening and initiated damage of the cellular compartmentalization of the chloroplast. In another study, damage detection can be evaluated by metal activities; for example, Cr regulates the metabolic functioning of the plant cell that leads to modification of photosynthetic pigmentation, and enhanced levels of production of ascorbate and glutathione-like metabolites, in response to severe metal concentration, and is predicted as an efficient parameter to analyse the damage to plants at the cellular level (Oves et al. 2016). Excessive dissolved metal stimulates alteration in fatty acid biosynthesis in the membrane, and such injuries in membranes may be increased via free acid radicals (Yadav and Mohanpuria 2009). The lipid peroxidation was accessed when plants were exposed to wastewater contaminations enriched with As, Cr, Ni, Zn, Cu, Al, and Cd. In *Bruguiera gymnorrhiza*, the application of various metal pollutants including Pb^{2+}, Cd^{2+}, and Hg^{2+} has accelerated lipid peroxidation rate. Indeed, lipid peroxidation causes an ill effect on the functionality of cell membranes, leading to cell death. Thus, altered lipid peroxidation, due to heavy exposure to metals, disorganizes membrane integrity, permeability, and fluidity and modulates the membrane-associated ATPase. Furthermore, evaluation of genotoxicity in plants caused due to interrupting the eventual process of DNA replication, processing, mutation, methylation, and repair, for instance, in *Cannabis sativa* and *Trifolium repens*, has been used to evaluate the metal-associated genotoxicity effect.

TABLE 6.1
Production of Free Radicals, Metabolites, and Antioxidants in Medicinal Plants under Water Pollution

Pollutant	Medicinal plant	Metabolites	Defence response	Changes	References
Copper and Zinc	*Pluchea lanceolata*	Quercetin, Tropane alkaloids	Increased quercetin and proline synthesis	Growth inhibition coupled with yellowing of leaves	Kumar et al. (2004)
Cadmium	*Catharanthus roseus*	Ajmalicine	Enhanced excretion of ajmalicine	Stimulated transcription of TDC and increased in cellular tryptamine concentration	Zheng and Wu (2004)
Silver, Copper, Zinc, Iron, and Manganese	*Salvia miltiorrhiza*	Tanshinone	Increased tanshinone content	Decreased biomass growth under the heavy metal condition	Zhang et al. (2004); Guo et al. (2005)
Cobalt and Nickel	*Hibiscus sabdariffa*	Anthocyanins and Flavons	Increase in the content of anthocyanins and flavons	Increased all the growth parameters	Eman et al. (2007)
Cadmium	*Phyllanthus amarus*	Hypophyllanthin	Enhanced phyllanthin and hypophyllanthin	Decreased protein, sugars, and chlorophyll concentration and increased starch level	Rai et al. (2005); Rai and Mehrorta (2008)
Cadmium	*Vetiveria zizanioides*	–	Increased antioxidant enzyme activity	Changes in root colour from white to brown	Aibibu et al. (2010)
Cadmium and Cobalt	*Trigonella foenum-graecum* L.	Diosgenin	Suppressed inflammation and proliferation	Increased diosgenin level in seedlings	De and De (2011)
Fertilizers	*Labisia pumila*	Phenolics, Flavonoids, and Saponin	Increased secondary metabolites production	Production of high-quality herbal plant	Ibrahim et al. (2013)

Treatment	Plant species	Compound	Effect	Effect	Reference
Phosphorus and potassium fertilizers	Salvia miltiorrhiza	Tanshinone	Accumulation of bioactive compounds	Increased biomass in response to fertilizer	Lu et al. (2013)
Salinity	Salvia miltiorrhiza	–	Increased CAT activity and decreased SOD activity	No changes at the morphological level	Gengmao et al. (2014)
Lead	Mentha crispa L.	Carvone	Tolerance to heavy metal	Reduction in fresh biomass, essential oil yield and enhanced carvone	Sá et al. (2015)
Lead, Copper, and Cadmium	Ocimum basilicum and Mentha spicata	Terpenoid	Increase in the percentage of linalool and essential oil yield and decrease in methyl chavicol	Plants showed normal growth with no symptoms of morphotoxicity	Kunwar et al. (2015)
Cadmium and Aluminium	Hypoxis hemerocallidea	Hypoxoside	Increased DPPH antioxidant scavenging	Reduction in hypoxoside accumulation	Okem et al. (2015a)
Cadmium and Aluminium	Drimia elata	Phenolics and flavonoids	Decreased secondary metabolites activities	Marked reduction in biomass	Okem et al. (2015b)
Cadmium	Monochoria hastata	–	Accumulating Cd, more in roots than shoot which acts as a tolerance mechanism	Changes in the chloroplast's shapes and accumulation of starch	Baruah et al. (2017)
Cadmium and Copper	Gynura procumbens	–	Promotes antioxidant and antimicrobial activity	Decrease in plant dry weight, total leaf area, and basal diameter	Ibrahim et al. (2017)
Cadmium and Lead	Eruca sativa L.	Osmolyte	Enhanced osmolyte accumulation	Upregulates enzymatic activities	Yildirim et al. (2019)

6.4   ACCLIMATION STRATEGIES TO OVERCOME POLLUTION STRESS

Plants develop various phytoremediation techniques, such as phytoextraction, phytostabilization, phytodegradation, photodegradation, and phytovolatilization, to survive in contaminant-polluted water. Plants absorb contaminants from water and store them in harvestable parts of roots and aerial parts of shoots, roots, and leaves, and they are volatilized through stomata where gas exchange occurs. Plant roots stimulate soil microbial communities in plant root zones to break down contaminants, both internally and through secreted enzymes. Roots and their exudates immobilize contaminants through adsorption, accumulation, and precipitation within the root zone, and thus prevent the spreading of contaminants (Figure 6.3).

Physiologically, phytohormones play a significant role in overcoming stress challenges caused by hyperaccumulation of heavy metals in wastewater. The regulation of phytohormones has been reported as fluctuating in accordance with the number of contaminants; for example, concentrations of metal ions affect the release and transportation of abscisic acid (ABA), brassinosteroids (BR), and indole acetic acid (IAA) (Wani et al. 2016). Brassinosteroids help in the detoxification of heavy metals by upregulating the productions of phytochelatins (PCs) that chelate accumulated ions in a plant cell. These steroids are also responsible for antioxidant activity and enhanced PC production to overcome the effect of Cu stress (Choudhary et al. 2011). According to Usha et al. (2009), the level of auxin and ABA is enhanced in response to the overproduction of PCs during the hyperaccumulation of Cu, Zn, and Cd in *Prosopis juliflora*. It was found that secondary messenger molecules combined or forming hormones-MAPK complex can cause efficient transcription and signalling during stress. However, ethylene is produced under senescence or age-related stresses. Besides these, the crosstalk between salicylic acid and jasmonic acid is a crucial defence response in stress acclimatization. Phytohormones like auxins, ABA, ethylenes, and BRs, etc. change their production or accumulation in a synergetic and antagonistic manner in mitigating stress issues (Komal et al. 2020).

6.5   DAMAGE REPAIR AND ROLE OF THE BIOCHEMICAL SYSTEM

In certain cases, the oxidative defence system is overcome by excessive contaminants and causes damage to intracellular machinery. The plants start suffering from oxidative stress in response to inhibition of reactive oxygen species (ROS) production, methylglyoxal (MG), and metal-dependent oxidative enzymes. This includes certain enzymes like superoxide dismutase (SOD) and catalase (CAT), along with the production of non-enzymatic free scavengers. A study on the leaves of *Nicotiana plumbaginifolia* revealed the induction of ascorbate peroxidase (APX) and CAT in response to excessive Fe exposure. Glutathione (GSH) is a well-known low molecular weight, non-enzymatic oxidant and redox buffer, and is found in cell compartments. GSH shields proteins from the deleterious effect of denaturation and indirectly protects membranes by maintaining the reduced state of α-tocopherol and zeaxanthin. GSH (Gly-Cys-Gly) acts as precursor molecules and helps in PC production, also forming a complex with metal ions (Komal et al. 2014) (Figure 6.3). GSH also controls

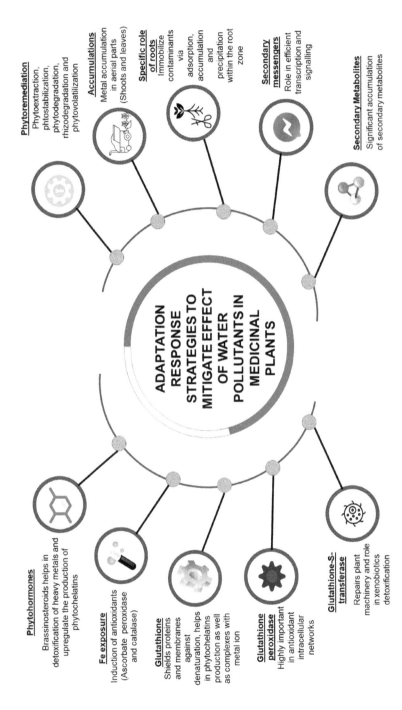

FIGURE 6.3 Physiological, biochemical, and molecular changes in medicinal plants as a response to water pollution.

the production of H_2O_2 (hydrogen peroxide); the enhanced level of GSH induces defence strategies against MG and ROS via mediating the expression and activation of enzyme activities concerning argininosuccinate (ASA) and GSH. Huang et al. (2010) reported an enhanced level of GSH in stems, roots, and leaves under different metal stress regimes (Pb, Cd, and Hg). In *Holcus lanatus*, an As-tolerant species, the level of GSH significantly enhanced as induced via As, as compared to As-sensitive species (Komal et al. 2014). Another, glutathione S transferase (GSTs), a metabolic isoenzyme, helps to repair plant machinery and is well recognized for xenobiotics detoxification. GSTs chiefly catalyse the conjunction of GSH in its reduced form with different compounds converted into soluble derivatives that can be sequestered into vacuoles or liberated from the infected cell. Enzymes like glutathione peroxidase (GPX) also exhibit peroxidase activity and constitute an important antioxidant intracellular network. It was previously reported that Ni, Hg, and Cu induce the enhanced activity of GPXs; however, there is no substantial effect on the exposure of Pb, Zn, and Co in plants. Similarly, in higher plants, the dehydroascorbate reductase (DHAR) enzyme is required in the ascorbate-glutathione (AsA-GSH) reaction. The oxidation of AsA (ascorbic acid) forms DHA (dehydroascorbic acid) via spontaneous disproportion (Yadav and Mohanpuria 2009). The localization of zinc in *Armeria maritima* ssp. Halleri (a zinc tolerant) was observed when exposed to high Zn contamination (Heumann 2002). Species like *Alyssum murale* shows hyperaccumulation strategies by forming a complex with a nitrogen-oxygen donor ligand for Ni transport and storage (Mcnear et al. 2010). Moreover, glutathione reductase (GR) catalyses the GSSG to GSH reduction through NADPH. A heavy metal like Cd causes a reduction of the GSSG/GSG ratio and initiates certain antioxidants, such as SOD and SR. The oxidation of reduced glutathione at Cys thiol group and activity of glutathione reductase catalyses the reverse reaction mediated through NADPH and performs a defence mechanism against oxidative stress caused by Cd generation. Glutathione reductase promotes cells to inhibit toxicity caused by reactive oxygen metabolites; thus it regulates the ascorbate and GSH in reduced form, helping cells to maintain sequestration of heavy metal in the cell.

Plant species remediate metal and are considered as good hyperaccumulators or considered as metal accumulators; for example, *Alyssuim wulfenianum* (Ni), *Arobiadopais hallerii* (Cd), *Pteris vittata* (Cu, Ni, Zn), *Astragulus bisukatus* (Se), *Pistia stratiotes* (Ag, Cd, Cr, Cu, Ni, Pb), *Thlaspi caerulescens* (Cd and Ni), *Sedum alfredii* (Pb), *Amanita muscaria* (Hg), *Crotalaria juncea* (Ni and Cr) *Rorippa globose* (Cd), and *Thlaspi caerulescens* (Zn, Pb, Zn, and Cd) (Sarma 2011) (Table 6.2).

6.6 SURVIVE OR DIE: MITIGATING THE CHALLENGE OF WATER POLLUTANTS IN MEDICINAL PLANTS

Moreover, dissolved heavy metals cause lipid oxidation and the production of oxylipins. Such oxylipins start signalling cascades as primary defence processes. The biosynthesis and accumulation of alkaloids, terpenoids, phenylpropanoids, and other primary, as well as secondary, metabolites are the chief adaptive response mechanisms of plants. Classically, medicinal plants also induce physiological and biochemical changes to cope with the challenge of heavy water contaminants. The concentration

TABLE 6.2
Defence Responses and Strategies in Medicinal Plants to Cope with the Impact of Water Pollutants

Plant name	Pollutant name	Metabolic response	Biochemical response	References
Pluchea lanceolata	Copper and Zinc	Glycinebetaine, soluble carbohydrates	Increased activity of superoxide dismutase enzyme	Kumar et al (2004)
Catharanthus roseus	Nickel and Cadmium	Serpentine	Induce active synthesis of secondary metabolites in plant	Zheng and Wu (2004)
Phyllanthus amarus	Chromium	Phyllanthin and hypophyllanthin	Function in heavy metal detoxification and enhance the growth of a plant	Rai and Mehrorta (2008)
Trigonella foenum-graecum	Arsenic and Zinc	Salicylic acid	Promote the activities of hydrolytic enzymes (α- and β-amylase), free amino acid contents	De and De (2011)
Hypoxis hemerocallidea	Cadmium and Aluminium	Phenolics and flavonoids	Induce active synthesis of secondary metabolites	Okem et al. (2015a)
Drimia elata	Cadmium and Aluminium	Proline	Promote heavy metal chelation	Okem et al. (2015b)
Mentha crispa L.	Lead	Proline	Enhance seedling length, leaf number, fresh biomass, and percentage of dry mass	Sá et al. (2015)
Gynura procumbens	Copper and Cadmium	Total phenolics, total flavonoids, and saponin content	Enhance the antibacterial activity	Ibrahim et al. (2017)
Ocimum basilicum	Nickel, Copper, and Zinc	Proline	Increase in NO intracellular level, upregulated p5CS enzymatic activity	Georgiadou et al. (2018)
Eruca sativa	Copper, Mercury, Chromium	Proline	Accumulation of antioxidant molecules (ascorbate, glutathione, α-tocopherol) and enzymes (SOD, POD, and CAT)	Yildirim et al. (2019)
Salvia miltiorrhiza	Silver, Copper, Zinc, Iron, and Manganese	Salicylic acid	Enhance antioxidant enzymes activity (SOD, POX)	Sharma et al. (2020)

of osmolytes (flavonoids and phenolics) is significantly enhanced along with the accumulation of active secondary metabolites upon exposure to Al and Cd in *Hypoxis hemerocallidea* (Okem et al. 2015a). Comparatively, Okem et al. 2015b suggested the metal chelation and proline accumulation route adopted by *Drimia elata*. In *Phyllanthus amarus*, the phyllanthin and hypophyllanthin contents are reported to be increased on Ch toxicity (Rai and Mehrotra 2008). According to Sá et al. 2015, proline accumulation, enhanced seedling length, fresh biomass, and leaf number per-centage dry mass were observed to be significantly elevated in high Pb concentration in *Mentha crispa*. Increased levels of Zn, Cu, and Ni also promote proline produc-tion, upregulation of P5CS enzymatic activity, and intracellular NO level in *Ocimum basilicum* (Georgiadou et al. 2018). Increased activities of hydrolytic enzymes (α- and β-amylase), free amino acid contents, and salicylic acid were observed in *Trigonella foenum-graecum* when exposed to Zn and As (De and De 2011). Accumulation of antioxidant molecules (ascorbate, glutathione, and α-tocopherol) and enzyme (SOD, POD, and CAT) activities were enhanced in *Eruca sativa* during overaccumulation of Cr, Hg, and Cu (Yildirim et al. 2019). Increased activity of superoxide dismutase enzyme and active synthesis of secondary metabolites were observed in *Pluchea lanceolata* and *Catharanthus roseus* respectively. The hyperaccumulation of Cd, Ni, Zn, and Cu also stimulates the production of serpentine, glycine betaine, and soluble carbohydrates (Zheng and Wu 2004; Kumar et al. 2004). Total phenolics, total flavonoids, and saponin content were increased in the presence of Cd and Cu in *Gynura procumbens* (Ibrahim et al. 2017). In *Salvia miltiorrhiza,* enhanced levels of Mn, Fe, Zn, Cu, and Ag cause enhanced antioxidant enzyme activity such as SOD, POX, and salicylic acid (Sharma et al. 2020). Medicinal plants show enhanced heavy metal accumulation potency; for instance, *Sambucus nigra* for Zn at 30.8–49.9 mg kg−1, *Aesculus hippocastanum* for Pb at 1480 μg g−1, *Bombax costatum* Zn for 67.1 at μg g−1, *Tilia sp.* for Zn at 13.8–32.5 mg kg−1, *Matricaria chamomilla* for Zn at 271 mg kg−1 dry wt, *Ocimum tenuiflorum* for Cr at 419 μg g−1 dry wt, *Costus speciosus* for Cd, Pb at 530 μg g−1 dry wt , *Cuminum cyminum* for Fe at 1.4 mg g−1 dry wt, *Matricaria chamomilla* for Zn at 271 mg kg−1 dry wt, *Agave amanuensis* for Cd at 900 μg g−1 dry wt, Pb 1390 μg g−1 dry wt, *Phyllanthus amarus* for Cd at 82 ppm, *Amaranthus dubius* for Cd at 150 ppm, *Hypericum sp.* for Cd at 0.5 mg kg−1 dry wt, *Bombax costatum* for Zn at 67.1 μg g−1, *Spilanthes oleracea* for Zn at 62.8 μg g−1 dry wt, *Amaranthus hybridus* for Hg at 336 ppm, and *Hibiscus sabdariffa* for Mn at 243 μg g−1 dry wt (Nasim and Dhir 2010).

Except for these strategies, such pollutants cause several types of damage in plant morphology, biochemistry, and physiology. In *Gynura procumbens*, exposure to Cu and Cd promotes the antioxidant system and simultaneously affects plant biomass, leaf area, and basal diameter. Similarly, tanshinone containing *Salvia miltiorrhiza* promotes tanshinone content upon elevating the concentration of Zn, Fe, Mn, Cu, and Ag and decreased biomass and growth parameters (Zhang et al. 2004; Guo et al. 2005). According to Ibrahim et al. (2013), contaminants like dissolved fertilizers enhance the production of saponins, flavonoids, phenolics, and other secondary metabolites *in Labisia pumila.* The levels of flavons, anthocyanins, and growth parameters show enhanced concentration when treated with Ni and Co in *Hibiscus sabdariffa* (Eman et al. 2007). In comparison, changes in root colour and elevated antioxidant enzyme activities were studied in *Vetiveria zizanioides* upon exposure to Cd stress (Aibibu et al. 2010).

6.7 CONCLUDING REMARKS

To date, more than 500 plant species are known to accumulate heavy metals. In medicinal plants, analytical results are the critical measures in evaluating the level of potential toxicity that could be dangerous to both humans and plants. *Acalypha wilkesiana, Achyranthes aspera, Acorus calamus, Aegle marmelos, Aframomum melegueta, Alchornea cordifolia, Aloe vera, Andrographis paniculata, Artemisia nilagirica, Bergemia liquilata, Boerhavia diffusa, Calotropis procera, Carcum carvi, Cassia alata, Cassia rhombifolia, Celastrus paniculata, Chromolaena odorata, Datura metal* and much more heavy metal accumulation by medicinal plants is chiefly caused via the dissolution of soluble metals in water which is contaminated and prone to contamination. However, few detection strategies exist that confirm heavy metal toxicity in medicinal plants. It also lacks standardization, techniques, and regulating governmental policies that detect permissible limits. Such a dearth restricts the development of herbal research and limits the immediate processing of new medicinal plants with improved traits. Authorities should estimate the safe values of metallic substances in medicinal plants. Moreover, extensive research focuses on such herbal plants or raw materials for marketing the pharmaceutical industries.

REFERENCES

Aibibu, N., Y.G. Liu, G.M. Zeng, X. Wang, B.B. Chen, H.X. Song, and L. Xu. 2010. Cadmium accumulation in *Vetiveria zizanioides* and its effects on growth, physiological and biochemical characters. *Bioresource Technology* 101: 6297–6307.

Ali, R.M., H.M. Abbas, and R.K. Kamal. 2007. The effects of treatment with polyamines on dry matter, oil and flavonoids contents in salinity stressed chamomile and sweet marjoram. *Plant, Soil and Environment* 53: 529–543.

Baruah, S., M.S. Bora, P. Sharma, P. Deb, and K.P. Sarma. 2017. Understanding of the distribution, translocation, bioaccumulation, and ultrastructural changes of *Monochoria hastata* plant exposed to cadmium. *Water, Air, and Soil Pollution* 228: 1–21.

Belaqziz, R., A. Romane, and A. Abbad. 2009. Salt stress effects on germination, growth and essential oil content of an endemic thyme species in Morocco (*Thymus maroccanus* Ball.). *Journal of Applied Sciences Research* 5: 858–863.

Choudhary, S.P., R. Bhardwaj, B. Gupta, P. Dutt, R. Gupta, M. Kanwar, and S. Biondi. 2011. Enhancing effects of 24-epibrassinolide and Putrescine on the antioxidant capacity and free radical scavenging activity of *Raphanus sativus* seedlings under Cu ion stress. *Acta Physiologiae Plantarum* 33: 1319–1333.

De, D., and B. De. 2011. Elicitation of diosgenin production in *Trigonella foenum-graecum* L. seedlings by heavy metals and signaling molecules. *Acta physiologiae plantarum* 33: 1585–1590.

Eman, A., N. Gad, and B. Nagam. 2007. Effect of cobalt and nickel on plant growth, yield and flavonoids content of *Hibiscus sabdariffa* L. *Australian Journal of Basic and Applied Science* 1: 73–78.

Fazal, H., B.H. Abbasi, N. Ahmad, and M. Ali. 2016. Elicitation of medicinally important anti-oxidant secondary metabolites with silver and gold nanoparticles in callus cultures of *Prunella vulgaris* L. *Applied Biochemistry and Biotechnology* 180: 1076–1092.

Furze, J.M., M.J.C. Rhodes, A.J. Parr, R.J. Robins, I.M. Whitehead, and D.R. Threlfall. 1991. Abiotic factors elicit sesquiterpenoid phytoalexin production but not alkaloid production in transformed root cultures of *Datura stramonium*. *Plant Cell Reports* 10: 111–114.

Gengmao, Z., S. Quanmei, H. Yu, L. Shihui, and W. Changhai. 2014. The physiological and biochemical responses of a medicinal plant (*Salvia miltiorrhiza* L.) to stress caused by various concentrations of NaCl. *PLOS ONE* 9: e89624.

Georgiadou, E.C., E. Kowalska, K. Patla, K. Kulbat, B. Smolińska, J. Leszczyńska, and V. Fotopoulos. 2018. Influence of heavy metals (Ni, Cu, and Zn) on nitro-oxidative stress responses, proteome regulation and allergen production in basil (*Ocimum basilicum* L.) plants. *Frontiers in Plant Science* 9: 862.

Guo, X.H., W.Y. Gao, H.X. Chen, and L.Q. Huang. 2005. Effects of mineral cations on the accumulation of tanshinone II A and protocatechuic aldehyde in the adventitious root culture of *Salvia niltiorrhiza*. *Zhongguo Zhong Yao Za Zhi* 30: 885–888.

Heumann, H.G. 2002. Ultrastructural localization of zinc in zinc tolerant *Armeria maritima ssp. halleri* by autometallography. *Plant Physiology* 159: 191–203. https://dst.gov.in/sites/default/files/Annual%20Report%202020-21%20in%20English.pdf (Ministry of Science and Technology, 2020)

Huang, W., X. He, and C. Liu. 2010. Effects of elevated carbon dioxide and ozone on foliar flavonoids of *Ginkgo biloba*. *Advanced Material Research* 113: 165–169.

Ibrahim, M.H., H.Z. Jaafar, E. Karimi, and A. Ghasemzadeh. 2013. Impact of organic and inorganic fertilizers application on the phytochemical and antioxidant activity of Kacip Fatimah (*Labisia pumila* Benth). *Molecules* 18: 10973–10988.

Ibrahim, M.H., Y. Chee Kong, and N.A. Mohd Zain. 2017. Effect of cadmium and copper exposure on growth, secondary metabolites and antioxidant activity in the medicinal plant Sambung Nyawa (*Gynura procumbens* (Lour.) Merr). *Molecules* 22: 1623.

Idso, S., B. Kimball, and G. Pettit. 2000. Effects of atmospheric CO_2 enrichment on the growth and development of *Hymenocallis littoralis* (amaryllidaceae) and the concentrations of several antineoplastic and antiviral constituents of its bulbs. *American Journal of Botany* 87: 769–773.

Jasim, B., R. Thomas, J. Mathew, and E.K. Radhakrishnan. 2017. Plant growth and diosgenin enhancement effect of silver nanoparticles in fenugreek (*Trigonella foenum-graecum* L.). *Saudi Pharmaceutical Journal* 25: 443–447.

Kartosentono, S., S. Suryawati, G. Indrayanto, and N. Zaini. 2002. Accumulation of Cd2+ and Pb2+ in the suspension cultures of *Agave amaniensis* and *Costus speciosus* and the determination of the culture's growth and phytosteroid content. *Biotechnology Letters* 24: 687–690.

Komal, T., M. Mustafa, Z. Ali, and A.G. Kazi. 2014. Heavy metal induced adaptation strategies and repair mechanisms in plants. *Journal of Endocytobiosis and Cell Research* 25: 33–41.

Kováčik, J., J. Grúz, M. Bačkor, J. Tomko, M. Strnad, and M. Repčák. 2008. Phenolic compounds composition and physiological attributes of *Matricaria chamomilla* grown in copper excess. *Environmental and Experimental Botany* 62 : 145–152.

Kumar V, and T. Khare. 2016. Differential growth and yield responses of salt-tolerant and sus-ceptible rice cultivars to individual (Na+ and Cl−) and additive stress effects of NaCl. *Acta Physiologiae Plantarum* 38 : 170.

Kumar, S., A. Narula, M.P. Sharma, and P.S. Srivastava. 2004. In vitro propagation of *Pluchea lanceolata*, a medicinal plant, and effect of heavy metals and different aminopurines on quercetin content. *In Vitro Cellular & Developmental Biology – Plant* 40: 171–176.

Kunwar, G., C. Pande, G. Tewari, C. Singh, and G.C. Kharkwal. 2015. Effect of heavy metals on terpenoid composition of *Ocimum basilicum* L. and *Mentha spicata* L. *Journal of Essential Oil Bearing Plants* 18: 818–825.

Lu, L., C.E. He, Y. Jin, X. Zhang, and J. Wei. 2013. Effects of the applications of phosphorus and potassium fertilizers at different growth stages on the root growth and bioactive compounds of '*Salvia miltiorrhiza*' Bunge. *Australian Journal of Crop Science* 7: 1533.

Luo, L., B. Wang, J. Jiang, M. Fitzgerald, Q. Huang, Z. Yu, H. Li, J. Zhang, J. Wei, C. Yang, H. Zhang, L. Dong, and S. Chen. 2021. Heavy metal contaminations in herbal medicines: determination, comprehensive risk assessments, and solutions. *Frontiers in Pharmacology* 11: 595335.

Mahalanabis, D., A.N. Alam, N. Rahman, and A. Hasnat. 1991. Prognostic indicators and risk factors for increased duration of acute diarrhoea and for persistent diarrhoea in children. *International Journal of Epidemiology* 20: 1064–1072.

McNear, D.H., R.L. Chaney, and D.L. Sparks. 2010. The hyperaccumulator *Alyssum murale* uses complexation with nitrogen and oxygen donor ligands for Ni transport and storage. *Phytochemistry* 71: 188-200.

Mishra, T. 2016. Climate change and production of secondary metabolites in medicinal plants: a review. *International Journal of Herbal Medicine* 4: 27–30.

Murch, S.J., and P.K. Saxena. 2002. Mammalian neurohormones: potential significance in reproductive physiology of St. John's wort (*Hypericum perforatum* L.)? *Naturwissenschaften* 89: 555–560.

Nasim, S.A., and B. Dhir. 2010. Heavy metals alter the potency of medicinal plants. *Reviews of Environmental Contamination and Toxicology* 203: 139–149.

Okem, A., W.A. Stirk, R.A. Street, C. Southway, J.F. Finnie, and J. Van Staden. 2015a. Effects of Cd and Al stress on secondary metabolites, antioxidant and antibacterial activity of *Hypoxis hemerocallidea* Fisch. & C.A. Mey. *Plant Physiology and Biochemistry* 97: 147–155.

Okem, A., C. Southway, W.A. Stirk, R.A. Street, J.F. Finnie, and J. Van Staden. 2015b. Effect of cadmium and aluminum on growth, metabolite content and biological activity in *Drimia elata* (Jacq.) Hyacinthaceae. *South African Journal of Botany* 98: 142–147.

Oves, M., M.K. Saghir, A. Q. Huda, M.F. Nadeen, and T. Almeelbi. 2016. Heavy Metals: biological importance and detoxification strategies. *Journal of Bioremediation and Biodegradation* 7: 334.

Preeti, M., C. Sukhmal, and V.K. Yadav. 2007. Optimal level of iron and zinc in relation to its influence on herb yield and production of essential oil in menthol mint. *Communications in Soil Science and Plant Analysis* 38: 561–578.

Rahimtoola, S. 2004. Digitalis therapy for patients in clinical heart failure. *Circulation* 109: 2942–2946.

Rai, V., and S. Mehrotra. 2008. Chromium-induced changes in ultramorphology and secondary metabolites of *Phyllanthus amarus* Schum & Thonn – an hepatoprotective plant. *Environmental Monitoring and Assessment* 147: 307–315.

Rai, V., S. Khatoon, S.S. Bisht, and S. Mehrotra. 2005. Effect of cadmium on growth, ultramorphology of leaf and secondary metabolites of *Phyllanthus amarus* Schum. and Thonn. *Chemosphere* 61: 1644–1650.

Rai, V., P. Vaypayee, S.N. Singh, and S. Mehrotra. 2004. Effect of chromium accumulation on photosynthetic pigments, oxidative stress defense system, nitrate reduction, proline level and eugenol content of *Ocimum tenuiflorum* L. *Plant Science* 167: 1159–1169.

Sá, R.A., O. Alberton, Z.C. Gazim, A. Laverde Jr, J. Caetano, and D.C. Dragunski. 2015. Phytoaccumulation and effect of lead on yield and chemical composition of *Mentha crispa* essential oil. *Desalination and Water Treatment* 53: 3007–3017.

Sarma, H. 2011. Metal hyperaccumulation in plants: a review focusing on phytoremediation technology. *International Journal of Environmental Science and Technology* 4: 118–138.

Sarma, H., S. Deka, H. Deka, and R.R. Saikia. 2011. Accumulation of Heavy Metals in Selected Medicinal Plants. *Reviews of Environmental Contamination and Toxicology*, 214: 63–86.

Sharma, A., G. Sidhu, F. Araniti, A.S. Bali, B. Shahzad, D.K. Tripathi, M. Brestic, M. Skalicky, and M. Landi. 2020. The role of salicylic acid in plants exposed to heavy metals. *Molecules* 25(3): 540.

Shriram, V., V. Kumar, R.M. Devarumath, T.S. Khare, and S.H. Wani. 2016. MicroRNAs as potential targets for abiotic stress tolerance in plants. *Frontiers in Plant Science* 7: 817.

Stuhlfauth, T., and H. Fock. 1990. Effect of whole season CO_2 enrichment on the cultivation of a medicinal plant, *Digitalis lanata. Journal of Agronomy and Crop Science* 164: 168–173.

Stuhlfauth, T., K. Klug, and H. Fock. 1987. The production of secondary metabolites by Digitalis lanata during CO_2 enrichment and water stress. *Phytochemistry* 26: 2735–2739.

Tirillini, B., A. Ricci, G. Pintore, M. Chessa, and S. Sighinolfi. 2006. Induction of hypericins in *Hypericum perforatum* in response to chromium. *Fitoterapia* 77: 164–70.

Tumova, L., J. Poustková, and J. Tuma. 2001. $CoCl_2$ and $NiCl_2$ elicitation and flavonoid production in *Ononis arvensis* l. culture in vitro. *Acta Pharmaceutica* 51: 159–162.

Usha, B., G. Venkataraman, and A. Parida. 2009. Heavy metal and abiotic stress inducible metallothionein isoforms from Prosopis juliflora (SW) DC show differences in binding to heavy metals in vitro. *Molecular Genetics and Genomics* 281: 99–108.

Wani, S.H., T. Dutta, N.R.R. Neelapu, and C. Surekha. 2017. Transgenic approaches to enhance salt and drought tolerance in plants. *Plant Gene* 11: 219–231. doi:10.1016/j.plgene.2017.05.006

Wani, S.H., V. Kumar, V. Shriram, and S. K. Sah. 2016. Phytohormones and their metabolic engineering for abiotic stress tolerance in crop plants. *Crop Journal* 4 : 162–176.

Yadav, S., and P. Mohanpuria. 2009. Responses of *Camellia sinensis* cultivars to Cu and Al stress. *Biologia Plantarum* 53: 737–740.

Yildirim, E., M. Ekinci, M. Turan, A.G.A.R. Güleray, O.R.S. Selda, A. Dursun, and T. Balci. 2019. Impact of cadmium and lead heavy metal stress on plant growth and physiology of rocket (*Eruca sativa* L.). *Kahramanmaraş Sütçü İmam Üniversitesi Tarım ve Doğa Dergisi* 22(6): 843–850.

Zhang, B., L.P. Zheng, W.Y. Li, and J.W. Wang. 2013. Stimulation of artemisinin production in *Artemisia annua* hairy roots by $Ag–SiO_2$ core-shell nanoparticles. *Current Nanoscience* 9: 363–370.

Zhang, C., Q. Yan, W. Cheuk, and J. Wu. 2004. Enhancement of tanshinone production in *Salvia miltiorrhiza* hairy root culture by Ag+ elicitation and nutrient feeding. *Planta Medica* 70: 147–151.

Zheng, Z., and M. Wu. 2004. Cadmium treatment enhances the production of alkaloid secondary metabolites in *Catharanthus roseus. Plant Science* 166: 507–514.

Zhou, L., G. Yang, H. Sun, J. Tang, J. Yang, Y. Wang, T.A. Garran, and L. Guo. 2016. Effects of different doses of cadmium on secondary metabolites and gene expression in *Artemisia annua* L. *Frontiers in Medicine* 11:137–146. doi:10.1007/s11684-016-0486-3.

7 Impact of Industrial Wastewater on Medicinal Plant Growth

Shipra Jha, Paras Porwal

CONTENTS

7.1 INTRODUCTION

Water insufficiency, along with the freshwater resources contamination through industrial wastewater, is an immense problem throughout the world. Industrial wastewater and sewage water have been utilized for plant irrigation in many developed and underdeveloped countries, due to the limited availability of fresh water. The utilization of sewage water for plant irrigation is mostly a hidden practice. Sewage water has improved soil fertility, organic matter, and soil microflora (Malla et al. 2007). In cities, this water is utilized by farmers because of its easy availability and also because it is a rich source of nutrients such as phosphate and nitrates (which act as fertilizer). In addition to this, it is a reliable and economic way to solve the water disposal problem (Rattan et al. 2005). The chlorophyll content, photosynthetic activity, carbohydrate content of leaves, and ultimately the yield of plants such as cauliflower, spinach, cabbage, groundnut, fodder, mulberry leaves, and leafy vegetables was increased on irrigating plants with sewage water (Ahmad et al. 2006; Zeid and Abou El Ghate 2007). Industrial wastewater contains plant growth nutrients, toxic heavy metals, and numerous pathogenic microorganisms. These plant growth nutrients improve soil fertility, water-holding capacity, and permeability (Kausar et al. 2017), whereas toxic heavy metals together with the pathogenic microorganisms are responsible for inducing serious health-related problems (Hoek et al. 2002). The effect of toxic contaminants and pathogen load can be minimized by diluting the wastewater and treating wastewater in ponds (avoiding direct contact with plants) before

DOI: 10.1201/9781003178866-7

133

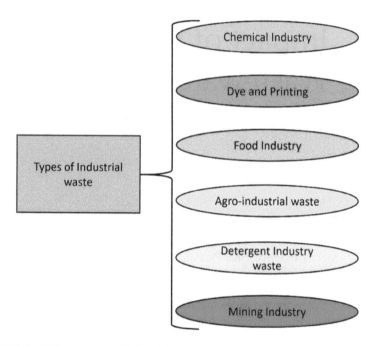

FIGURE 7.1   Different types of industrial waste.

irrigation, respectively (Constantino et al. 2005). Sometimes, fertilizers are used along with the industrial wastewater to fulfil the plant nutrient requirement (Akhtar et al. 2012; Chalkoo 2014). The continuous and massive use of fertilizers is responsible for various types of environmental pollution and is a big threat to biodiversity by inflowing groundwater and eutrophication (Iqbal 2012). Many research studies have confirmed that short-term utilization of sewage water is not harmful to plants but may cause phytotoxicity in the long term (Malla et al. 2007). Farmers residing close to cities and industries are utilizing declining water quality (industrial effluents) for irrigating the plant crops which exerts serious health issues (CPCB 2009). Taken together, this chapter summarizes the effects of various industrial effluents on the growth parameters of medicinal plants.

7.2   HEAVY METALS AND MEDICINAL PLANTS

The concentration of heavy metals, such as arsenic (As), cadmium (Cd), chromium (Cr), aluminium (Al), and manganese (Mn), in wastewater is increasing, due to various anthropogenic activities like rapid industrialization (industrial waste) and urbanization, agricultural practices (overuse of chemical fertilizers), municipal waste, and so on. In addition to this, plants are exposed to these heavy metals via atmospheric dust, rainfall, and plant safeguard agents (Maobe et al. 2012). These heavy metals are the principal contaminants in medicinal plants and leafy vegetables and their uptake mechanism in plants depends on innumerable factors, such as plant species, soil characteristics (pH), and organic matter content (Ajasa et al. 2004; Shaban et al. 2016). However, its transfer and accumulation in soil and plants affect plant growth negatively. This also leads to food chain exposure and is responsible for causing severe health disorders in humans, such as skin eruptions, neurological and gastro-intestinal diseases, lung cancer, kidney damage, and reproductive effects (Aery and Rana 2007; Singh et al. 2011; Gharebaghi et al. 2017). The herb named basil (*Tulsi*) is known to be utilized for various medicinal properties, such as improving appetite, acidity, kidney function, and other infections (Hocking et al. 2000). A reduction in plant height, number of leaves, and seed germination was reported when *Ocimum basilicum* and *Ocimum purpuresens* were grown under excess Cd (Islam 2007). Ibrahim et al. (2017) also observed a significant decrease in basal diameter, leaf area, plant dry weight, and chlorophyll concentration when the medicinal plant *Gynura procumbens* was grown over heavy metal contaminated soil with Cu and Cd (stressed conditions). Valerian (*Valeriana officinalis* L.) is a perennial herb and its root extract is utilized for reducing anxiety, nervous agitation, depression, and insomnia (Weeks 2009). The valeric acid found in this herb is known to reduce abdominal cramps during menstruation by relaxing gastrointestinal smooth muscles (Wielgosz et al. 2012). The pH of soil plays a decisive role in heavy metal uptake by plants, although the uptake mechanism is a complex process and depends on diverse factors, such as soil characteristics, plant species and genotype, microbial activity (including rhizosphere), metal motility in soil, and so on (Adamczyk-Szabela et al. 2015). The research studies reveal that valerian seeds do not germinate at pH 4.5 so it has proved difficult to examine heavy metal uptake while at alkaline pH (up to 10) low zinc uptake and heavy metal motility are shown, depending upon organic constituents of soil (Violantel et al. 2010; Adamczyk-Szabela et al. 2015). Thus, soil pH has a strong impact on the upregulation of heavy metals by the plants.

7.3   THE DETERGENT INDUSTRY AND MEDICINAL PLANTS

Tomato (*Solanum lycopersicum* L.) is a vegetable that belongs to the family *Solanaceae* that is cultivated (outdoors or indoors) and consumed throughout the world. It contributes a good source of several vitamins, such as vitamins C, A, and B, and minerals, such as iron, potassium, and phosphorus, to the diet (Charchar et al. 2003). Tomato is eaten fresh or in processed form, such as juice, puree, powder, sauce, ketchup, soup, paste, canned product, etc. Currently, due to the presence of various bioactive compounds such as lycopene, it is being utilized for the development and formulation of functional foods and nutraceuticals (Friedman 2013). The medicinal benefits of tomato, especially due to the presence of carotene, phenolics, and lycopene (an antioxidant), include prevention or treatment of eye-related diseases (cataract, macular degeneration), risk of cancer (prostate and colon), cardiovascular diseases, asthma, and bone metabolism (Burton-Freeman and Reimers 2011). The effect of various detergent effluent concentrations (5%, 10%, 15%, 20%, and 25%) conducted on medicinal plant tomato in Nigeria reveals a reduction in foliar anatomy, yield and growth at higher concentrations of detergent effluent exceeding 25%, whereas at lower effluent concentrations from 5% to 20% a progressive increase in plant growth parameters was observed, such as increased plant height, fruit, shoot and root weight (wet/dry basis), fruit number, biomass, stem girth, germination time, flowering, and ultimately overall growth and yield as compared to the control (Okanume et al. 2017).

Okra (*Abelmoschus esculentus*) is a vegetable that belongs to the family Malvaceae, cultivated as a garden crop or on a commercial scale in several countries, such as India, Iran, Pakistan, Bangladesh, Japan, Thailand, Afghanistan, Turkey, Ethiopia, and Western Africa. Okra is not a staple food, consumed as diet food and its seeds are rich in protein (lysine, tryptophan), unsaturated fatty acid (linoleic acid), and phenolic compounds. Okra contains folate, ascorbic acid, vitamin B, fibre, calcium, iron, zinc, manganese, and potassium (Moyin-Jesu 2007; Gosslau and Chen 2004). The medicinal benefits of okra include prevention or treatment of chronic diseases such as cancer, heart disease (hypocholesterolemic effect), regulating the absorption of blood sugar, gastro-protective, anti-ulcerogenic, and diuretic, chemo-preventive agents, and many more (Kendall and Jenkins 2004; Gemede et al. 2015). The effect of various soap and detergent effluent concentrations (1%, 05%, 10%, 20%, 25%, 50, 75%, and 100%) conducted on the medicinal plant okra in Nigeria reveals that wastewater below 50% only supports germination, flowering after 37 days (treated with 1%, 5%, 10% concentration), and greener colour (with 1%, 5%, 10% concentration). Thus, for irrigation purposes, detergent and soap effluent below 5% can be used, due to lower concentrations of nitrogen and phosphate (Fatoba et al. 2011).

7.4   THE TEXTILE INDUSTRY AND MEDICINAL PLANTS

On a global scale, the textile/garment industry contributes significantly to the economy of a country. The industry uses synthetic chemicals with diverse compositions, ranging from organic to inorganic substances for dyeing and printing, and continuously adding toxic substances in groundwater reservoirs and running water (Islam et al. 2015). The effluent obtained from the textile industry is highly coloured (impedes light penetration and photosynthetic activities), contains excessive salts (saline soil), various micronutrients, and non-biodegradable substances with high biological and chemical oxygen demands (Wynne et al. 2001). The utilization of this untreated wastewater for plant irrigation leads to the transfer and accumulation of these toxic substances in plant tissues and has a detrimental effect on plant and human growth (Garg and Kaushik 2008; Hai 2007). But at a lower concentration of effluent, a positive impact on germination and growth of *Pisum sativum* and, at 100 % effluent concentration, a reduction in plant root and shoot was noticed (Malaviya et al., 2012). *Medicago sativa* L. (alfalfa) is a leguminous plant with high protein and nutrition content and also affects the fertility of soils. The medicinal properties of alfalfa are utilized for the treatment of elevated cholesterol, diabetes, asthma, osteoarthritis, kidney, bladder, and prostate conditions, and menopausal symptoms (Bora and Sharma 2011; Mahawar and Akhtar 2016). It also contains various vitaminss such as vitamins A, C, E, and K, and minerals, such as iron, phosphorous, potassium, and calcium, and is a good source for commercial chlorophyll (Choi et al. 2013). The effect of various effluent concentrations (20%, 40%, 60%, 80%, and 100%) conducted on the medicinal plant *Medicago sativa* reveals a reduction in root and shoot length, along with the chlorophyll concentration at a higher concentration of effluent exceeding 40%, which may be due to higher total solids and increased salinity (Mahawar and Akhtar 2016). In a study conducted on *Vigna munga* L. (black gram), shoot and root length growth was promoted at 25% effluent concentration together with growth-promoting effect, but above 25% this reduced significantly as compared with the control sample (Wins et al. 2010). Mohammad and Khan (1985) find no detrimental effect of textile industry effluents on soil and plants below 50% effluent concentration. *Sesamum indicum* L. (sesame) finds its medicinal value in antibacterial activity (against *Staphylococcus* and *Streptococcus*), lower blood cholesterol in humans due to the presence of sesamolin and sesamin compounds, and is also utilized in the cosmetic industry (Anilakumar et al. 2010). Lenin et al. (2014) also noticed similar results in the medicine plant sesame that, at 20% effluent concentration (sago factory), promotes seed germination, weight (fresh and dry), and root and shoot length whereas at above 20% effluent concentration significantly reduced. When textile and dye effluent is diluted with water (1:3 ratio), this has a positive impact on the growth and germination of groundnut and no detrimental effect on the growth and germination of field crops (Parameswari 2014).

7.5 THE DAIRY INDUSTRY AND MEDICINAL PLANTS

Increased modernization and consumer health consciousness has forced industry people to develop a diverse range of milk products. The milk-processing industries undertake different unit operations to convert raw milk into processed milk products, such as the production of cream, butter, sour milk, different varieties of cheese, milk (condensed or dried), yoghurt, and ice cream including frozen desserts, and many more. A large amount (depending on plant size and processing) of wastewater is produced in many dairy operations, such as pasteurization, homogenization, and cleaning of dairy tankers and plant. The milk industries produce effluent approximately three times greater than milk volume (Manu et al. 2012; Kaur and Sharma 2017). Dairy industry effluent is rich in total dissolved solids, and organic and inorganic nutrients (Verma and Singh 2017). *Vigna radiata* L. (mung bean), also known as green gram, contains carbohydrates, proteins, vitamins, minerals, dietary fibres, and bioactive compounds (for nutraceuticals formulation). It is consumed as a pulse or in sprouted form as salad. The medicinal properties of mung bean include antidiabetic, antihyperlipidemic, antimicrobial, anti-inflammatory, antitumor, anticancerous, antisepsis, and many more (Tang et al. 2014). The effect of various dairy effluent concentrations (20%, 40%, 60%, 80%, and 100%) conducted on medicinal plants in mung bean and mustard reveals maximum seed germination of 84% and 100% at 20% effluent concentration, while only 25% and 32 % seed germination at 100% effluent was observed in mung bean and mustard seeds respectively when compared with the control sample (Kapil and Mathur, 2020). The lower concentration of effluent (below 20%) upholding seed germination and overall plant growth parameters may be due to low concentration of dairy effluent but, at higher concentrations, a decrease in seed germination was observed which may be due to increased osmotic pressure. Similar findings were also reported on green and black gram (Prasannakumar et al. 1997), soybeans (Gaikar et al. 2010) on irrigating with industry effluent and Ragi (Lakshmi and Sundaramoorthy 2003), wheat (Singh et al. 2002; Koushik et al. 2005) and on irrigating with other industrial effluents. Therefore, the dairy industrial effluent (eco-friendly, cost-effective) at low concentration can be best utilized as liquid fertilizer for irrigation of medicinal crop plants.

7.6   THE OIL INDUSTRY AND MEDICINAL PLANTS

Crude oil is the unrefined oil located below ground that can only be used after refining, such as kerosene, gasoline, etc. The accidental spillage of crude oil during mining affects the plant ecosystem adversely. Oil spillage hinders germination and other nutrient uptake of plants by building a layer that prevents the availability of oxygen to seed (Oyem and Oyem, 2013). Oil spillage hampers the normal physical operations of soil that includes blocking of soil pores, restricting root extension, limiting aeration, and rate of infiltration; increased soil acidity, nitrogen, heavy metals, and organic content matters (Abosede 2013). The plants grown in oil spillage areas are reported for impaired physiological functioning, such as dryness of plant, leaf chlorosis, biomass reduction, root-shoot length reduction, and brownish or yellowish colour of vegetation (Agbogidi 2011; Osuagwu et al. 2013). Maize or corn (*Zea mays* L.) is consumed as a staple food, after wheat and rice. It is a good source of various nutrients and phytochemical compounds, such as carotenoids, phytosterols, and phenolic compounds. Phytochemicals help to prevent or treat many diseases, like nausea, vomiting, stomach upset, and bladder irregularity. The zein protein and maize oil are extensively used in the nutraceutical and pharmaceutical industries. Additionally, the resistant starch helps to retard atherosclerosis, risk of cecal cancer, and obesity complications (Shah et al. 2016). The impact of crude oil contamination at treatment rate (0, 10, 20, 30, 40, 50, and 60 mL) was conducted (on field and pot experiment) on medicinal plant maize to analyse different growth parameters after 4, 8, and 12 weeks of planting. It reveals a progressive decrease in plant height, leaf area index, and dry matter yield, with an increased level of crude oil contamination in both the experiments (field and pot) with an exception that, in pot experiment at 10 and 20 mL treatment, increase in the height of the plant was observed, as compared to the control plant (Agim et al. 2021). Thus, cultivation of plants should be avoided in oil spillage fields, and polluted soil can be remediated quickly and hydrocarbons can be degraded through remediation.

7.7   OTHER INDUSTRIES' EFFLUENT AND MEDICINAL PLANTS

Cement kiln dust is obtained as the main by-product from the factories of the cement industries, which interferes with the photosynthesis, seed production, and pollen germination of plant leaves such as *Gentiana kurroo* (a herb used for digestive disorders). Effluent from these industries dumped directly into nearby rivers without treatment is responsible for causing negative effects on the flora and fauna of rivers (Kunal et al. 2016). Ipeaiyeda et al. (2017), in their research study conducted on cement industry effluent on river (Onyi, Nigeria) water quality, states that river water was alkaline (pH 9), with high turbidity (7.5–804 FTU) that interferes with light penetration: chloride (if higher can corrode metal) and sulphate (within regulatory limit), excessive phosphate and nitrate (accelerated eutrophication), BOD below 6 mg/L (below regulatory limit), and COD above 30 mg/L (above regulatory limit). The alkaline pH and chemical conformation of cement kiln dust upsets carbohydrate metabolism harshly, and this causes surface plasmolysis and ultimately arrests starch formation in plants (Czaja 1960; Lawler et al. 2002).

7.8 CONCLUSION

The external morphology of medicinal plants does not guarantee safety from pollutants, especially when the plant is cultivated on or around contaminated sites. Thus, it creates a quality concern over medicinal products that require meticulous control and risk analysis before use. The prolonged accumulation of essential or non-essential metals above the threshold causes various health-related problems in humans and also negatively impacts plant growth parameters. So, controlled growth with the adaptation of Good Agricultural Practices (GAP) and Good Manufacturing Practices (GMP) minimizes medicinal plant contamination. Therefore, industrial effluent treatment with a cost-effective approach must be performed before discarding into natural water bodies (rivers/lakes) to minimize water pollution and simultaneously increase the reuse of wastewater for irrigation purposes. The research work conducted on the effect of wastewater from various industries on medicinal plants can be utilized as a yardstick to predict a multitude of toxic effects of pollutants in human beings. In conclusion, many research studies have been carried out on a controlled environment, but now field trials are needed to perform risk analysis and develop practical guidelines for medicinal plant producers.

REFERENCES

Abosede, E.E. 2013. Effect of crude oil pollution on some soil physical properties. *Journal of Agriculture and Veterinary Science* 6(3): 14–17.

Adamczyk-Szabela, D., J. Markiewicz, and W.M. Wolf. 2015. Heavy metal uptake by herbs. IV. Influence of soil pH on the content of heavy metals in *Valeriana officinalis* L. *Water, Air, & Soil Pollution* 226(4).

Aery, N.C., and D.K. Rana. 2007. Interactive effects of Zn, Pb and Cd in Barley. *Journal of Environmental Science & Engineering* 49(1): 71–76.

Agbogidi, O.M. 2011. Effects of crude oil contaminated soil on biomass accumulation in *Jatropha curcas* L. Seedlings. *Journal of Ornamental and Horticultural* Plants 1(1): 43–49.

Agim, L.C. 2021. Lethality of crude oil contamination on properties of soil and growth parameters of maize (*Zea mays* L.). *International Journal of Environmental Sciences & Natural Resources* 28(3): 1–11.

Ahmad, B., K. Bakhsh, and S. Hassan. 2006. Effect of sewage water on spinach yield. *International Journal of Agriculture and Biology* 3: 423–25.

Ajasa, A.M.O., M.O. Bello, A.O. Ibrahim, et al. 2004. Heavy trace metals and macronutrients status in herbal plants of Nigeria. *Food Chemistry* 85(1): 67–71.

Akhtar, N., A. Inam, and N. Khan. 2012. Effects of city wastewater on the characteristics of wheat with varying doses of nitrogen, phosphorus and potassium. *Recent Research in Science and Technology* 4: 18–29.

Anilakumar, K., A. Pal, F. Khanum, and A.S. Bawa. 2010. Nutritional, medicinal and industrial uses of sesame (*Sesamum indicum* L.) seeds – an overview. *Agriculturae Conspectus Scientificus* 75(4): 159–168.

Bora, K.S., and A. Sharma. 2011. Phytochemical and pharmacological potential of *Medicago sativa*: A review. *Pharmaceutical Biology* 49(2): 211–220.

Burton-Freeman, B., and K. Reimers. 2011. Tomato consumption and health: emerging benefits. *American Journal of Lifestyle Medicine* 5(2): 182–191.

Chalkoo, S., S. Sahay, A. Inam, and S. Iqbal. 2014. Application of wastewater irrigation on growth and yield of chilli under nitrogen and phosphorus fertilization. *Journal of Plant Nutrition* 37(7): 1139–1147.

Charchar, A.U., J.M. Gonzaga, V. Giordano, et al. 2003. Reaction of tomato cultivars to infection by a mixed population of *m. Incognita* Race and *M. Javanica* in the field. *Nematologia Brasileira* 27: 49–54.

Choi, K.C., J.M. Hwang, S.J. Bang et al. 2013. Chloroform extract of alfalfa (*Medicago sativa*) inhibits lipopolysaccharide-induced inflammation by downregulating ERK/NF-KB signaling and cytokine production. *Journal of Medicinal Food* 16(5): 410–420.

CPCB. 2009. Status of water supply, wastewater generation and treatment in class-I cities and class-II towns of India. Control of Urban Pollution series: CUPS/70/2009-10. New Delhi: Central Pollution Control Board, Ministry of Environment and Forests.

Czaja, A.T. 1960. Die Wirkung von verstäubtem Kalk und Zement auf Pflanzen. *Plant Foods for Human Nutrition* 7: 184–212. https://doi.org/10.1007/BF01099766

Fatoba, P.O., K.S. Olorunmaiye, and A.O. Adepoju. 2011. Effects of soaps and detergents wastes on seed germination, flowering and fruiting of tomato (*Lycopersicon esculentum*) and okra (*Abelmoschus Esculentus*) plants. *Ecology Environment and Conservation* 17(1): 7–11.

Friedman, M. 2013. Anticarcinogenic, cardioprotective, and other health benefits of tomato compounds lycopene, α-tomatine, and tomatidine in pure form and in fresh and processed tomatoes. *Journal of Agricultural and Food Chemistry* 61(40): 9534–9550.

Gaikar, R.B., Uphade, A. Gadhave, and S. Kuchekar. 2010. Effect of dairy effluent on seed germination & early seedling growth of soyabeans. *Rasayan Journal Chemistry* 3: 137–139.

Garg, V.K., and P. Kaushik. 2008. Influence of textile mill wastewater irrigation on the growth of sorghum cultivars. *Applied Ecology and Environmental Research* 6(1): 1–12.

Gemede, H.F., N. Ratta, G.D. Haki, et al. 2015. Nutritional quality and health benefits of okra (*Abelmoschus esculentus*): a review. *Journal of Food Processing Technology* 6(458): 2.

Gharebaghi, A., M.H.A. Haghighi, and H. Arouiee. 2017. Effect of cadmium on seed germination and earlier basil (*Ocimum basilicum* L. and *Ocimum basilicum* Var. *Purpurescens*) seedling growth. *Trakia Journal of Sciences* 1(1): 1–4.

Gosslau, A., and K.Y. Chen. 2004. Nutraceuticals, apoptosis, and disease prevention. *Nutrition* 20(1): 95.

Hai, F.I., K. Yamamoto, and K. Fukushi. 2007. Hybrid treatment systems for dye wastewater. *Critical Reviews in Environmental Science and Technology* 37(4): 315–377.

Hocking, P.J., and M.J. McLaughlin. 2000. Genotypic variation in cadmium accumulation by seed of linseed and comparison with seeds of some other crop species. *Australian Journal of Agricultural Research* 51(4): 427–433.

Hoek, W. van der, M. U. Hassan, J. H. J. Ensink, et al. 2002. Urban wastewater: a valuable resource for agriculture. A case study from Haroonabad, Pakistan. Research report 63, *International Water Management Institute (IWMI)*, Colombo, Sri Lanka.

Ibrahim, M.H., Y.C. Kong, and N.A.M. Zain. 2017. Effect of cadmium and copper exposure on growth, secondary metabolites and antioxidant activity in the medicinal plant *Sambung nyawa* (*Gynura Procumbens* (Lour.) Merr). *Molecules* 22(10): 1623.

Ipeaiyeda, A.R., and G.M. Obaje. 2017. Impact of cement effluent on water quality of rivers: a case study of Onyi river at Obajana, Nigeria. *Cogent Environmental Science* 3: 1.

Iqbal S., A. Inam, S. Sahay, and S. Chalkoo. 2012. Balance use of inorganic fertilizers on chilli (*Capsicum annuum* L.) irrigated with wastewater. *Biosciences International* 1(3): 82–89.

Islam, E.U., X.E. Yang, Z.L. He, Q. Mahmood. 2007. Assessing potential dietary toxicity of heavy metals in selected vegetables and food crops. *Journal of Zhejiang University. Science. B* 8(1): 1–13.

Islam, J.B., M. Sarkar, A.K.M.L. Rahman, and K.S. Ahmed. 2015. Quantitative assessment of toxicity in the Shitalakkhya river, Bangladesh. *The Egyptian Journal of Aquatic Research* 41(1): 25–30.

Kapil, J., and N. Mathur. 2020. The impact of dairy effluent on germination parameters of seeds of mung bean (*Vigna Radiata*) and mustard (*Brassica nigra*). *International Journal of Economic Plants* 7(4): 170–175.

Kaur, V., and G. Sharma. 2017. Impact of dairy industrial effluent of Punjab (India) on seed germination and early growth of *Triticum aestivum*. *Indian Journal of Science and Technology* 10(16): 1–9.

Kausar, S., S. Faizan, and I. Haneef. 2017. Nitrogen level affects growth and reactive oxygen scavenging of fenugreek irrigated with wastewater. *Tropical Plant Research* 4(2): 210–224.

Kendall, C.W.C., and D.J.A. Jenkins. 2004. A dietary portfolio: maximal reduction of low-density lipoprotein cholesterol with diet. *Current Atherosclerosis Reports* 6(6): 492–498.

Koushik, P., V.K. Garg, and B. Singh. 2005. Effect of textile effluents on growth performance of wheat cultivars. *Bioresource Technology* 96(10): 1189–1193.

Kunal, A.R., and R. Siddique. 2016. Bacterial treatment of alkaline cement kiln dust using bacillus *Halodurans* strain KG1. *Brazilian Journal of Microbiology* 47(1): 1–9.

Lakshmi, S., and P. Sundaramoorthy. 2003. Effect of chromium on germination and biochemical changes in blackgram. *Journal of Ecobiology* 15: 7–11.

Lawler, J.J., S.P. Campbell, A.D. Guerry, et al. 2002. The scope and treatment of threats in endangered species recovery plans. *Ecological Applications* 12(3): 663–667.

Lenin, M., K.S. Mariyappan, and M.R. Thamarikannan. 2014. Effect of Sago Factory Effluent on Seed Germination and Seedling Growth Of gingelly (*Sesamum Indicum* L.) varieties. *International Journal of Life Sciences Biotechnology and Pharma Research* 3(1): 2250–3137.

Lucho-Constantino, C.A., M. Alvarez-Suarez, R.I. Beltran-Hernandez, et al. 2005. A multivariate analysis of the accumulation and fractionation of major and trace elements in agricultural soils in Hidalgo State, Mexico irrigated with raw wastewater. *Environment International* 31(3): 313–323.

Mahawar, P., and A. Akhtar. 2016. Impact of dye effluent on growth and chlorophyll content of alfalfa (*Medicago sativa* L.). *Annals of Plant Sciences* 5(10): 1432–1435.

Malaviya, P., R. Hali, and N. Sharma. 2012. Impact of Dyeing Industry Effluent on Germination and Growth of Pea (*Pisum sativum*). *Journal of Environmental Biology* 33(6): 1075–1078.

Malla, R., Y. Tanaka, K. Mori, and K.L. Totawat. 2007. Effect of short-term sewage irrigation on chemical build up in soils and vegetables. *The Agricultural Engineering International: The CIGRE Ejournal* IX.

Manu, K.J., M.V.M. Kumar, and V.S. Mohana. 2012. Effect of dairy effluent (treated and untreated) on seed germination, seedling growth and biochemical parameters of maize (*Zea Mays* L.). *International Journal of Research in Chemistry and Environment (IJRCE)* 2(1): 62–69.

Maobe, M.A.G., E. Gatebe, L. Gitu, and H. Rotich. 2012. Profile of heavy metals in selected medicinal plants used for the treatment of diabetes, malaria and pneumonia in Kisii region, Southwest Kenya. *Global Journal of Pharmacology* 6(3): 245–251.

Mohammad, A., and A.U. Khan. 1985. Effects of a Textile Factory Effluent on Soil and Crop Plants. *Environmental Pollution Series A, Ecological and Biological* 37(2): 131–148.

Moyin-Jesu, E.I. 2007. Use of plant residues for improving soil fertility, pod nutrients, root growth and pod weight of okra (*Abelmoschus esculentum* L). *Bioresource Technology* 98(11): 2057–2064.

Okanume, O.E., O.M. Joseph, O.A. Agaba, et al. 2017. Effect of industrial effluent on the growth, yield and foliar epidermal features of tomato (*Solanum lycopersicum* L.) in Jos, Plateau State, Nigeria. *Notulae Scientia Biologicae* 9(4): 549–556.

Osuagwu, A.N., A.U. Okeigbo, I.A. Ekpo, et al. 2013. Effect of crude oil pollution on growth parameters, chlorophyll content and bulbils yield in air potato (*Dioscorea bulbifera* L.). *International Journal of Applied Sciences and Technology* 3(4): 37–42.

Oyem, I.L.R., and I.L. Oyem. 2013. Effects of crude oil spillage on soil physico-chemical properties in Ugborodo Community. *International Journal of Modern Engineering Research* 3(6): 3336–3342.

Parameswari, M. 2014. Effect of textile and dye effluent irrigation on germination and its growth parameters of green gram, black gram and red gram. *International Journal of Environmental Science and Toxicology Research* 2(1): 6–10.

Prasannakumar, P.-G., P.-R Pandit, and R. Kumar. 1997. Effect of dairy effluent on seed germination bioassays to assess toxicity of seed germination, seedling growth and pigment of green gram (*Phaseolus areus* L.) and black gram effluent (*Phaseolus mungo* L.). *Advances in Plant Sciences* 2: 146–149.

Rattan, R.K., S.P. Datta, P.K. Chenkar, et al. 2005. Long-term impact of irrigation with sewage effluents on heavy metal content in soils, crops and groundwater – a case study. *Agriculture, Ecosystems & Environment* 109(3): 310–322.

Shaban, N.S., A.A. Khaled, and N.Y.H. El-Houda. 2016. Impact of toxic heavy metals and pesticide residues in herbal products. *Journal of Basic and Applied Sciences* 5(1): 102–106.

Shah, T.R., K. Prasad, and P. Kumar. 2016. Maize – A potential source of human nutrition and health: a review. *Cogent Food & Agriculture* 2(1): 1166995.

Singh, A., S.B. Agrawal, J.P.N. Rai, and P. Singh. 2002. Assessment of the pulp and paper mill effluent on growth, yield and nutrient quality of wheat (*Triticum aestivum* L.). *Journal of Environmental Biology* 23(3): 283–288.

Singh, R., N. Gautam, A. Mishra, and R. Gupta. 2011. Heavy metals and living systems: an overview. *Indian Journal of Pharmacology* 43(3): 246–253.

Tang, D., Y. Dong, H. Ren, L. Li, and C. He. 2014. A review of phytochemistry, metabolite changes, and medicinal uses of the common food mung bean and its sprouts (*Vigna Radiata*). *Chemistry Central Journal* 8(1): 4.

Verma, A., and A. Singh. 2017. Physico-chemical analysis of dairy industrial effluent. *International Journal of Current Microbiology and Applied Sciences* 6(7): 1769–1775.

Violante, A., V. Cozzolino, L. Perelomov, et al. 2010. Mobility and bioavailability of heavy metals and metalloids in soil environments. *Journal of Soil Science and Plant Nutrition* 10(3): 268–292.

Weeks, S.B. 2009. Formulations of dietary supplements and herbal extracts for relaxation and anxiolytic action: relarian. *Medical Science Monitor: International Medical Journal of Experimental and Clinical Research* 15(11): RA256–262.

Wielgosz, T., J. Kozłowski, and J. Cis. 2012. Herbs from the Pharmacy of Nature. Publicat.

Wins, J.A., and M. Murugan. 2010. Effect of textile mill effluent on growth and germination of black gram–*Vigna Mungo* (L.) Hepper. *International Journal of Pharma and Bio Sciences* 1(1): 1–7.

Wynne, G., D. Maharaj, and C. Buckley. 2001. Cleaner production in the textile industry. In Lessons from the Danish Experience. School of Chemical Engineering, University of Natal, Durban, pp. 3–17.

Zeid, I.M., and H.M Abou El Ghate. 2007. Effect of sewage water on growth, metabolism and yield of Bean. *Journal of Biological Sciences* 7(1): 34–40.

8 Heavy Metal Pollution and Medicinal Plants
An Overview

Allah Ditta, Naseer Ullah, Xiaomin Li,
Ghulam Sarwar Soomro, Muhammad Imtiaz,
Sajid Mehmood, Amin Ullah Jan,
Muhammad Shahid Rizwan,
Muhammad Rizwan, Iftikhar Ahmad

CONTENTS

8.1 INTRODUCTION

Natural and anthropogenic components could influence the spatiotemporal heterogeneity of heavy metals (HMs) in the biosphere and earth. The comprehensive utilization of fertilizers and chemicals, and the disposal of certain industrial wastes through sewage systems have substantially contributed to the degradation of our ecosystems (Ullah et al., 2020). A drastic spike of HMs in certain situations is caused by human activities in the environment (Kanwal et al., 2021; Naveed et al., 2021). Farming techniques entail synthetic chemical contaminants, including metal-containing pest-management strategies, the use of hazardous wastewater, and chemical fertilizers (Irshad et al., 2021; Mehmood et al., 2021; Murtaza et al., 2021). In addition, industrial growth with the discharge of effluents rich in toxic chemicals has also contributed to the pollution of various water and soil reserves with HMs. Over 50 elements are being classified as 'heavy metals', having a density greater than 5 g cm^{-3}. Whilst most of these are necessary at a specific concentration for the completion of multiple metabolic processes and ultimately for optimum plant growth and productivity, several of these are listed as toxic for the plants when exposed to higher levels (Ahmad et al., 2020). Certain metals, like chromium (Cr), manganese (Mn), molybdenum (Mo), zinc (Zn), nickel (Ni), iron (Fe), and copper (Cu) are important HMs for several bio-chemical and physiological functions within crops, like seed formation, endothermic chemical transformations, electron transport, and several other cellular metabolic processes (Jan et al., 2021; Rizwan et al., 2021; Thind et al., 2021). However, at small doses, non-essential elements like arsenic (As), mercury (Hg), cadmium (Cd), Ni, and lead (Pb) initiate certain biochemical processes that are harmful to crops (Niamat et al., 2019; Ijaz et al., 2020; Liu et al., 2020; Naveed et al., 2020; Sabir et al., 2020).

Millions of people across diverse ethnic perspectives usually believe that syn-thetic medicines are successful in healing various types of infections. Nevertheless, medication has already been evolving quickly, and various synthetic medicines are being suggested to heal various types of medical complications, but consumers prefer herbal plants with fewer complications. In the present era, because of lower med-ical expenses, consumers are much more proactive in consuming natural remedies to cure disorders. As per the recommendation of the World Health Organization (WHO) study, approximately 80 per cent of individuals prefer to adopt indigenous medicinal plants to cure certain illnesses and this has been the main phase of healthcare services. Medicinal plants could be eaten in the context of vegetables, fruit, oils, or pharma-ceutical items towards therapeutic applications as well as for medical treatment.

The plants need a variety of macro- and micronutrients as well as certain naturally occurring substances from the soil for optimum growth and development. The plants can even absorb the HMs in the soil environment. In certain crops and agricultural products, residues of Cd and Pb have been found and identified between the various classes of minerals. In the last century, HM deposition in the atmosphere seems to be an enduring feature, mostly influenced by external intervention in the natural environ-ment. As a result, medicinal herbs sprouting in their natural habitat are subjected to various sources of HM exposure. Molecular and morphological reactions including oxidative stress due to the production of secondary metabolites such as reactive oxygen species (ROS) in different medicinal plants occur under various pollutants stress. In this chapter, comprehensive exposure to HMs and the subsequent impact on plants, especially medicinal ones, have been discussed. In each section, the research gaps have been highlighted and future perspectives regarding research have been discussed.

8.2   MECHANISMS OF HM UPTAKE

Owing to the potential aerosol accumulation through the air, the introduction of metals within the living cells of medicinal plants is of special importance from the emission perspective. Whether by diffusion together with water or through ion exchange among soil particles and the root surface, HMs in soil may pass through the surface of the root. The migration of the metal atoms further into the roots develops either through passive cell membrane-mediated diffusion or through active trans-port via potential electrochemical concentration gradients. Within the cell, metals can travel across an electrochemical gradient via signalling pathways inside the cel-lular membranes. Other pathways of metal penetration which may take place through cell membranes include via stomata or leaf cuticles, or maybe both during foliar uptake. Through the exudates of rhizomes and processes inside the root system, the emergence of metals through particles of soil could be encouraged. Efficient accumulation was being modified through crops to consume critical metabolites; however, many additional metals present are often picked up at the same time.

8.3 HEAVY METAL STRESS, AND THE GROWTH AND DEVELOPMENT OF MEDICINAL PLANTS

Given the consumption of medicinal plants as traditional medicine, nutrition, and dietary sources, these may be contaminated with various organic or inorganic contaminants. HMs are considered chronic inorganic toxins with a category of considerable importance, having potentially harmful effects on the exposed entity including medicinal plants. A summary of the consequences of HMs on the growth and development of medicinal plants is given in Table 8.1. Unlike certain environmental chemicals, which dissolve in water and carbon dioxide as decomposable materials, HMs could persist in the atmosphere and soil for long periods and, as described above, can cause a variety of detrimental consequences for the environment and public health (Asiminicesei et al., 2020).

The formation of hazardous ROS is induced through HM susceptibility – i.e. the circumstance under which crops trigger certain cellular pathways or hormonal factors, which counteract the consequences of disturbance caused under HM stress. In this fashion, the plants defend themselves against the formation of free radicals, stopping certain biochemical compounds and macromolecules from being damaged. The accumulation of bioactive compounds, like enzymatic antioxidants, proline, glutathione, and phenolic compounds, and flavonoids, are techniques for defending plants from HM intervention. Plants make numerous chemical materials, which are functionally distinct and may rarely be engaged in growth and development or propagation, yet are important to protect crops against abiotic stresses. Bioactive compounds had several functionalities and excellent biocompatibility and include more than several metabolic categories.

Under HM stress, the main metabolic activities influencing plant growth result through the formation of peripheral compounds. However, the climate significantly influences the cellular metabolism of crops. Such external variables may cause the development of bioactive compounds to activate immediately. With physiological and biochemical conditions, plant secondary enzymes could offer security to medicinal plants. To combat biological influences impacting the quantity or structure of bioactive compounds in medicinal plants, polyphenols and certain bioactive components having diverse medicinal features provide a beneficial effect on the medicinal capacity of the crop. Anthocyanidins, phenolic compounds, lignin, phenolic acids, alkaloids, and glycosides are among the bioactive compounds produced under HM stress. The deposition of HMs in various parts of medicinal plants has been seen in several studies (Lajayer et al., 2017; Maleki et al., 2017; Mafakheri and Kordrostami, 2021). Strong HM concentrations could negatively affect the growth and production of medicinal plants (Shen et al., 2020; Kulal et al., 2020; Azab and Hegazy, 2020; Bempah et al., 2012). A substantial reduction in the concentration of carotenoids, starch, cellulose, carbohydrate, and length of shoot and root, as well as biomass production

TABLE 8.1
Consequences of Heavy Metals on the Growth and Development of Medicinal Plants

Species	Experiment type	Heavy metals stress	Impact	References
Catnip, thyme and Fine leaf *Schizonepeta* herb	Pot	1.0–2.5 mg Cd kg⁻¹ and 500–1500 mg Pb kg⁻¹	The plant heights, dry weights of the control groups during Cd stress were less than those of the categories of catnip, thyme, and Fine leaf *Schizonepeta* Herb	Zhou et al., 2020
Tartary buckwheat	Pot	Time (15 and 30 days) and Hg concentration (0, 25, 50, and 75 μM)	Hg was readily consumed by seedlings with higher root content, resulting in a decrease in the length of the root and shoot. The root and shoot Hg uptakes were associated with each other strongly and specifically	Pirzadah et al., 2018
Ligusticum chuanxiong	Pot	5 mg Cd kg⁻¹, 300 and 500 mg Pb kg⁻¹	The net photosynthetic rate, dry-matter mass of roots, rhizomes, leaves, intercellular CO_2 concentrations, transpiration rate, and stomatal conductance were substantially reduced by Cd and Pb exposure	Zeng et al., 2020
Narcissus tazetta	Pot	0.5, 1.0 mM cadmium chloride	Peroxidase activity increased under Cd stress	Soleimani et al., 2020
Panax notoginseng	Pot	0, 0.6, 6.0, 12.0 mg Cd kg⁻¹ using $CdCl_2 \cdot 2.5H_2O$	Under different Cd treatment levels, the yields of Rg1, Rb1 and total saponins of *P. notoginseng* decreased	Zu et al., 2020
Menthol mint (*Mentha arvensis* L.)	Pot	50 μM Cd	Reduced growth, photosynthetic parameters, mineral nutrient concentration, and increased oxidative stress biomarkers like electrolyte leakage, malondialdehyde, and hydrogen peroxide contents under Cd stress	Zaid et al., 2020
Malabar spinach (*Basella alba*)	Pot-culture factorial	0, 25, 50, and 100 mg Cr kg⁻¹	Cr significantly decreased vine length, leaf number, leaf area, fresh-, and dry biomass, root length, and dry-matter ratio	Zewail et al., 2020
Cucurbita pepo	Pot	Cd-concentrations (0, 100, 300, and 500 μM)	The production of biomass decreased under the Cd-stress	Labidi et al., 2021

of *Phyllanthus amarus* can result from an increase in Cr concentration (Jaison and Muthukumar, 2017). Related findings had been observed previously in Cd-exposed crops of *P. amarus* where the concentration of starch was raised. In another study, Shen et al. (2017) found that smelting and mining activities have contributed to soil pollution with HMs and this has resulted in HM deposition in different parts of medicinal plants like fengdan. Elemental Zn, Pb, Cu, and Cd were among the main metal contaminants accumulated in different parts of the *Fengdan moutan* cortex. Similarly, Sharmila et al. (2016) reported that *Brassica juncea* (Indian mustard) adapted to the concentration of Cd^{2+} via intensity variation, a substantial decrease in primary productivity of cells in photosystem II with quite a subsequent improvement in amounts of proline.

Shen et al. (2020) reported a strong association among Zn, Fe, Mg, and Cu uptake and translocation processes between soil to plants, which was confirmed through strong elemental correlations among the atomic radii of the fundamental elements Zn, Fe, Mg, and Cu, due to the similarities among their atomic radii as well as octahedral configuration. Zhou et al. (2020) studied the variation in the concentration of amino acid, malondialdehyde (MDA), and antioxidant enzymes in the fine leaves of the *Schizonepeta* herb, thyme, and catnip exposed to Pb and Cd stress. The results showed that concentrations of HMs and species of plants were correlated to the overall amount of amino acids synthesized by the root system as well as the amount of each amino acid. *Monochoria hastata* plants were subjected to Cd stress in another experiment. The plants displayed some signs of noticeable toxicity like leaf fall, withering, and chlorosis when the Cd level was 15 mg L^{-1} (Baruah et al., 2016). Similarly, (Brima, 2017) reported that, among everything in the medicinal plants, Cd occurred at the minimum amounts while *Vitex agnus-castus* displayed such features at the maximum concentration and the concentrations of Cd ranged from 0.01 to 0.10 mg kg^{-1}.

Physiological factors of crops enable harmful chemicals to accumulate in stems, shoots, or leaves. Most species have noticeable oxidative damage like necrosis, while others do not have visual symptoms and need biochemical investigation. Some experiments have shown that certain varieties of medicinal plants have a distinct capacity for bioaccumulation of HMs (Galal and Shehata, 2015). For instance, *Plantago major* is a hyperaccumulator of Fe and aluminium (Al) since the shoot can accumulate concentrations greater than 1,000 mg kg-1. A further experiment by Angelova et al. (2015) indicated that lavender (*Lavandula vera* L.) is capable of tolerating higher concentrations of different HMs (Pb, Cd, and Zn) and hence can be classified as a hyperaccumulator. The plant species could be effectively utilized for the phytoremediation of toxic HMs in soils. Research by Lajayer et al. (2019) demonstrated that low concentrations of phytonutrients or HMs would accumulate in edible portions of the basil plant, whereas the study by Fattahi et al. (2019) demonstrated that sweet basil production was severely impaired by polluted soil containing Cd and Pb. Furthermore, HMs could inflict detrimental consequences on germinating seeds as well as on plant physicochemical characteristics in polluted soils and lands irrigated with wastewater (Mazhar et al., 2019; Mehmood et al., 2019).

The comparative accumulation of HMs depends on soil properties, the species and age of the plant under study, and the nature, form, and concentration of HM stress (Mehmood et al., 2018a, 2018b). Valuable modulation in plant height was shown by *Phaseolus aureus* planted over Hoagland's solution containing Cd or Hg, although modulation was more noticeable in seeds sprouted/developed over the solution of nutrient containing Hg = 3 µM as compared to seeds planted as well as developed in a solution of nutrient containing Cd = 13.2 µM. Once five-day-old seedlings were subjected to Hg or Cd-containing solutions, many seedlings expired at lower concentrations (3 µM Cd) during twenty-four hours of treatment. However, no decline was reported with concentrations greater than 200 µM Hg in young plants. Among perfectly healthy young plants, the demonstrated analgesic effect of Cd was observed to be only according to the specificity of species (Shaw and Rout, 2002).

HMs absorption in medicinal plants can interact with the accumulation of NPK in the plant body and may result in the deficiency of certain macronutrients. In another experiment conducted by Mishra et al. (2014) it was found that supplementing Cd to the medicinal plant (*Withania somnifera*) resulted in NPK deficiency along with interruption in growth and development, and necrosis, as well as necrotizing at higher concentrations of Cd. Moreover, photosynthetic pigments' concentration improved compared to the control (1.75-fold higher than control) at 50 µM and decreased by 0.13-fold at 300 µM. In comparison to the control, total chlorophyll contents (chl a + chl b), chl a and b, and carotenoids declined by 0.1- to 0.2-fold under higher concentrations of Cd, i.e. 200 to 300 µM.

The harmful effects of certain HMs can be minimized by the application of certain other HMs. For example, the higher concentrations of Cd significantly reduced growth, productivity, and pigment production as well as relative water content (RWC) in mustard (*Brassica juncea*) plants, while the application of selenium (Se) reversed this effect and improved vegetation, pigment production, and RWC (Ahmad et al., 2016; Zhang et al., 2018). Under Cd stress, the number of osmolytes (proline and glycine betaine), sugars, and nutritional profile were significantly improved, while Cd contents were reduced with the addition of Se. The disruption of small microstructures of cells seems to be the result of the deposition of ROS induced by HMs; in parallel with the apparent toxicological consequences over increase in the concentration of Pb^{2+} (from 5 to 50 mg L^{-1}), the photosynthesis rate of *Potamogeton crispus* dropped significantly. Simultaneously, excessive amounts of Pb^{2+} exacerbated phenotypic disruption of cellular structures like fracturing of chloroplast, loss, and breakdown of the envelope around chloroplast, cramping of mitochondrial cristae, vacuolation and deflection of mitochondria, nuclear condensation, and nucleoli deformation, as well as disturbance of the nuclear membrane (Hu et al., 2007).

Depending upon the type, form, and concentration, HMs might impede the germination of medicinal plant seeds. After soaking in KNO_3 and GA_3, a reduced ripening of seeds of *Catharanthus roseus* exposed to various concentrations of $CdCl_2$ and $PbCl_2$ for 24 hours was recorded. Both the protease and alpha-amylase were greatly diminished in seeds under different levels of $CdCl_2$ and $PbCl_2$. Similarly, Zn and Cu substantially ($p < 0.05$) lowered the seed germination of *Eucomis autumnalis* at 1 mg L^{-1}. The growth of root in three species of *Eucomis autumnalis*, *Bowiea volubilis* and

Merwilla natalensis was adversely affected under low concentrations of Cu and Zn (1–2 mg L^{-1}), while Hg doses of 0.5 and 1 mg L^{-1} greatly reduced the germination percentage of *E. autumnalis* and *B. Volubilis*, respectively. When the concentration of Hg and Cd was 2 mg L^{-1}, some deleterious impacts were demonstrated in the root system of *B. volubilis*. About 0.5 mg L^{-1} HMs significantly reduced the shoot length of *Merwilla natalensis*. Jeliazkova and Craker (2003) studied the effect of different concentrations of Zn, Pb, Cd, and Cu on vegetative and root growth in fennel (*Foeniculum vulgare* L.), anise (*Pimpinella anisum* L.), and caraway (*Carum carvi* L.). Findings demonstrated that, relative to the control, Cd promoted seed germination and growth parameters of caraway by around 20 per cent at 6 mg L^{-1}. Subsequently, at around 6 and 10 mg L^{-1} Cd and 100 and 500 mg L^{-1} Pb, seedling growth and growth parameters of anise plants decreased. Yet, at 100 mg L^{-1} Pb, seedling growth in caraway was accelerated. In all three species of plant, Zn and Cu greatly decreased seed germination and root development. In addition, it indicated that the toxic effects on vegetative growth and root growth of medicinal plants are dependent upon species of HM and the plant being exposed.

8.4   IMPACT OF HEAVY METALS ON MEDICINAL PLANTS' METABOLIC PATHWAYS

Under HM stress, the formation of bioactive compounds in medicinal plants is caused and it depends on the nature, form, and concentration of both the HM under consideration and the plant being exposed. This helps better to undertake the required critical analysis to establish the required composition for a specific medicinal plant and species of HM. Medicinal plants resist HM stress by maintaining bioactive constituents and keeping them constant. Rai et al. (2005) reported that in *Phyllanthus amarus*, the chemotherapeutic substances like molecules of hypophyllanthin and phyllanthin increased owing to the abiotic stress induced by Cd. In another study, Rai and Mehrotra (2008) demonstrated that concentrations of hypophyllanthin and phyllanthin have risen under Cr stress. The results of Cr on *Ocimum tenuiflorum* were examined by Rai et al. (2004). In contrast to the controls, 14.46, 24.61, 16.80, and 3.83 per cent more eugenol in *Ocimum tenuiflorum* was recorded under 10, 20, 50, and 100 µM Cr stress, respectively; Cr-treated plants produced greater eugenol. Manquian-Cerda et al. (2016) studied the development of phenolic compounds and antioxidant reaction in plantlets of *Vaccinium corymbosum* L exposed to 50 and 100 µM Cd^{2+} for 7, 14, and 21 days. In the results, Cd^{2+} stress significantly boosted the amounts of malondialdehyde (MDA), with a maximum rise over 14 days. Through the incorporation of Cd2+ in the chemical solution, a boost in the concentration of chlorogenic acid, the principal phenolic compound present in the blueberry plant species, was observed. Under higher concentrations of HMs, medicinal plants can reduce their capacity to synthesize bioactive constituents. Murch et al. (2003) experimented on seeds of plants cultivated in sterilized and controlled conditions and augmented by 25 or 50 mM of Ni. The results showed that Ni stress significantly impaired the ability to generate or absorb hyperforin as well as displaying a 15–20-fold reduction in levels of pseudohypericin and hypericin.

In some medicinal plants, bioactive constituents are resistant to HM stress and can stay intact. Upon ten days of exposure to low (3 μM) and strong (60 and 120 μM) Cd concentrations in *Matricaria chamomilla* L. that were about four weeks old, the intensification of dry mass and contents of N were not substantially altered by either Cd concentration (Kováčik et al., 2006). Here appeared, indeed, a substantial decrease in leaf chlorophyll as well as water quality. Herniarin amongst those molecules related to coumarin was not impaired by Cd, whereas certain (E)-2-β-d-glucopyranosyloxy-4-methoxycinnamic acids (GMCAs) and precursors (Z) greatly improved across each Cd concentration exposed. Kim et al. (1991) reported the highest increase in berberine production with the addition of $CuSO_4$ (200 and 500 μM). Through enhancing the formation of bioactive compounds, $CuSO_4$ polymeric network layers delivered a substantial portion of berberine into the solution.

Higher concentrations of HMs may also affect the molecular structure of active ingredients. The influence of Pb under various concentrations, i.e. 900, 1800, 3600, 7200, and 9000 mg/kg, on the production and chemical composition of the essential oil and phytoaccumulation of mint was studied by Sá et al. (2015). The application of Pb had not greatly affected the length of the root and aerial elements. The prevalence of Pb strongly affected the abundance of leaves and flowering, as well as producing weight, indicating an exceptional tolerance. The strong Pb concentration greatly increased the yield of essential oil to about ten times greater than that of the control. The strong Pb concentrations also influenced the molecular framework of essential oil in *M. crispa* but the major ingredient's amount ranged from 39.3 to 90 per cent in plants under control, i.e. cultivated in uncontaminated soil, and those plants grown under Pb stress. Kunwar et al. (2015) conducted a greenhouse experiment in plastic pots containing soils spiked with different concentrations of Cu, Cd, and Pb using *M. spicata* L. and *O. basilicum* L. as test plants. The critical oil proportion in aerial parts was determined via GC and GC-MS. The contents of essential oils in *O. basilicum* L. were significantly higher and there was a quantitative difference in cultivated seeds throughout each soil modified with HM. For *M. spicata*, no major improvements in the essential oil contents under HM stress were recorded, representing a common indication of unequal reactions to metals. The authors suggested that there is a need to select the best medicinal plant with maximum medicinal contents while interacting with a particular metal in the soil.

8.5   PRODUCTION AND SCAVENGING OF ROS

Through the production of free radicals or ROS, HMs can cause cell death (Singh et al. 2006). ROS can interfere with fatty acids, enzymes, and biomolecules as well as pigments and result in protein denaturation and degradation of membranes around various organelles through lipid peroxidation, thereby influencing the efficiency and feasibility of the cells (Imtiaz et al., 2018). A summary of the interaction of HMs with medicinal plants, their impacts and defence mechanisms is provided in Figure 8.1. Mechanisms such as enzymatic and non-enzymatic antioxidants in medicinal plants could mitigate the detrimental effects arising from the oxidative damage at the cellular and sub-cellular levels, caused by the production of ROS under HM stress. To

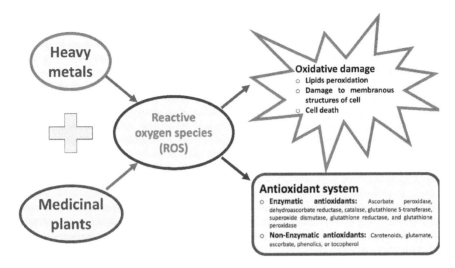

FIGURE 8.1 Interaction of HMs with medicinal plants, their impacts and defence mechanisms.

scavenge various types of ROS, plant species utilize several enzymatic molecules like catalase (CAT), superoxide dismutase (SOD), guaiacol peroxidase (GPX), and ascorbate peroxidase (APX), or monomeric antioxidants like ascorbic acid, cysteine, and non-protein thiols, which defend against the probable deleterious effects as well as tissue malfunction under HM stress.

HM stress induces depletion of molecular oxygen to create intermediates like hydrogen peroxide (H_2O_2), hydroxyl radicals (OH), and superoxide radicals (O_2) that are significantly hazardous or toxic for the plants' metabolic activities. Radicals of superoxide could suppress metallic ions (Mn^+) in organelles like Fe^{3+} and Cu^{2+}. The synthesis of superoxide radicals through H_2O_2 and O_2 could be catalysed by superoxide dismutase (Heldt, 2005). Hydroxyl radicals are created via the interaction of H_2O_2 with metal ions by a reduction reaction (Heldt, 2005). Hydroxyl radicals are quite dangerous and can degrade proteins or lipoproteins. No defence antibodies against OH radicals is found in the plant cell. Consequently, the avoidance of the reduction process by removing superoxide radicals utilizing SOD is important for cells. Hydrogen peroxide negatively affects multiple metabolites and its deleterious effects can be controlled by APX and CAT (Heldt, 2005). Catalase is a crucial enzyme, which catalyses oxygen and water to break down hydrogen peroxide and defends plants against ROS (Chelikani et al., 2004). Ascorbate peroxidase (APX) is an additional significant enzyme for the reduction of H_2O_2. Ascorbate, an essential plant antioxidant, is oxidized and transformed into a monodehydroascorbate radical that is instinctively converted to ascorbate via ferredoxin reduction. In a selective manner, dehydroascorbate and ascorbate could be integrated into two molecules of monodehydroascorbate. Through the redox process catalysed via dehydroascorbate reductase, dehydroascorbate may also have been transformed into ascorbate via glutathione (GSH). Glutathione, consisting of three amino acids, i.e. glycine, glutamate, and cysteine, is indeed an essential antioxidant within plant tissues. The oxidation of GSH contributes to the production of a disulfide (GSSG) bond of two molecules of glutathione among the compounds of cysteine. GSSG reduction is catalysed by the glutathione reductase enzyme and regulated by NADPH as the reductant (Heldt, 2005).

A typical crop response to metal toxicity is the development of H_2O_2. Cuypers et al. (2011) demonstrated a high accumulation of H_2O_2 in leaves and roots of *A. thaliana* seedlings subjected to Cd stress. In either leaves or roots, plasma bridges are used as a leading tool against stress conditions (Cuypers et al., 2011). Depending on the plant species, the quality of tissue as well as the HMs associated, the degree of lipid peroxidation may vary. Susceptibility to Cu boosted lipid peroxidation in root system *A. thaliana* seedlings higher than that of the Cd (Cuypers et al., 2011). Under HM stress, lipid peroxidation results in the production of malondialdehyde, which is known to be a potential stress biomarker. In shoots of *Withania somnifera*, an enhanced amount of MDA showed extreme oxidative stress via triggering extreme ROS generation, which destroyed cell walls (Mishra et al., 2014). Pandey et al. (2007) reported a rise in MDA contents after treatment of $CdCl_2$ and $PbCl_2$ to *Catharanthus roseus*. Oxylipins constitute bioactive molecules responsible for signalling, generated from polyunsaturated oxidized fatty acids, and implicated in genomic control in circumstances of abiotic stress. The key proteins for the biosynthesis of oxylipins are lipoxygenases (LOXes). Cytosolic LOX1 was highly triggered in seedlings of *A. thaliana* under Cd and Cu treatments, while LOX2 expression decreased in Cu-treated plants and plants treated with 5 M Cd. During HM treatments, each LOX mRNAs expanded in leaf tissue; however, greater expression was noted for LOX22 (Cuypers et al., 2011). The above illustrates that, depending on the plant tissue and HM type, the genomic sequence will vary. In medicinal plants under HM stress, numerous non-enzymatic agents are generated which are actively involved in mechanisms of rejuvenation of HMs and ROS scavenging (Mishra et al., 2014).

Mishra et al. (2014) demonstrated that non-enzymatic antioxidants like non-protein thiol (NPSH), phenolics, and AsA play an important role in the biotransformation of Cd under different concentrations of Cd in medicinal plants. In addition, many other non-enzymatic antioxidants like total ascorbate, tocopherol, proline, DHA, GSH, and flavonoid were also involved in the biotransformation of Cd up to some concentrations, i.e. 150–200 µM except proline. DHA and ascorbate (AsA) have been the main effective species of low molecular weight antioxidants that perform a crucial function in the protection against oxidative stress with increased ROS concentrations triggered through HM stress (Mishra et al., 2014). The molecules of oxygen radicals produced by NADPH oxidase are an essential factor for the generation of ROS in cellular membranes, the mechanism which preferentially affects the HMs distributed (Cuypers et al., 2011). NADPH oxidases (RBOHC/F) rely on cellular compartments as well as the HM involved. Suppression of NADPH oxidases (RBOHC/F) resulted in higher genomic activity in the roots of *A. thaliana* seedlings exposed to Cu stress. In comparison, Cd exposure resulted in a major amplification of RBOHD expression of genes. For the RBOHC genome, the strongest amplification triggered by Cd stress was observed in leaves, and indeed RBOHF genotype was highly mediated.

Susceptibility to Cu greatly raised the frequency of transcription of RBOHD genes (Cuypers et al., 2011). The findings revealed that NADPH oxidase transcription is reliant on HMs in foliage tissues as well as genetic expression forms.

In *Potamogeton crispus*, an increase in the activity of peroxidase as well as malondialdehyde concentration appeared with an improvement in Pb2+ production, ranging from 5 to 50 mg/L, while the activity of CAT and SOD declined initially and then increased (Hu et al., 2007). The effects of Cr on *Ocimum tenuiflorum* were investigated by Rai et al. (2004). The results showed that lipid peroxidation caused by Cr combined with the release of K as well as adverse consequences reduces the concentration of non-protein thiol, chlorophyll activity, ascorbic acid, cysteine, and protein. The leaves had an improved concentration of proline. SOD, GPX, and CAT impulsivity suggested that endogenous antioxidants performed some important functions throughout defending crops from the detrimental effects of Cr.

The results of various Cd concentrations on *Bacopa monnieri* demonstrated an elevated lipid peroxidation as well as decreased photosynthesis and protein quality; detrimental amounts of Cd generated oxidative stress (Singh et al., 2006). Nevertheless, *B. Monnieri* was able to counter oxidative damage by triggering separate enzymatic and non-enzymatic antioxidants under HM stress. Bagheri et al. (2017) investigated the effect of short- (2 days) and long-term (4 days) Cd treatments (40 mg kg^{-1} soil) on spinach grown on two soils with normal (300 μM SO$_4^{2-}$) and deficient (30 μM SO$_4^{2-}$) sulphur contents. The oxidative stress was enhanced by S deficiency (− S and − Cd) and Cd stress (+ S and + Cd) including respective mixture (− S and + Cd) relative to the control (+ S and − Cd). The severity of the stress prototype was as follows: (S deficiency) < (Cd stress) < (deficiency of S + Cd stress). Deficiency of sulphur was identified with a cumulative decrease in total ascorbate (AsA + DHA), dehydroascorbate (DHA), total glutathione (GSH + GSSG), ascorbate (AsA), and glutathione (GSH). The concentration of total glutathione was raised dramatically once Cd was added in the presence of adequate S, in addition to a small improvement in its oxidized and reduced kinds. The S availability increased the development of glutathione reductase (GR), CAT, APX, and SOD, against Cd stress while an opposite effect was noted under Cd stress with S deficiency. Soluble protein and total carotenoid production decreased during all S and Cd stresses, with S deficit as well as combined stress treatments showing the largest decrease, while a significantly negative effect on fresh and dry biomass was noted under both S and Cd stress. The results suggested that, for improved production of CAT, GR, APX, and SOD against Cd stress, an effective concentration of S is necessary.

Rahoui et al. (2017) had also reported that Cd stress stimulates the protective mechanisms in organically generated *Medicago truncatula* seedling roots. Across several genetic variants examined excluding TN1, 11, and A17, Cd stress (100 μM) immediately enhanced ROS generation, improved antioxidants such as total CAT, PRX, SOD, and as well as metabolites related to the transformation of ascorbate and glutathione like APX and MDA. The production of soluble phenolics was also enhanced under Cd treatment along with peroxidase improvement. Consequently, the lignification presence was confined to the newly formed protoxylem products developed in the region of the root tip, typically the area of elongation. Developmental modifications such as lignocellulosic accumulation increased the content of cellulose as well as pectin and endothelial meatus, and compact and mutilated hair was observed in the roots of plants treated with Cd.

In addition, GPX, SOD, CAT amounts along with antioxidant enzymes improved, and variations in the isoenzyme expression trends were also identified in plants with Ni treatment. Enhanced lipid peroxidation, H_2O_2 production and, electrolyte leakage, together with an excessive volume of antioxidant enzymes like GSH, APX, and SOD, were seen in mustard (*Brassica juncea*) subjected to higher concentrations of Cd (Ahmad et al., 2016). Se fortification not only lowered the lipid peroxidation and H_2O_2 production but also enhanced the production of antioxidant enzymes at higher amounts as well as maintaining the concentration of Cd in shoot and root. Higher doses of Cd led to diminished flavonoids and ascorbic acid, although total phenols and tocopherol were enhanced during similar concentrations of Cd. The application of Se maintained the flavonoid and AsA contents as well as subsequent changes in total phenols and tocopherol. The findings suggested that, through modulation of anti-oxidant enzymes and phytohormones as well as bioactive compounds, foliar appli-cation of Se would minimize the detrimental consequences of Cd stress throughout mustard plants.

In plants treated with Cd, the Cd concentration within crops was boosted. The elevated rates of lipid peroxidation throughout the leaves and the concentration of H_2O_2 in the root tissues are due to oxidative stress induced through Cd. The oxidative stress caused due to the accumulation of Cd in leaves elevated the amount of DNA damage. Similarly, Kumar et al. (2013) exposed 0, 0.25, 0.5, 0.75, 1.0, and 1.25 mM of Pb to *Talinum triangulare* (Jacq.) Willd for seven days and investigated their effect on stress, concentrating on the biochemistry of root cells and DNA damage through functional as well as molecular analysis. The findings showed that Pb stress inten-sified the development of ROS, oxidation of protein, and DNA damage, ultimately resulting in decreased concentration of proteins, depending on the dose, cell death, and lipid peroxidation.

The ROS induced via HM stress could trigger the pathway of transcriptional regu-lation like the channels of mitogen-activated protein kinase (MAPK) (Jonak et al., 2004; Lin et al., 2005; Liu et al., 2010; Panda and Achary, 2014). The pathway of MAPK contains the module called MAPKKK-MAPKK-MAPK, known to be one of the key pathways mediating the proliferation and differentiation of intracellular inputs to intracellular interactions (Panda and Achary, 2014). Via phosphorylation, MAPKKK stimulates MAPKK, and MAPKK could phosphorylate MAPK.

MAPKs, p38 MAPK, c-Jun NH2-terminal kinase (JNK), and extracellular signal-regulated protein kinase (ERK) transfer intracellular stimuli towards the nucleus, and a wide variety of molecular and cellular processes, segmentation, and cell death have been reported as being involved. The function of MAPK in the stress of HMs has been observed in many studies, like with the treatment of tributyltin, inorganic Hg or Cd triggers p38 MAPK, ERK, and JNK (Matsuoka and Igisu, 2002). In manipu-lating the genome, the sequence of MAPK indeed performs an essential character, which stimulates enzymes, proteins, or transcription factors (TFs) (Yang et al., 2003; Luo et al., 2016). Yeh et al. (2004) observed that, in response to 100–400 µM Cd, $OsMAPK_2$ expressed itself in growth media. Four separate MAPKs (SAMK, MMK_3, MMK_2, and SIMK) were triggered by the introduction of alfalfa (*Medicago sativa*) seeds to higher concentrations of Cu or Cd ions. A comparative analysis of the energetics of MAPK stimulation showed that Cu ions triggered SAMK, MMK_3, MMK_2, and SIMK quite swiftly, whereas Cd ions stimulated slow production of MAPK (Jonak et al., 2004). Susceptibility to 10 μM CdCl2 for 8 h contributed to sustained stimulation and formation of c-fos mRNA and ERK kinase (Ding and Templeton, 2000). Cd2+ stress triggered the c-Jun kinase/stress-activated protein kinase (SAPK) as well. The analyses described previously suggest that multiple HMs may be focused on particular cascades and enzymes.

8.6 CONCLUSIONS AND FUTURE PERSPECTIVES

Within their native habitat, medicinal plants may be subject to various sorts of abiotic stresses like frost, drought, salinity, and HM toxicity. The presence of HMs in the biosphere is attributable to industrialization and urbanization-related geological disturbances as well as anthropogenic activities. In the soil and marine environments, substantial amounts of HMs are contained, whereas a comparatively lower fraction is detected in vapour in the surrounding air. The toxicity of HMs influences the output of different medicinal plants and plant populations and further inhibits the proper functioning and integrity of the entire environment. The deposition of HMs in soil is of considerable concern since these are being swallowed up or absorbed and assimilated into plant biomass. The drugs extracted from these medicinal plants are being bio-accumulated in humans.

A series of diverse biochemical and physiological alterations accompanied by transformations in the metabolic processes of plants occurs due to the absorption of HMs. Overall, excessive quantities of HMs may severely influence the vegetative growth, development, and production of medicinal plants. The potential inhibitory effect of HMs, furthermore, relies upon its extent as well as the composition of soil's physical properties, species of plants, growth conditions, and HMs. The generation of bioactive molecules is also influenced by the nature, form, and concentration of HMs while these molecules are of major importance for the optimum growth and productivity of medicinal plants. The beneficial implications of HMs on the development of useful active compounds in medicinal plants have also been reported by several researchers. The accomplishment of the desired impact on metabolic pathways relies on the species of medicinal plants and the species and concentration of HMs, which suggests determining the required HM species and optimum concentration for a given medicinal plant and/or active ingredient.

The preliminary indication of stress is accompanied by the production of ROS, cascades of MAPK, or stimulation of several TFs, as well as enzymes like proteins and antioxidant enzymes in the medicinal plants against HM stress. In common, the enzymatic and non-enzymatic antioxidant system is triggered for the excavation of ROS produced under metal stress. Secondary metabolites are also produced like sugars, phenols, and others. The harvesting of viable medicinal plants for the development of useful bioactive compounds, like AsA, flavonoids, phenolics, tocopherol, essential oils, and non-protein thiol, could perhaps be a useful replacement while working with soils contaminated with HMs and not suitable for agricultural crop production. There are also certain HMs which have the potential to reduce the harmful consequences of several other HMs. This aspect requires further research to understand the mechanisms behind the interaction of HMs and the metal being used to treat HM stress. Today's understanding regarding the foliar absorption of HMs and the metabolic processes associated, as well as the biosynthetic mechanisms and consequences of various metabolic enzymes in medicinal plants that are essential for the management of ambient emissions of HMs, is minimal and requires further research. In addition, among specific types of medicinal plants, the involvement of enzymatic/non-enzymatic antioxidants or transporter genomes in HM biotransformation is also not completely understood and needs to be investigated across both root and foliar pathways for HM absorption.

REFERENCES

Ahmad, I., M. Tahir, U. Daraz, A. Ditta, M.B. Hussain and Z.U.H. Khan. 2020. Responses and tolerance of cereal crops to metals and metalloids toxicity. In: Mirza Hassanuzzaman (ed.). Agronomic Crops. Springer, Singapore, pp. 235–264. https://doi.org/10.1007/978-981-15-0025-1_14

Ahmad, P., E.A. Allah, A. Hashem, M. Sarwat, and S. Gucel. 2016. Exogenous application of selenium mitigates cadmium toxicity in *Brassica juncea* L. (Czern & Cross) by up-regulating antioxidative system and secondary metabolites. *Journal of Plant Growth Regulation* 35: 936–950.

Angelova, V.R., Grekov, D.F., Kisyov, V.K., and Ivanov, K.I. 2015. Potential of lavender (*Lavandula vera* L.) for phytoremediation of soils contaminated with heavy metals. *International Journal of Biological, Biomolecular, Agricultural, Food and Biotechnological Engineering* 9: 522–529.

Asiminicesei, D.-M., Vasilachi, I.C., and Gavrilescu, M. 2020. Heavy metal contamination of medicinal plants and potential implications on human health. *Revista de Chimie* 71: 16–36.

Azab, E. and Hegazy, A.K. 2020. Monitoring the efficiency of *Rhazya stricta* L. plants in phytoremediation of heavy metal-contaminated soil. *Plants* 9: 1057.

Bagheri, R., J. Ahmad, H. Bashir, M. Iqbal, and M.I. Qureshi. 2017. Changes in rubisco, cysteine-rich proteins and antioxidant system of spinach (*Spinaciaoleracea* L.) due to sulphur deficiency, cadmium stress and their combination. *Protoplasma* 254(2): 1031–1043.

Baruah, S., M.S. Bora, P. Sharma, P. Deb, and K.P. Sarma. 2016. Understanding of the distribution, translocation, bioaccumulation, and ultrastructural changes of monochorea hastata plant exposed to cadmium. *Water, Air, & Soil Pollution* 228: 1–21.

Bempah, C.K., Boateng, J., Asomaning, J., and Asabere, S.B. 2012. Heavy metals contamination in herbal plants from some Ghanaian markets. *Journal of Microbiology, Biotechnology and Food Sciences* 2: 886–896.

Brima, E.I. 2017. Toxic elements in different medicinal plants and the impact on human health. *International Journal of Environmental Research and Public Health* 14(10): 1209.

Chelikani, P., I. Fita, and P.C. Loewen. 2004. Diversity of structures and properties among catalases. *Cellular and Molecular Life Sciences* 61(2): 192–208.

Cuypers, A., S. Karen, R. Jos, O. Kelly, K. Els, R. Tony, H. Nele, V. Nathalie, G. Yves, C. Jan, and V. Jaco. 2011. The cellular redox state as a modulator in cadmium and copper responses in *Arabidopsis thaliana* seedlings. *Journal of Plant Physiology* 168(4): 309–316.

Ding, W., and D.M. Templeton. 2000. Activation of parallel mitogen-activated protein kinase cascades and induction of c-fos by cadmium. *Toxicology and Applied Pharmacology* 162(2): 93–99.

Fattahi, B., K. Arzani, M.K. Souri, and M. Barzegar. 2019. Effects of cadmium and lead on seed germination, morphological traits, and essential oil composition of sweet basil (*Ocimum basilicum* L.). *Industrial Crops and Products* 138: 111584.

Galal, T.M., and H.S. Shehata. 2015. Bioaccumulation and translocation of heavy metals by *Plantago major* L. grown in contaminated soils under the effect of traffic pollution. *Ecological Indicators* 48: 244–251.

Heldt, H.W. 2005. Plant Biochemistry. Elsevier Academic Press, London; San Diego, CA.

Hu, J.Z., G.X. Shi, Q.S. Xu, X. Wang, Q.H. Yuan, and K.H. Du. 2007. Effects of Pb2 + on the active oxygen-scavenging enzyme activities and ultrastructure in *Potamogeton crispus* leaves. *Russian Journal of Plant Physiology* 54(3): 414–419.

Ijaz M., M.S. Rizwan, M. Sarfraz, S. Ul-Allah, A. Sher, A Sattar, L. Ali, A. Ditta, and B. Yousaf. 2020. Biochar reduced cadmium uptake and enhanced wheat productivity in alkaline contaminated soil. *International Journal of Agriculture and Biology* 24: 1633–1640. https://doi.org/10.17957/IJAB/15.1605

Imtiaz, M., M. Ashraf, M.S. Rizwan, M.A. Nawaz, M. Rizwan, S. Mehmood, B. Yousaf, Y. Yuan, M.A. Mumtaz, A. Ditta, M. Ali, S. Mahmood, and S. Tu. 2018. Vanadium toxicity in chickpea (*Cicer arietinum* L.) grown in red soil: effects on cell death, ROS and antioxidative systems. *Ecotoxicology and Environmental Safety* 158: 139–144. https://doi.org/10.1016/j.ecoenv.2018.04.022

Irshad, S., Z. Xie, S. Mehmood, A. Nawaz, A. Ditta, and Q. Mahmood. 2021. Insights into conventional and recent technologies for arsenic bioremediation: a systematic review. *Environmental Science and Pollution Research* 28(15): 18870–18892. https://doi.org/10.1007/s11356-021-12487-8

Jaison, S., and T. Muthukumar. 2017. Chromium accumulation in medicinal plants growing naturally on tannery contaminated and non-contaminated soils. *Biological Trace Element Research* 175: 223–235.

Jan, A.U., F. Hadi, A. Shah, A. Ditta, M.A. Nawaz, and M. Tariq. 2021. Plant growth regulators and EDTA improve phytoremediation potential and antioxidant response of *Dysphania ambrosioides* (L.) Mosyakin & Clemants in a Cd-spiked soil. *Environmental Science and Pollution Research* 28: 43417–43430. https://doi.org/10.1007/s11356-021-13772-2

Jeliazkova, E.A., and L.E. Craker. 2003. Seed germination of some medicinal and aromatic plants in heavy metal environment. *Journal of Herbs, Spices & Medicinal Plants* 10(2): 105–112.

Jonak, C., H. Nakagami, and H. Hirt. 2004. Heavy metal stress. Activation of distinct mitogen-activated protein kinase pathways by copper and cadmium. *Plant Physiology* 136(2): 3276–3283.

Kanwal, U., M. Ibrahim, F. Abbas, M. Yamin, F. Jabeen, A. Shahzadi, A.A. Farooque, M. Imtiaz, A. Ditta, and S. Ali. 2021. Phytoextraction of Lead using a hedge plant [*Alternanthera bettzickiana* (Regel) G. Nicholson]: Physiological and bio-chemical alterations through bioresource management. *MDPI-Sustainability* 13(9): 5074. https://doi.org/10.3390/su13095074

Kim, D.I., H. Pedersen, and C.K. Chin. 1991. Stimulation of berberine production in *Thalictrum rugosum* suspension cultures in response to addition of cupric sulfate. *Biotechnology Letters* 13(3): 213–216.

Kováčik, J., J. Tomko, M. Bačkor, and M. Repčák. 2006. Matricaria chamomilla is not a hyperaccumulator, but tolerant to cadmium stress. *Plant Growth Regulation* 50(2–3): 239–247.

Kulal, C., R.K. Padhi, K. Venkatraj, K.K. Satpathy, and S.H. Mallaya 2020. Study on trace elements concentration in medicinal plants using EDXRF technique. *Biological Trace Element Research* 198(1): 293–302.

Kumar, A., M.N.V. Prasad, V.M.M. Achary, and B.B. Panda. 2013. Elucidation of lead-induced oxidative stress in talinum triangulare roots by analysis of antioxidant responses and DNA damage at cellular level. *Environmental Science and Pollution Research* 20(7): 4551–4561.

Kunwar, G., C. Pande, G. Tewari, C. Singh, and G.C. Kharkwal. 2015. Effect of heavy metals on terpenoid composition of *Ocimumbasilicum* L. and *Menthaspicata* L. *Journal of Essential Oil Bearing. Plants* 18(4): 818–825.

Labidi, O., V. Vives-Peris, A. Gómez-Cadenas, R.M. Perez-Clemente, and N. Sleimi 2021. Assessing of growth, antioxidant enzymes, and phytohormone regulation in *Cucurbita pepo* under cadmium stress. *Food Science & Nutrition* 9(4): 2021–2031.

Lajayer, B.A., M. Ghorbanpour, and S. Nikabadi. 2017. Heavy metals in contaminated environment: destiny of secondary metabolite biosynthesis, oxidative status and phytoextraction in medicinal plants. *Ecotoxicology and Environmental Safety* 145: 377–390.

Lajayer, B.A., N. Najafi, E. Moghiseh, M. Mosaferi, and J. Hadian. 2019. Micronutrient and heavy metal concentrations in basil plant cultivated on irradiated and non-irradiated sewage sludge-treated soil and evaluation of human health risk. *Regulatory Toxicology and Pharmacology* 104: 141–150.

Lin, C.W., H.B. Chang, and H.J. Huang. 2005. Zinc induces mitogen-activated protein kinase activation mediated by reactive oxygen species in rice roots. *Plant Physiology and Biochemistry* 43: 963–968.

Liu, X.M., K.E. Kim, K.C. Kim, X.C. Nguyen, H.J. Han, M.S. Jung, et al. 2010. Cadmium activates *Arabidopsis* MPK3 and MPK6 via accumulation of reactive oxygen species. *Phytochemistry* 71: 614–618.

Liu, Y.Z., M. Imtiaz, A. Ditta, M.S. Rizwan, M. Ashraf, S. Mehmood, O. Aziz, F. Mubeen, M. Ali, N.N. Elahi, R. Ijaz, S, Lelel, C. Shuang, and S. Tu. 2020. Response of growth, antioxidant enzymes and root exudates production towards As stress in *Pteris vittata* and *Astragalus sinicus* colonized by arbuscular mycorrhizal fungi. *Environmental Science and Pollution Research* 27: 2340–2352. https://doi.org/10.1007/s11356-019-06785-5

Luo, Z.B., J. He, A. Polle, and H. Rennenberg. 2016. Heavy metal accumulation and signal transduction in herbaceous and woody plants: paving the way for enhancing phytoremediation efficiency. *Biotechnology Advances* 34(6): 1131–1148.

Mafakheri, M., and M. Kordrostami, 2021. Recent advances toward exploiting medicinal plants as phytoremediators. In: Handbook of Bioremediation. Elsevier, pp. 371–383.

Maleki, M., M. Ghorbanpour, and K. Kariman. 2017. Physiological and antioxidative responses of medicinal plants exposed to heavy metals stress. *Plant Gene* 11: 247–254.

Manquian-Cerda, K., M. Escudey, G. Zuniga, N. Arancibia-Miranda, M. Molina, and E. Cruces. 2016. Effect of cadmium on phenolic compounds, antioxidant enzyme activity and oxidative stress in blueberry (*Vaccinium corymbosum* L.) plantlets grown in vitro. *Ecotoxicology and Environmental Safety* 133: 316–326.

Matsuoka, M., and H. Igisu. 2002. Effects of heavy metals on mitogen-activated protein kinase pathways. *Environmental Health and Preventive Medicine* 6(4): 210–217.

Mazhar, S., A. Ditta, L. Bulgariu, I. Ahmad, M. Ahmed, and A.A. Nadiri. 2019. Sequential treatment of paper and pulp industrial wastewater: prediction of water quality parameters by Mamdani Fuzzy Logic Model and phytotoxicity assessment. *Chemosphere* 227: 256–268. https://doi.org/10.1016/j.chemosphere.2019.04.022

Mehmood, S., M. Imtiaz, S. Bashir, M. Rizwan, S. Irshad, G. Yuvaraja, M. Ikram, O. Aziz, A. Ditta, S.U. Rehman, Q. Shakeel, M.A. Mumtaz, W. Ahmed, S. Mahmood, D. Chen, S. Tu. 2019. Leaching behavior of Pb and Cd and transformation of their speciation in co-contaminated soil receiving different passivators. *Environmental Engineering Science* 36(6): 749–759. https://doi.org/10.1089/ees.2018.0503

Mehmood S., M. Rizwan, S. Bashir, A. Ditta, O. Aziz, L.Z. Yong, Z. Dai, M. Akmal, W. Ahmed, M. Adeel, M. Imtiaz, and S. Tu. 2018a. Comparative effects of biochar, slag and Ferrous-Mn ore on lead and cadmium immobilization in soil. *Bulletin of Environmental Contamination and Toxicology* 100(2): 286–292. https://doi.org/10.1007/s00128-017-2222-3

Mehmood, S., D.A. Saeed, M. Rizwan, M.N. Khan, O. Aziz, S. Bashir, M. Ibrahim, A. Ditta, M. Akmal, M.A. Mumtaz, W. Ahmed, S. Irshad, M. Imtiaz, S. Tu, and A. Shaheen. 2018b. Impact of different amendments on biochemical responses of sesame (*Sesamum Indicum* L.) plants grown in lead-cadmium contaminated soil. *Plant Physiology and Biochemistry* 132: 345–355. https://doi.org/10.1016/j.plaphy.2018.09.019

Mehmood, S., X. Wang, W. Ahmed, M. Imtiaz, A. Ditta, M. Rizwan, S. Irshad, S. Bashir, Q. Saeed, A. Mustafa, and W. Li. 2021. Removal mechanisms of slag against potentially toxic elements in soil and plants for sustainable agriculture development: a critical review. *MDPI-Sustainability* 13(9): 5255. https://doi.org/10.3390/su13095255

Mishra, B., R.S. Sangwan, S. Mishra, J.S. Jadaun, F. Sabir, and N.S. Sangwan. 2014. Effect of cadmium stress on inductive enzymatic and nonenzymatic responses of ROS and sugar metabolism in multiple shoot cultures of Ashwagandha (*Withania somnifera* Dunal). *Protoplasma* 251(5): 1031–1045.

Murch, S.J., K. Haq, H.V. Rupasinghe, and P.K. Saxena. 2003. Nickel contamination affects growth and secondary metabolite composition of St. John's wort (*Hypericum perforatum* L.). *Environmental and Experimental Botany* 49(3): 251–257.

Murtaza, G., A. Ditta, N. Ullah, M. Usman, and Z. Ahmed. 2021. Biochar for the management of nutrient impoverished and metal contaminated soils: preparation, applications, and prospects. *Journal of Soil Science and Plant Nutrition* 21: 2191–2213.

Naveed, M., A. Ditta, M. Ahmad, A. Mustafa, Z. Ahmad, M. Conde-Cid, S. Tahir, S.A.A. Shah, M.M. Abrar, and S. Fahad. 2021. Processed animal manure improves morpho-physiological and biochemical characteristics of *Brassica napus* L. under nickel and salinity stress. *Environmental Science and Pollution Research* 28: 45629–45645. https://doi.org/10.1007/s11356-021-14004-3

Naveed, M., S.S. Bukhari, A. Mustafa, A. Ditta, S. Alamri, M.A. El-Esawi, M. Rafique, S. Ashraf, and M.H. Siddiqui. 2020. Mitigation of nickel toxicity and growth promotion in sesame through the application of a bacterial endophyte and zeolite in nickel contaminated soil. *International Journal of Environmental Research and Public Health* 17(23): 8859. https://doi.org/10.3390/ijerph17238859

Niamat, B., M. Naveed, Z. Ahmad, M. Yaseen, A. Ditta, A. Mustafa, M. Rafique, R. Bibi, and X. Minggang. 2019. Calcium-enriched animal manure alleviates the adverse effects of salt stress on growth, physiology and nutrients homeostasis of *Zea mays* L. *MDPI-Plants* 8(11): 480. https://doi.org/10.3390/plants8110480

Panda, B.B., and V.M.M. Achary. 2014. Mitogen-activated protein kinase signal transduction and DNA repair network are involved in aluminum-induced DNA damage and adaptive response in root cells of *Allium cepa* L. *Frontiers in Plant Science* 5: 256.

Pandey, S., K. Gupta, and A.K. Mukherjee. 2007. Impact of cadmium and lead on *Catharanthus roseus* – A phytoremediation study. *Journal of Environmental Biology* 28: 655–662.

Pirzadah, T.B., B. Malik, I. Tahir, Q.M. Irfan, and R.U. Rehman. 2018. Characterization of mercury-induced stress biomarkers in *Fagopyrum tataricum* plants. *International Journal of Phytoremediation* 20: 225–236.

Rahoui, S., Y. Martinez, L. Sakouhi, C. Ben, M. Rickauer, E. El Ferjani, L. Gentzbittel, and A. Chaoui. 2017. Cadmium-induced changes in antioxidative systems and differentiation in roots of contrasted *Medicago truncatula* lines. *Protoplasma* 254(1): 473–489.

Rai, V., and S. Mehrotra. 2008. Chromium-induced changes in ultramorphology and secondary metabolites of *Phyllanthus amarus* Schum and Thonn. – An hepatoprotective plant. *Environmental Monitoring and Assessment* 147(1–3): 307–315.

Rai, V., S. Khatoon, S.S. Bisht, and S. Mehrotra. 2005. Effect of cadmium on growth, ultramorphology of leaf and secondary metabolites of *Phyllanthus amarus* Schum. and Thonn. *Chemosphere* 61(11): 1644–1650.

Rai, V., P. Vajpayee, S.N. Singh, and S. Mehrotra. 2004. Effect of chromium accumulation on photosynthetic pigments, oxidative stress defense system, nitrate reduction, proline level and eugenol content of *Ocimum tenuiflorum* L. *Plant Science* 167(5): 1159–1169.

Rizwan, M.S., M. Imtiaz, J. Zhu, B. Yousaf, M. Hussain, L. Ali, A. Ditta, M.Z. Ihsan, G. Huang, M. Ashraf, and H. Hu. 2021. Immobilization of Pb and Cu by organic and inorganic amendments in contaminated soil. *Geoderma* 385: 114803. https://doi.org/10.1016/j.geoderma.2020.114803

Sá, R.A., O. Alberton, Z.C. Gazim, A. Laverde Jr, J. Caetano, A.C. Amorin, and D.C. Dragunski. 2015. Phytoaccumulation and effect of lead on yield and chemical composition of *Mentha crispa* essential oil. *Desalination and Water Treatment* 53(11): 3007–3017.

Sabir, A., M. Naveed, M.A. Bashir, A. Hussain, A. Mustafa, Z.A. Zahir, M. Kamran, A. Ditta, A. Núñez-Delgado, Q. Saeed, and A. Qadeer. 2020. Cadmium mediated phytotoxic impacts in *Brassica napus*: managing growth, physiological and oxidative disturbances through combined use of biochar and *Enterobacter* sp. MN17. *Journal of Environmental Management* 265: 110522. https://doi.org/10.1016/j.jenvman.2020.110522

Sharmila, P., P.K. Kumari, K. Singh, N.V.S.R.K. Prasad, and P. Pardha-Saradhi. 2016. Cadmium toxicity-induced proline accumulation is coupled to iron depletion. *Protoplasma* 254: 763–770.

Shaw, B.P., and N.P. Rout. 2002. Hg and Cd induced changes in the level of proline and the activity of proline bio synthesizing enzymes in *Phaseolus aureus* Roxb. and *Triticum aestivum* L. *Biologia Plantarum* 45 : 267–271.

Shen, Z., Y. Chen, D. Xu, L. Li, and Y. Zhu. 2020. Interactions between heavy metals and other mineral elements from soil to medicinal plant Fengdan (*Paeonia ostii*) in a copper mining area, China. *Environmental Science and Pollution Research* 27: 33743–33752.

Shen, Z.J., D.C. Xu, Y.S. Chen, and Z. Zhang. 2017. Heavy metals translocation and accumulation from the rhizosphere soils to the edible parts of the medicinal plant Fengdan (*Paeonia ostii*) grown on a metal mining area, China. *Ecotoxicology and Environmental Safety* 143: 19–27.

Singh, S., S. Eapen, and S.F. D'Souza. 2006. Cadmium accumulation and its influence on lipid peroxidation and antioxidative system in an aquatic plant, *Bacopa monnieri* L. *Chemosphere* 62(2): 233–246.

Soleimani, S.H., F. Bernard, M. Amini, and R.-A. Khavari-Nezhad. 2020. Cadmium accumulation and alkaloid production of *Narcissus tazetta* plants grown under in vitro condition with cadmium stress. *Plant Physiology Reports* 25: 51–57.

Thind, S., I. Hussain, A. Ditta, A. Perveen, R. Rasheed, M.A. Ashraf, S. Hussain, and Q. Mahmood. 2021. Alleviation of Cd stress by silicon nanoparticles during different phenological stages of Ujala wheat variety. *Arabian Journal of Geosciences* 14:1028.

Ullah, I., A. Ditta, M. Imtiaz, S. Mehmood, M. Rizwan, M.S. Rizwan, A.U. Jan, and I. Ahmad. 2020. Assessment of health and ecological risks of heavy metal contamination: a case study of agricultural soils in Thall, Dir-Kohistan. *Environmental Monitoring and Assessment* 192: 786. https://doi.org/10.1007/s10661-020-08722-3

Yang, S.H., A.D. Sharrocks, and A.J. Whitmarsh. 2003, Transcriptional regulation by the MAP kinase signaling cascades. *Gene* 320: 3–21.

Yeh, C.M., L.J. Hsiao, and H.J. Huang. 2004. Cadmium activates a mitogen-activated protein kinase gene and MBP kinases in rice. *Plant and Cell Physiology* 45(9): 1306–1312.

Zaid, A., F. Mohammad, and Q. Fariduddin. 2020. Plant growth regulators improve growth, photosynthesis, mineral nutrient and antioxidant system under cadmium stress in menthol mint (*Mentha arvensis* L.). *Physiology and Molecular Biology of Plants* 26: 25–39.

Zeng, J., X. Li, X. Wang, K. Zhang, Y. Wang, H. Kang, G. Chen, T. Lan, Z. Zhang, S. Yuan, C. Wang, and Y. Zhou. 2020. Cadmium and lead mixtures are less toxic to the Chinese medicinal plant *Ligusticum chuanxiong* Hort. than either metal alone. *Ecotoxicology and Environmental Safety* 193: 110342.

Zewail, R.M., H.S. El-Desoukey, and K.R. Islam. 2020. Chromium stress alleviation by salicylic acid in Malabar spinach (*Basella alba*). *Journal of Plant Nutrition* 43: 1268–1285.

Zhang, E, Y. Yuan, Z. Qian, G. Fei, A. Ditta, S. Mehmood, M.S. Rizwan, M.A. Mustaq, M. Rizwan, O. Aziz, R. Ijaz, J. Afzal, M. Imtiaz, and S. Tu. 2018. Seed priming with selenium to affect seed germination, seedling growth, and electrolyte leakage in rice under vanadium and cadmium stress. *Journal of Environment and Agriculture* 3(1): 262–273.

Zhou, M., Y. Zhi, Y. Dai, J. Lv, Y. Li, and Z. Wu. 2020. The detoxification mechanisms of low-accumulating and non-low-accumulating medicinal plants under Cd and Pb stress. *RSC Advances* 10: 43882–43893.

Zu, Y., X. Mei, B. Li, T. Li, Q. Li, L. Qin, and Z. Yang. 2020. Effects of calcium application on the yields of flavonoids and saponins in *Panax notoginseng* under cadmium stress. *International Journal of Environmental Analytical Chemistry*. DOI: 10.1080/03067319.2020.1781835

9 The Influence of Environmental Pollution on Secondary Metabolite Production in Medicinal Plants

Swati T. Gurme, Mahendra L. Ahire,
Jaykumar J. Chavan, Pankaj S. Mundada

CONTENTS

9.1 INTRODUCTION

Biodiversity is the very basis of human survival and economic well-being and includes all forms of life and ecosystems (McNeely et al. 1990). There are more than 50 million plant species on Earth, out of which about 50,000 to 80,000 are considered across the world to be medicinal plants (Schippmann et al. 2002). India is one of the major centres of origin and diversity of crops and medicinal plants. India contains 2 of the 14 mega-biodiversity hotspots of the world, with a diversity of about 20,000 species of higher plants.

Medicinal plants have been known as the primary source of drugs from ancient times. The widespread use of herbal remedies and healthcare preparations, from traditional herbs to medicinal plants, has been traced for their medicinal properties (Sharma et al. 2020). Though humans have developed modern sophisticated pharmaceutical chemicals to treat illness, medicinal plants remain an important tool for treating illness. Human beings have been utilizing medicinal plants for basic preventive and curative healthcare properties. According to a survey carried out by the

World Health Organization (WHO), 80 per cent of the population of developing countries is still relying on traditional medicines, mostly plant-based drugs (Sahito et al. 2003). *Ayurveda* is an old, well-documented, mostly plant-based system of medicine. It has many positive points over the allopathic system of medicine, such as fewer side effects, low cost, and long-lasting effect, and generally suits the temperament of individuals. The principles of positive health and therapeutic measures were related to the physical, mental, social, and spiritual welfare of human beings. It was estimated that there were approximately 100,000 different plant-derived compounds, with a large number of new ones being added to the list every year (Verpoorte et al. 1998). As a drug, mostly all the parts of medicinal plants are useful for treatment – viz. fruit, vegetables, and extracts of plant parts (Sahito et al. 2003). Various compounds produced by plants can be broadly classified as primary metabolites and secondary metabolites. The balance between the activities of the primary and secondary metabolism is a dynamic one, which will be largely affected by the growth, tissue differentiation, and development of the plant body.

In modern medicine, plants are used as sources of direct therapeutic agents and as a taxonomic marker for the discovery of new compounds (Nalawade and Tsay 2004). Although modern medicines may be available in developed countries, herbal medicines (phytopharmaceuticals) have often been maintained popularly for historical and cultural reasons. The chemical synthesis of bioactive compounds is difficult because of their complex structure and high cost (Shimomura et al. 1997). These are influenced mainly by cultivation period, season of collection, plant-to-plant variability in the medicinal content, lack of adequate methods for the production and standardization of the crop, and lack of understanding of the unique plant physiology.

The growth and development of any plant is the result of plant–environment interactions in which several climatic and edaphic factors play an important role. All these factors synchronously drive the biochemical, physiological, and molecular responses of plants. Plants are always exposed to environmental extremities. These environmental extremities are either natural or synthetic, which exerts extreme pressure on the growth and development of plants. To withstand such pressure, plants diverted their metabolism towards the synthesis of certain secondary metabolites.

9.2   SECONDARY METABOLITES

Secondary metabolites (SMs) are a group of chemical compounds produced by plants from primary metabolic pathways. The concept of secondary metabolites was first introduced by Albrecht Kossel who received the Nobel Prize in Physiology and Medicine in 1910. The concentration of different SMs is mainly exaggerated by environmental stress and pollution (Khan 2007; Uddin 2019). SMs, the trademark of plants, have a wide variety of applications, not only for the plant itself but also for humans. They serve as weapons against microbes (antimicrobial), parasites, bugs, etc., act as metal transport machinery, sexual hormones, and are used mainly for the treatment of human diseases (Uddin 2019). SMs present in plants are categorized as alkaloids, terpenoids, phenolics, steroids, and flavonoids. These metabolites have a wide diversity in structure and size and are found in very large numbers throughout the plant kingdom. Along with the SMs, plants also produce antioxidant enzymes and non-enzymatic antioxidants like ascorbic acid, alpha-tocopherol, carotenoids, gluta-thione, and proline (Mittler 2002; Lajayer 2017). Some of the secondary compounds, also referred to as natural products, include drugs such as morphine, codeine, cocaine, and quinine; anticancerous alkaloids (vinblastine, vincristine, ajmalicine), belladonna alkaloids, colchicines, physostigmine, pilocarpine, and reserpine, and steroids like diosgenin, digoxin, and digitoxin.

Plants produce SMs as a response to physiological stimuli and stress (Elzaawely et al. 2007). Plants are subjected to a multitude of biotic and abiotic stresses during their entire life cycle. These stresses affect almost every aspect of the plant at the mor-phological, biochemical, physiological, and molecular level (Lokhande et al. 2011; Ahire et al. 2013). Changing environmental conditions alter the metabolism of both primary and secondary metabolites in plants. Plants respond to these altered envir-onmental conditions by regulation of stomata, altering the levels of osmoprotectants, antioxidant enzymes, and certain phytohormones at transcriptional and translational levels (Li et al. 2020; Mundada et al. 2020; Mundada et al. 2021). Earlier, many authors reported that environmental stresses resulted in the enhanced production of certain secondary metabolites in several medicinal plants (Wahid and Ghazanfar 2006; Cheng et al. 2007; Ghorpade et al. 2011; Ahire et al. 2013; Ahire et al. 2014).

9.3 ENVIRONMENTAL POLLUTION AND ITS CONSEQUENCES

Living things and the non-living environment are inseparably interactive with each other and make changes accordingly. The major causes of change in the medicinal properties or effectiveness of plants are rapid change in climate, development of the urban and industrial sector, population explosion, shrinking forest cover, destructive harvesting, and floods (Figure 9.1). These types of human-made obstruction in the environment are the reasons for water, soil, and air pollution. Several researchers have illustrated the adverse impacts of environmental pollutants on growth and develop-ment in plants and their primary, as well as secondary, metabolism. However, very few researchers have investigated the impact of environmental pollution on the pro-duction of secondary metabolites in medicinal plants. We provide a brief overview of this issue in this chapter.

9.3.1 SOIL AND WATER POLLUTION

Soil and water pollution results from the accumulation of pollutants, chemicals, and contaminants in high quantities which are toxic to microbes, plants, animals, humans, and the soil itself. Due to soil and water pollution, agricultural land is turning into non-agricultural or non-fertile and to deserts (Thurston 1992). The United Nations Food and Agricultural Organization revealed that, annually, 75 billion tonnes of soil from agricultural land is lost due to erosion, high salt concentration, and water-logging. The reasons for soil pollution are ever increasing, in the use of fertilizers, pesticides, insecticides, herbicides, high salinity, discharge of radioactive elements from industries, percolation of contaminated water in fertile land, unfavourable and harmful irrigation practices, sanitary sewage leakage, acid rain, fuel leakage from automobiles, and so on (Freedman and Hutchinson 1981; El-Ramady et al. 2015).

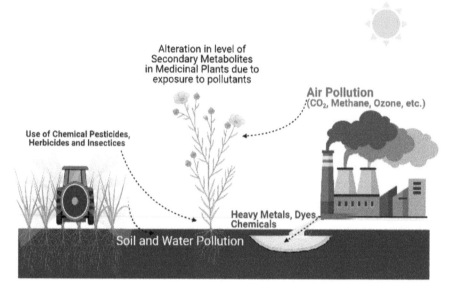

FIGURE 9.1   Outline of environmental pollution sources and their impact on plants.

Soil pollution not only affects the soil itself and its constituents but is also responsible for air pollution by releasing volatile molecules into the atmosphere. This ultimately suggests that soil and air pollution increase proportionally. The leaching of toxic
material into water through rain and sewage leads to water pollution in streams, lakes,
rivers, and oceans. These environmental pollutions directly or indirectly affect life
on the planet. The soil which is significantly contaminated also affects human life
through the accumulation of toxic chemicals in plant bodies. As plants cannot get
rid of these pollutants, they accumulate them and change their behaviour according
to the environmental hazards. To accomplish this, plants change and produce a huge
variety of chemical compounds in the form of hormones, enzymes, proteins, primary
and secondary metabolites, according to their defence strategy.

The main causative agent for soil pollution is heavy metal. They may accumulate in
the human body directly or indirectly. The uptake of heavy metals beyond the permissible limit can cause metabolic disruption to plants as well as humans (Khan 2007).
Heavy metals are categorized into the following types as described in Table 9.1.

Metals are released due to rhizosphere activities and root/microbial symbiotic
association, while their accumulation in plants is facilitated through water diffusion
from soil to root surface and to plant body along with ion exchange between clay
particles and the root surface (Maleki et al. 2017). Once the particles of heavy metals
are attached to the root surface their movement in the root occurs either by active or
passive transport and/or via the electrochemical potential gradients mediated by the
cell membrane (Clarkson and Luttge 1989). In addition to these paths of entry, foliar
uptake is another way of entry for heavy metals into plant cells, and this happens
through stomata, leaf cuticle, or both organelles (Maleki et al. 2017). Uptake and

TABLE 9.1
Non-Critical and Toxic Metals

Non-critical	Toxic but rare	Toxic and accessible
Na, K, Mg, Ca, Fe, Li, Sr and A	Ti, Hf, Zr, W, Nb, Ta, Re, Ga, La, Os, Rh, Ir, Ru, Ba	Be, Co, Ni, Cu, Zn, Sn, As, Se, Te, Pd, Ag, Cd, Pt, Au, Hg, Tl, Pb, Sb, Bi, Cr

Source: Wood (1974).

translocation of heavy metals through root to leaves are mainly dependent on factors like type of metal, environmental condition, plant species, growth stage, constituents of cell, and plant detoxification method (Lajayer 2017).

Mobility of heavy metals within the plants may affect adversely or favourably the biosynthesis of active constituents in different plant species. This may be due to the inactivation of essential enzymes or interruption of biosynthetic pathways which are implicated in the production of secondary metabolites (Thangavel et al. 1999; Murch et al. 2003; Pandey et al. 2007; Nasim et al. 2010). Specifically, this adverse or favourable effect of heavy metals on medicinal plants shows an impact directly or indirectly on pharmacologically active substances or by manipulating the pharmaco-kinetics. This may occur because of accumulation in harvestable plant parts. The concentration of heavy metals or affected biosynthetic medicinal components changes the overall performance (maturity and harvesting time) and the importance of particular medicinal plants (Sharma 2020).

Medicinal plants exposed to different levels of heavy metals show different results. Exposure of the *Phyllanthus amarus* to chromium leads not only to alteration in secondary metabolite production but also affects plant biomass, root and shoot length, and decrease in the level of photosynthetic capability due to loss in chlorophyll, carotenoids, and proteins (Rai et al. 2005; Rai and Mehrotra 2008). Stunted growth, chlorosis, and necrosis are observed in *Withania somnifera* upon exposure to Cd, leading to deficiency of NPK (Mishra et al. 2014) and a decrease in relative water content in *Brassica juncea* (Ahmad et al. 2016), while the addition of another heavy metal (selenium) diminishes the effect of Cd and support for the biomass growth, pigment content, and relative water content of the plant (Ahmad et al. 2016). The effect of Ni (25–50 mM) on seedlings of *Hypericum perforatum* (St John's wort plants) shows a negative effect on the production of hypericin and pseudohypericin (Murch et al. 2003) compared with under controlled conditions.

Exposure of Cr in *Ocimum tenuiflorum* shows high production of eugenol as compared to control plants (Rai et al. 2004). Similarly, a favourable effect of Cd supplementation was observed in *Phyllanthus amarus* leading to the production of phyllanthin and hypophyllanthin (Rai et al. 2005; Rai and Mehrotra 2008). A positive effect of Cd supplementation (50 and 100 µM) was also reported in the case of *Vaccinium corymbosum*. It showed enhanced production of malondialdehyde (MDA) and chlorogenic acid (phenolic compound) after 14 days of exposure (Manquian-Cerda et al. 2016). This suggested the positive interaction of heavy metals for the production of phenolic and antioxidant responses in medicinal plants. Another example of a medicinal plant, i.e., *Matricaria chamomilla*, showed steady behaviour under Cd exposure at low concentration (3 μM), while at a higher concentration of Cd (60–120 μM) enhanced production of herniarin (the coumarin-related product) was observed. The enhanced level of herniarin resulted from an increase in its precursor's quantity (Z)- and (E)-2-β-d-glucopyranosyloxy-4-methoxycinnamic acids (GMCAs) at the time of Cd exposure (Maleki et al. 2017).

The cell suspension study in *Thalictrum rugosum* showed a significant increase in berberine in the presence of $CuSO_4$ (Kim et al. 1991). A similar result of elevation in phenolics and flavonoids was shown by Radwan et al. (2018) in *Ficus carica* and *Schinus molle* in lead (Pb)-polluted and non-polluted sites. Identically the effect of $ZnSO_4$ (200μM) and $CuSO_4$ (150μM) in *Pluchealan ceolata* shows a high level of quercetin in regenerants (Kumar et al. 2004).

Exposure of medicinal plants to heavy metals not only influences the concentration of secondary metabolites but also manipulates the content of associated antioxidants, anti-inflammatory compounds, and some essential oils too. The decrease in the level of essential oil and growth of the plant was assessed in the case of *Mentha piperita* (cv Tundza and Clone No 1) (Lamiaceae) and *Mentha arvensis* var. *piperascens* (cv Mentolna-14) on exposure to Pb (Zheljazkov and Nielsen 1996). A study carried out in a greenhouse by Kunwar et al. (2015) shows that exposure of *O. basilicum* to varying concentrations of Pb, Cu, and Cd-treated soil resulted in quantitative rather than qualitative change in essential oil. The stimulation of jasmonic acid and ethylene upon application of heavy metals was also illustrated in certain plants (Maksymiec 2007). Ethylene regulates the production of alkaloids, hyoscyamine, and scopolamines. A change in the concentration of ethylene ultimately changes the production of these compounds (Pitta-Alvarez et al. 2000). This result was observed in *in vitro* hairy root culture of *Brugmansia candida* exposed to Ag. The release of scopolamines is elevated compared to hyoscyamine, and this may be due to ethylene-mediated down-regulation of hyoscyamine-6-β-hydroxylase (H6H) which is responsible for the synthesis of scopolamines from hyoscyamine (Pitta-Alvarez et al. 2000). The combined and individual effects of Pb and Cd were studied on *Ligusticum chuanxiong* by Zeng et al. (2020). As compared to individual metal exposure, combined exposure conditions resulted in a superior antioxidant defence strategy in *L. chuanxiong*. This condition affects mainly the content of ferulic acid, tetramethylpyrazine, and ligustilide, along with the total weight of the plant (Zeng et al. 2020). Heavy metal exposure also results in lipid peroxidation and the production of highly active signalling molecules (Gratão et al. 2005).

9.3.2 Air Pollution

In recent times, problems of air pollution have been increasing due to an uncontrolled rise in population, rapid industrialization, and deforestation. There is an alarming rise in the concentration of air-polluting gases such as carbon dioxide (CO_2), methane (CH_4), nitrous oxide (N_2O), nitrogen oxides (NOx), and sulfur dioxide (SO_2), along with secondary pollutants like ozone (O_3). Among all the greenhouse gases, CO_2 emission is more major than other gases. The global average atmospheric carbon dioxide in 2019 was 409.8 parts per million which is higher than at any point in at least the past 800,000 years (Lindsay 2020).

Medicinal plants respond to the altered levels of CO_2 and show variable responses in terms of accumulation of secondary metabolites which are the basis of medicinal activity in plants (Mishra 2016). Digoxin is a cardenolide glycoside used in cardiac failure treatment (Rahimtoola 2004). An elevation in the level of digoxin was reported in *Digitalis lanata* under elevated CO_2 concentrations. However, in a similar experiment the level of other glycosidic derivatives, namely digitoxin, digitoxigenin and digoxin-mono-digitoxoside, were reduced (Stuhlfauth and Fock 1990). Along with concentration, the duration of exposure to these gases is also a crucial factor in the regulation of the level of plant secondary metabolites. In a study carried out in *Hymenocallis littoralis*, a plant with antineoplastic and antiviral properties showed an increase in the concentration of three types of alkaloids, namely pancratistatin, 7-deoxynarciclasine, and 7-deoxy-trans dihydronarciclasin, in the first year, but this concentration was reduced in subsequent years (Idso et al. 2000). In *Papaver setigerum*, an elevation in the level of CO_2 from 300 µmol mol^{-1} to 600 µmol mol^{-1} resulted in the elevation of four alkaloids, viz. morphine, codeine, papaverine, and noscapine (Ziska et al. 2008). Similarly, in *Hypericum perforatum*, a neuroprotective plant showed elevation of phenolics in the presence of an excess of CO_2 (Zobayed and Saxena 2004). Schonhof et al. (2007) reported elevation in the level of glucosinolate derivatives in *Brassica oleracea* var. italic Plenck at elevated levels of CO_2. An increase in the level of tannins and phenolic compounds was reported in *Zingiber officinale* by Ghasemzadeh et al. (2010). A differential response in the accumulation of different terpenoids was observed in the case of *Ginkgo biloba* under excess of CO_2 and O_3. Around 15 per cent more accumulation of quercetin aglycon and reduction of kaempferol aglycon by 10 per cent was reported in studies carried out by Huang et al. (2010). By contrast, elevation in the level of almost all types of secondary metabolites was reported in *Catharanthus roseus* (Saravanan and Karthi 2014). An elevation in the level of CO_2 from 400µmol mol^{-1} to 1,200 µmol mol^{-1} led to an increase in the content of flavonoids and phenolic compounds in *Elaeis guineensis* (Ibrahim and Jaafar 2012).

Ozone is one of the trace gases found in Earth's atmosphere, especially in the stratosphere, and is known for its beneficial property of absorption of UV radiation. Increasing the concentration of ozone is known to decrease the yield in field crops (Kumar et al. 2017). Similar adverse effects of ozone are also reported in medicinal plants. Bortolin et al. (2016) exposed *Capsicum baccatum* to chronic ozone concentrations. The level of capsaicin and dihydrocapsaicin was reduced by more than half in the pericarp of the treated plants compared with the control. On the other hand, the content of total carotenoids and phenolic compounds was increased. In seeds, the level of capsaicin was reduced significantly compared with dihydrocapsaicin. Exposure to ozone in *Salvia officinalis* at a concentration of 120 ± 13 ppb for 90 consecutive days led to an increase in phenolic content, namely gallic acids, catechinic acid, caffeic acid, and rosmarinic acid (Pellegrini et al. 2015). In the case of *Melissa officinalis*, increased ozone concentrations led to an elevation in anthocyanin content significantly, along with phenolics and tannins (Pellegrini et al. 2011; Shakeri et al. 2016). Exposure of *Betula pendula* to ozone led to an increase in flavonoid (hyperoside) content and decrease in terpenoid (papyriferic acid) and phenolic (betuloside) compounds (Lavola et al. 1994). A field study on *Sida cordifolia* L. exposed to elevated levels of ozone (ambient + 20 ppb) showed an increased level of steroid content in leaves and roots. However, the terpene content showed a variable response (Ansari et al. 2021). In the medicinal herb *Hypericum perforatum,* exposure to ozone (110 ppb, 5 hours) led to an increase in the content of phenolic compounds by 45 per cent, flavonoids by 58 per cent, and anthocyanins by 2 per cent (Pellegrini et al. 2018).

9.4   CONCLUSION

Plants exhibit diversity among the different species and the array of secondary metabolites produced in them. These secondary metabolites essentially contribute to their medicinal properties. Over the past centuries, these biologically active compounds have played an important role in the welfare of human beings. These medicinal plants serve as a reservoir of several pharmaceutically active ingredients. Environmental pollution affects almost every aspect of plant growth, affecting normal growth and metabolism, imposing stress conditions in plants. Plants respond to these stresses by overproduction and accumulation of antioxidants and alteration of secondary metabolite profile. Looking at the paramount productive threshold of such metabolic alterations, appropriate conservatory practices are needed before these plants lose their bioactive components in the long run. Furthermore, there is a need to understand the mechanism of stress alleviation in medicinal plants at physiological and molecular levels.

ACKNOWLEDGEMENTS

The authors are grateful to Yashavantrao Chavan Institute of Science, Satara for financial assistance. Financial assistance to the faculty under the self-funded project is gratefully acknowledged.

REFERENCES

Ahire, M.L., S. Laxmi, P.R. Walunj, P.B. Kavi Kishor, and T.D. Nikam. 2014. Effect of potassium chloride and calcium chloride induced stress on in vitro cultures of *Bacopa monnieri* (L.) Pennell. *Journal of Plant Biochemistry and Biotechnology* 23: 366–378.

Ahire, M.L., P.R. Walunj, P.B. Kavi Kishor, and T.D. Nikam. 2013. Effect of sodium chloride induced stress on growth, proline, glycine betaine accumulation, antioxidative defense and bacoside A content in in vitro regenerated shoots of *Bacopa monnieri* (L.) Pennell. *Acta Physiologiae Plantarum* 35: 1943–1953.

Ahmad, P., E.A. Allah, A. Hashem, M. Sarwat, and S. Gucel. 2016. Exogenous application of selenium mitigates cadmium toxicity in *Brassica juncea* L. (Czern & Cross) by up-regulating antioxidative system and secondary metabolites. *Journal of Plant Growth Regulation* 35: 936–950.

Ansari, N., M. Agrawal, and S.B. Agrawal. 2021. An assessment of growth, floral morphology, and metabolites of a medicinal plant *Sida cordifolia* L. under the influence of elevated ozone. *Environmental Science and Pollution Research* 28: 832–845.

Bortolin, R.C., F.F. Caregnato, A.M.D. Junior, A. Zanotto-Filho, K.S. Moresco, A. de Oliveira Rios, and J.C.F. Moreira. 2016. Chronic ozone exposure alters the secondary metabolite profile, antioxidant potential, anti-inflammatory property, and quality of red pepper fruit from *Capsicum baccatum*. *Ecotoxicology and Environmental Safety* 129: 16–24.

Cheng, A.X., Y.G. Lou, Y.B. Mao, S. Lu, L.J. Wang, and X.Y. Chen. 2007. Plants terpenoids: biosynthesis and ecological functions. *Journal of Integrative Plant Biology* 49: 179–186.

Clarkson, D.T., and U. Luttge. 1989. Mineral nutrition: divalent cations, transport and compartmentation. *Progress in Botany* 51: 93–112.

El-Ramady, H., N. Abdalla, T. Alshaal, É. Domokos-Szabolcsy, N. Elhawat, J. Prokisch, A. Sztrik, M. Fári, S. El-Marsafawy, and M.S. Shams. 2015. Selenium in soils under climate change, implication for human health. *Environmental Chemistry Letters* 13: 1–19.

Elzaawely, A.A., T.D. Xuan, and S. Tawata. 2007. Changes in essential oil, kava pyrones and total phenolics of *Alpinia zerumbet* (Pers.) BL Burtt. & RM Sm. leaves exposed to copper sulphate. *Environmental and Experimental Botany* 59: 347–353.

Freedman, B., and T.C. Hutchinson. 1981. Sources of metal and elemental contamination of terrestrial environments. In: N.W. Lepp (ed.) Effect of heavy metal pollution on plants. Springer, Netherlands, pp. 35–94.

Ghasemzadeh, A., H.Z. Jaafar, and A. Rahmat. 2010. Antioxidant activities, total phenolics and flavonoids content in two varieties of Malaysia young ginger (*Zingiber officinale* Roscoe). *Molecules* 15: 4324–4333.

Ghorpade, R.P., A. Chopra, and T.D. Nikam. 2011. Influence of biotic and abiotic elicitors on four major isomers of boswellic acid in callus culture of *Boswellia serrata* Roxb. *Plant Omics Journal* 4: 16–176.

Gratão, P.L., A. Polle, P.J. Lea, and R.A. Azevedo. 2005. Making the life of heavy metal-stressed plants a little easier. *Functional Plant Biology* 32: 481–494.

Huang, W., Q. Deng, B. Xie, J. Shi, F.H. Huang, B. Tian, Q. Huang, and S. Xue. 2010. Purification and characterization of an antioxidant protein from *Ginkgo biloba* seeds. *Food Research International* 43: 86–94.

Ibrahim, M.H., and H.Z. Jaafar. 2012. Primary, secondary metabolites, H2O$_2$, malondialdehyde and photosynthetic responses of *Orthosiphon stimaneus* Benth. to different irradiance levels. *Molecules* 17: 1159–1176.

Idso, S.B., B.A. Kimball, G.R. Pettit III, L.C. Garner, G.R. Pettit, and R.A. Backhaus. 2000. Effects of atmospheric CO$_2$ enrichment on the growth and development of *Hymenocallis littoralis* (Amaryllidaceae) and the concentrations of several antineoplastic and antiviral constituents of its bulbs. *American Journal of Botany* 87: 769–773.

Khan, M.A., I. Ahmad, and I. Rahman. 2007. Effect of environmental pollution on heavy metals content of *Withania somnifera*. *Journal of Chinese Chemical Society* 54: 339–343.

Kim, D.I., H. Pedersen, and C.K. Chin. 1991. Stimulation of berberine production in *Thalictrum rugosum* suspension cultures in response to addition of cupric sulfate. *Biotechnology Letters* 13: 213–216.

Kumar, S., A. Narula, M.P. Sharma, and P.S. Srivastava. 2004. *In vitro* propagation of *Pluchea lanceolata*, a medicinal plant, and effect of heavy metals and different aminopurines on quercetin content. *In Vitro Cellular and Developmental Biology – Plant* 40: 171–176.

Kumar, V., T. Khare, S. Arya, V. Shriram, and S.H. Wani. 2017. Effects of toxic gases, ozone, carbon dioxide, and wastes on plant secondary metabolism. In: M. Ghorbanpour, and A. Varma (eds). Medicinal plants and environmental challenges. Springer, Cham, pp. 81–96.

Kunwar, G., C. Pande, G. Tewari, C. Singh, and G.C. Kharkwal. 2015. Effect of heavy metals on terpenoid composition of *Ocimum basilicum* L. and *Mentha spicata* L. *Journal of Essential Oil Bearing Plants* 18: 818–825.

Lajayera, B.A., M. Ghorbanpourb, and S. Nikabadi. 2017. Heavy metals in contaminated environment: Destiny of secondary metabolite biosynthesis, oxidative status and phytoextraction in medicinal plants. *Ecotoxicology and Environmental Safety* 145: 377–390.

Lavola, A., R. Julkunen-Tiitto, and E. Pääkkönen. 1994. Does ozone stress change the primary or secondary metabolites of birch (*Betula pendula* Roth.)? *New Phytologist* 126: 637–642.

Li, Y., D. Kong, Y. Fu, M.R. Sussman, and H. Wu. 2020. The effect of developmental and environmental factors on secondary metabolites in medicinal plants. *Plant Physiology and Biochemistry* 148: 80–89.

Lindsay, R. 2020. Climate Change: Atmospheric Carbon Dioxide. www.climate.gov/news-features/understanding-climate/climate-change-atmospheric-carbon-dioxide

Lokhande, V.H., T.D. Nikam, V.Y. Patade, M.L. Ahire, and P. Suprasanna. 2011. Effects of optimal and supra-optimal salinity stress on antioxidative defence, osmolytes and in vitro growth responses in *Sesuvium portulacastrum* L. *Plant Cell Tissue Organ and Culture* 104: 41–49.

Maksymiec, W. 2007. Signaling response in plants to heavy metal stress. *Acta Physiologiae Plantarum* 29: 177 –187.

Maleki, M., M. Ghorbanpour, and K. Kariman. 2017. Physiological and antioxidative responses of medicinal plants exposed to heavy metals stress. *Plant Gene* 11: 247–254.

Manquian-Cerda, K., M. Escudey, G. Zuniga, N. Arancibia-Miranda, M. Molina, and E. Cruces. 2016. Effect of cadmium on phenolic compounds, antioxidant enzyme activity and oxidative stress in blueberry (*Vaccinium corymbosum* L.) plantlets grown *in vitro*. *Ecotoxicology and Environmental Safety* 133: 316–326.

McNeely, J.A., K.R. Miller, W.V. Reid, R.A. Mittermeier, and T.B. Werner. 1990. Conserving the World's Biological Diversity. IUCN, World Resources Institute, Conservation International, WWF-US and the World Bank, Washington, DC.

Mishra, B., R S. Sangwan, S. Mishra, J.S. Jadaun, F. Sabir, and N.S. Sangwan. 2014. Effect of cadmium stress on inductive enzymatic and nonenzymatic responses of ROS and sugar metabolism in multiple shoot cultures of Ashwagandha (*Withania somnifera* Dunal). *Protoplasma* 251: 1031–1045.

Mishra, T. 2016. Climate change and production of secondary metabolites in medicinal plants: A review. *International Journal of Herbal Medicine* 4: 27–30.

Mittler, R. 2002. Oxidative stress, antioxidants and stress tolerance. *Trends in Plant Science* 7: 405–410.

Mundada, P.S., V.T. Barvkar, S.D. Umdale, S. Anil Kumar, T.D. Nikam, and M.L. Ahire. 2021. An insight into the role of silicon on retaliation to osmotic stress in finger millet (*Eleusine coracana* (L.) Gaertn). *Journal of Hazardous Materials* 403: 124078.

Mundada, P.S., T.D. Nikam, S. Anil Kumar, S.D. Umdale, and M.L. Ahire. 2020. Morpho-physiological and biochemical responses of finger millet (*Eleusine coracana* (L.) Gaertn.) genotypes to PEG-induced osmotic stress. *Biocatalysis and Agricultural Biotechnology* 23: 101488.

Murch, S.J., K. Haq, H.V. Rupasinghe, and P.K. Saxena. 2003. Nickel contamination affects growth and secondary metabolite composition of St. John's wort (*Hypericum perforatum* L.). *Environmental and Experimental Botany* 49: 251–257.

Nalawade, S.M., and H.S. Tsay. 2004. *In vitro* propagation of some important Chinese medicinal plants and their sustainable usage. *In Vitro Cellular and Developmental Biology Plant* 40: 143–154.

Nasim, S.A., and B. Dhir. 2010. Heavy metals alter the potency of medicinal plants. *Reviews of Environmental Contamination and Toxicology* 203: 139–149.

Pandey, S., K. Gupta, and A.K. Mukherjee. 2007. Impact of cadmium and lead on *Catharanthus roseus* – A phytoremediation study. *Journal of Environmental Biology* 28: 655–662.

Pellegrini, E., G. Carucci, A. Campanella, G. Lorenzini, and C. Nali. 2011. Ozone stress in *Melissa officinalis* plants assessed by photosynthetic function. *Environmental and Experimental Botany* 73: 94–101.

Pellegrini, E., A. Francini, G. Lorenzini, and C. Nali. 2015. Ecophysiological and antioxidant traits of *Salvia officinalis* under ozone stress. *Environmental Science and Pollution Research* 22: 13083–13093.

Pellegrini, E., A. Campanella, L. Cotrozzi, M. Tonelli, C. Nali, and G. Lorenzini. 2018. Ozone primes changes in phytochemical parameters in the medicinal herb *Hypericum perforatum* (St. John's wort). *Industrial Crops and Products* 126: 119–128.

Pitta-Alvarez, S.I., T.C. Spollansky, and A.M. Giulietti. 2000. The influence of different biotic and abiotic elicitors on the production and profile of tropane alkaloids in hairy root cultures of *Brugmansia candida*. *Enzyme and Microbial Technology* 26: 252–258.

Radwan, A.M., N.F. Reyad, A.E.R.M. Donia, and M.A. Ganaie. 2018. Comparative studies on the effect of environmental pollution on secondary metabolite contents and genotoxicity of two plants in Asir area, Saudi Arabia. *Tropical Journal of Pharmaceutical Research* 17: 1599–1605.

Rahimtoola, S. 2004. Digitalis therapy for patients in clinical heart failure. *Circulation* 109: 2942 –2946.

Rai, V., and S. Mehrotra. 2008. Chromium-induced changes in ultramorphology and secondary metabolites of *Phyllanthus amarus* Schum & Thonn. – An hepatoprotective plant. *Environmental Monitoring and Assessment* 147: 307–315.

Rai, V., S. Khatoon, S.S. Bisht, and S. Mehrotra. 2005. Effect of cadmium on growth, ultramorphology of leaf and secondary metabolites of *Phyllanthus amarus* Schum and Thonn. *Chemosphere* 61: 1644–1650.

Rai, V., P. Vajpayee, S.N. Singh, and S. Mehrotra. 2004. Effect of chromium accumulation on photosynthetic pigments, oxidative stress defense system, nitrate reduction, proline level and eugenol content of *Ocimum tenuiflorum* L. *Plant Science* 167: 1159–1169.

Sahito, S.R., M.A. Memon, T.G. Kazi, and G.H. Kazi. 2003. Evaluation of mineral contents in medicinal plant *Azadirachta indica* (Neem). *Journal of the Chemical Society of Pakistan* 25: 139–143.

Saravanan, S. and S. Karthi. 2014. Effect of elevated CO_2 on growth and biochemical changes in *Catharanthus roseus* – A valuable medicinal herb. *World Journal of Pharmacy and Pharmaceutical Sciences* 3: 411–422.

Schippmann, U., D.J. Leaman, and A.B. Cunningham. 2002. Impact of cultivation and gathering of medicinal plants on biodiversity: Global trends and issues. In: FAO, Biodiversity and the ecosystem approach in agriculture, forestry and fisheries, Rome, pp. 1–21.

Schonhof, I., H. Klaring, A. Krumbein, M. Schreiner. 2007. Interaction between atmospheric CO_2 and glucosinolates in Broccoli. *Journal of Chemical Ecology* 33: 105–114.

Shakeri, A., A. Sahebkar, and B. Javadi. 2016. *Melissa officinalis* L. A review of its traditional uses, phytochemistry and pharmacology. *Journal of Ethnopharmacology* 188: 204–228.

Sharma, M., R. Thakur, M. Sharma, and A.K. Sharma. 2020. Changing scenario of medicinal plants diversity in relation to climate change: A review. *Plant Archives* 20: 4389–4400.

Shimomura, K., K. Yoshimatsu, M. Jaziri, and K. Ishimaru. 1997. Traditional medicinal plant genetic resources and biotechnology applications. In: K. Watanabe, and E.R.G. Pehu (eds). Plant biotechnology and plant genetic resources for sustainability and productivity. R.G. Landes Company and Academic Press, Austin, TX, pp. 209–225.

Stuhlfauth, T., and H. Fock. 1990. Effect of whole season CO_2 enrichment on the cultivation of a medicinal plant, *Digitalis lanata. Journal of Agronomy and Crop Science* 164: 168–173.

Thangavel, P., A.S. Sulthana, and V. Subburam. 1999. Interactive effects of selenium and mercury on the restoration potential of leaves of the medicinal plant, *Portulaca oleracea* Linn. *Science of the Total Environment* 243: 1–8.

Thurston, H.D. 1992. Sustainable practices for plant disease management in traditional farming systems. Westview Press, Boulder, Oxford and IBH publishing Co., New Delhi.

Uddin, M. 2019. Environmental factors on secondary metabolism of medicinal plants. *Acta Scientific Pharmaceutical Sciences* 3: 34–46.

Verpoorte, R., R. van der Heijden, H.J.G. ten Hoopen, and J. Memelink. 1999. Metabolic engineering of plant secondary metabolite pathways for the production of fine chemicals. *Biotechnology Letters* 21: 467–479.

Wahid, A., and A. Ghazanfar. 2006. Possible involvement of some secondary metabolites in salt tolerance of sugarcane. *Journal of Plant Physiology* 163: 723–730.

Wood, J.M. 1974. Biological cycles for toxic elements in the environment. *Science* 183: 1049–1052

Zenga, J., X. Lia, X. Wang, K. Zhang, Y. Wang, H. Kang, G. Chen, T. Lan, Z. Zhang, S. Yuan, C. Wang, and Y. Zhou. 2020. Cadmium and lead mixtures are less toxic to the Chinese medicinal plant *Ligusticum chuanxiong* Hort. than either metal alone. *Ecotoxicology and Environmental Safety* 193: 110342.

Zheljazkov, V.D., and N.E. Nielsen. 1996. Effect of heavy metals on peppermint and cornmint. *Plant and Soil* 178: 59–66.

Ziska, L., S. Panicker, and H. Wojno. 2008. Recent and projected increases in atmospheric carbon dioxide and the potential impacts on growth and alkaloid production in wild poppy (*Papaver setigerum* DC.). *Climate Change* 91: 395–403.

Zobayed, S., and P. Saxena. 2004. Production of St. John's wort plants under controlled environment for maximizing biomass and secondary metabolites. *In Vitro Cellular and Developmental Biology – Plant* 40: 108–114.

10 Biotechnological Studies of Medicinal Plants to Enhance Production of Secondary Metabolites under Environmental Pollution

Saumya Pandey

CONTENTS

10.1 INTRODUCTION

Plants are natural chemical factories of a range of bioactive compounds synthesized during secondary metabolism. These secondary metabolites form the basis of the traditional medicine system and have been used since the prehistoric era due to their therapeutic potency, low cost, availability, and absence of side effects (Dias et al. 2012). Various bioactive compounds, such as alkaloids, terpenoids, and phenylpropanoids, are used as raw materials for the development of various modern drugs. In addition to having immense application in the pharmaceutical and food industries, these

DOI: 10.1201/9781003178866-10

177

secondary metabolites have also been used in the perfume, cosmetic, and agrochemical industries (Chandran et al. 2020).

Secondary metabolites play a significant role in survival and plant growth under stressful environmental conditions, such as temperature, salinity, and light intensity. These stresses induce various complex biochemical pathways depending mainly on the developmental stages, cell type, and environmental factors. The initiation and differentiation of the cellular structure specifically involved in the biosynthesis and accumulation of secondary metabolites are controlled by developmental factors (Bartwal et al. 2013). Thus, the medicinal properties may vary in different plant parts (root, stems, leaves, and flowers) at different developmental stages (Broun et al. 2006). For example, the essential oil content in the stem bark of *Cinnamomum cassia* increased with the increase of growth year and was found at its maximum (2.61% w/w) in a 12-year-old plant (Geng et al. 2011). Furthermore, environmental stresses cause modification in the expression of genes involved in the regulation of biosynthetic pathways of secondary metabolites (Borges et al. 2017; Sharma 2018). For example, drought stress enhances the biosynthesis of plant glycoside glycyrrhizin in the liquorice (*Glycyrrhiza glabra*) root. Drought stress conditions led to modulation of the expression of genes (SQS, bAS, LUS, and CAS) involved in glycyrrhizin biosynthesis (Nasrollahi et al. 2014).

Environmental pollution and climate change are interrelated and influence each other through complex interactions. The major environmental stresses likely to be increased by climate change are drought, heat, salinity, and waterlogging. The quality and quantity of secondary metabolites of medicinal plants are influenced by environmental factors, and thus by controlling and optimizing these factors high-quality drugs with novel therapeutic efficacy can be produced (Ncube et al. 2012; Yang et al. 2018). Several biotechnological approaches, such as screening and selection of cell lines with the highest production, plant cell and tissue culture, cell immobilization, hairy root culture, induction by elicitors, media components and plant growth regulator optimization, recombinant DNA technology together with modern techniques of analysis like NMR, HPLC, GC-MS, and LC-MS, have been exploited for the production of various valuable secondary metabolites.

10.2 ENVIRONMENTAL FACTORS AND THEIR EFFECT ON THE CONTENT OF SECONDARY METABOLITES

The basic components of the environment, such as the biosphere, atmosphere, water, and land, act as a repository for all pollutants. Environmental pollution is one of the major concerns at a global scale caused by urbanization, industrialization, population growth, agricultural activities, exploration, and mining, etc. It imposes a very serious impact on human health, terrestrial and aquatic animals, microorganisms, and plants (Kazemi and Ghorbanpour 2017). Environmental pollution poses a negative impact on both biotic and abiotic factors of the environment.

Being immobile, plants constantly interact with the rapidly changing and potentially damaging climate change. Thus, to counteract the effect, plants have evolved various complex defence mechanisms involving the synthesis of a diverse range of chemical metabolites which play a major role in the adaptation of plants (Holopainen and Gershenzon 2010). Although secondary metabolites have different structures and functions, they originate from the intermediate products of primary metabolism. For example, phenylalanine, the precursor of phenylpropanoid metabolism, is produced from the intermediate product (erythrose-4-phosphate) of the Calvin cycle and pentose phosphate pathway (Caretto et al. 2015).

Several reports have suggested that biotic and abiotic stress conditions, such as temperature, cold, salinity, light intensity, microbial attack, and many more, can result in changes at the gene or protein level of affected plants, thus altering the metabolite pool of plants (Szathmáry et al. 2001; Loreto and Schnitzler 2010). This results in increased activity of enzymes that play an important role in the secondary metabolism in plants. For example, the enzyme activity of phenylalanine ammonia-lyase (PAL) and chalcone synthase (CHS) involved in flavonoid synthesis is affected during environmental stress. The synthesis of specific secondary metabolites in the plants is highly regulated and produced in either a tissue-specific or developmental phase-specific or environmental factor-specific or species-specific manner (Osbourn et al. 2003).

The divergence from optimum environmental factors leads to an evident effect on the rate of biochemical and physiological processes and thus causes stress to the plant. To minimize the negative effect of particular stresses, the plants synthesize secondary metabolites of different compositions and quantities depending on the duration and magnitude of stress as well as the physiological status of the plant (Niinemets 2010). Consequently, understanding plant response to particular stress conditions and its effect on the quantity and quality of secondary metabolites is necessary for the exploitation of desired metabolites in pharmaceutical drug development (Ncube et al. 2012).

Increased biosynthesis of phenolic compounds in plants was reported at higher temperatures and reduced water availability (Glynn et al. 2004; Alonso-Amelot et al. 2007). Similarly, Zhang et al. (2011) reported higher levels of alkaloids in *Achnatherum inebrians* under salinity stress conditions. He et al. (2009) studied the effect of ozone level on the level of secondary metabolism of *Ginkgo biloba* and reported higher levels of terpenes and lower levels of phenolics under increased ozone stress. Furthermore, the level of salidroside in the medicinal herb *Rhodiola sachalinensis* can be increased by enhancing organic matter, exchangeable nitrogen, and total nitrogen in the soil, whereas soil with exchangeable potassium and phosphorous content greater than 180 mg/ml and 5 mg/ml respectively results in reduced salidroside yields (Yan et al. 2004). The advent of several advanced tools and techniques has enabled plants to be used as 'green factories' for developing desired metabolites in specific concentrations and tissue by inducing different stress conditions.

10.3 THERAPEUTIC IMPORTANCE OF MEDICINAL PLANT-DERIVED SECONDARY METABOLITES

In addition to providing food, plants act as repositories of various chemical compounds exhibiting therapeutic potential. The World Health Organization has listed 21,000 plants having medicinal properties, of which India accounts for almost 2,500 plant varieties (Shukla et al. 2019; Pundarikakshudu and Kanaki 2019). Secondary metabolites are synthesized in plants via various metabolic pathways, such as acetate-malonate, phenylpropanoid, 2-C-methyl-d-erythritol-4-phosphate, mevalonate, glucose, and amino acid pathways, and environmental factors can also act as inducers for their production in plants (Gonçalves and Romano 2018; Shitan 2016) They are responsible for the medicinal properties of plants and also have a significant role in the survival of plants during environmental stress (Kliebenstein 2013). These plant-derived secondary metabolites have been used for the prevention and treatment of diseases for ages, due to their antifungal, antiviral, antidiabetic, antioxidant, immune-modulatory, hepato-protective, and reno-protective properties (Rasool Hassan 2012; Ghorbanpour et al. 2017).

Nearly 80 per cent of human populations in developing countries are still dependent on plant-based traditional herbal medicines for their primary health care (WHO 2019). Despite the tremendous development in the field of synthetic chemistry, nearly 25 to 28 per cent of modern medicines are developed from bioactive constituents of plant origin (Samuelsson 2004). Recently plant-derived drugs have been regaining their popularity as they are cheap, easily available, naturally healing, show no side effects, and have a long-term therapeutic effect as compared to allopathic drugs (Narula et al. 2004). The global market for botanical and plant-derived drugs should reach $ 39.6 billion by 2022 from $ 29.4 billion in 2017 at a compound annual growth rate of 6.1 per cent (BBC Research 2017).

Secondary metabolites can be categorized into three chemically distinct groups: (1) phenolics (examples: daidzein, quercetin) which have antibacterial, anticancer, cardio-protective, anti-inflammatory, immune-protective, and skin protection from UV properties (Wang et al. 2018); (2) terpenes (examples: limonene, phyllaembicilins) which have antidepressant, antidiabetic, anticancer, antiviral, and antiplasmodial activity (Cox-Georgian et al. 2019); (3) alkaloids (examples: morphine, atropine) which demonstrate antibacterial, antitumoor, anti-inflammatory, hypnotic, local anaesthetic, antimitotic, and psychotropic activity (Masci et al. 2019). The diverse range of bioactive components derived from medicinal plants has been exploited in several traditional and modern medicines, such as taxol and artemisinin from *Taxus baccata* and *Artemisia annua*, which show anticancer activity. Similarly, digoxin and lantoside C from *Digitalis lanata* have been used for the treatment of congestive heart failure (Wawer 2008).

10.4 BIOTECHNOLOGICAL APPROACHES TO ENHANCE SECONDARY METABOLITE PRODUCTION

Biotechnological approaches have been extensively exploited for the conservation of plants and enhanced production of secondary metabolites as well as enabling engineering of novel metabolites. Furthermore, totipotency and *in vitro* transformation efficiency of plant cells have facilitated the application of these approaches to boost the production of secondary metabolites (Figure 10.1 and Table 10.1).

10.4.1 PLANT TISSUE CULTURE

Plant tissue culture is at the core of medicinal biotechnology, facilitating higher production of metabolites even in low-yielding and biotic/abiotic stress-susceptible

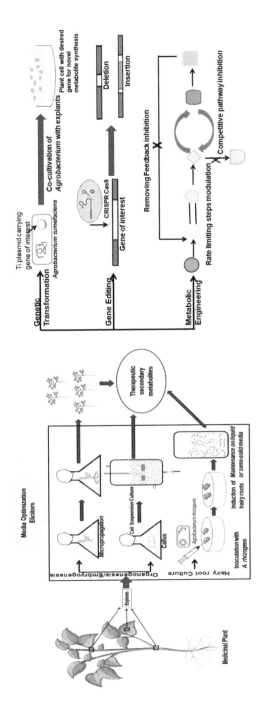

FIGURE 10.1 Biotechnological approaches for the production of secondary metabolites.

TABLE 10.1

Biotechnological Approaches to Improve Secondary Metabolite Production in Plants

Medicinal plant	Secondary metabolite	Biotechnological approaches	Medicinal properties	References
Argemone mexicana L.	Sanguinarine	Methyl jasmonate and fungal elicitors	Antimicrobial, antioxidant, anti-inflammatory, and antitumour	Trujillo-Villanueva et al. 2010
Camptotheca acuminate Decne.	Camptothecin	Callus Culture Overexpression of *ORCA3* gene in transgenic hairy root lines	Antitumour	Thengane et al. 2003; Ni et al. 2011
Datura stramonium L.	Sesquiterpenoid	Elicitation with Cu and Cd	Antiepileptic, anti-asthmatic, analgesic, antioxidant, antimicrobial, insecticidal	Furze et al. 1991
Eleutheroccous senticosus	β-sitosterol and stigmasterol	Overexpression of squalene synthase gene PgSS1	Antiosteoarthritic, antihypercholesterolemic, cytotoxicity, antitumour, hypoglycaemic, antimutagenic, antioxidant, anti-inflammatory	Seo et al. 2005
Eschscholzia californica Cham.	Columbamine	Overexpression of *C. japonica*-SMT gene	Anticancer	Rastegari et al. 2019
Glycyrrhiza glabra L.	Isoprenylated flavonoids	Hairy root culture	Antimicrobial, antioxidant, anti-inflammatory, antitussive, antidiabetic, antiviral, anticancer, antimutagenic, antiulcer, and hepatoprotective.	Asada et al. 1998
Mucuna pruriens (L.) DC	L-DOPA (L-3, 4-dihydroxy phenylalanine	Cell suspension culture	Anti-parkinsonian	Wichers et al. 1985
Ocimum basilicum L.	Rosmarinic acid, Lithospermic A acid, Lithospermic B Acid	Hairy root culture	Anti-microbial, immune-modulatory, antidiabetic, anti-allergic, anti-inflammatory, hepato- and renal-protectant	Tada et al. 1996

Species	Compound	Method	Activity	Reference
Panax ginseng C.A. Mey.	Ginsenoside	Hairy root culture	Antioxidant, Anti-inflammatory, vasorelaxation, antiallergic, antidiabetic, and anticancer	Mallol et al. 2001
Papaver somniferum L.	Morphine Codeine	*Agrobacterium*-mediated transformation via shoot organogenesis	Analgesic, antitussive	Park and Facchini 2000
Salvia cinnabarina M. Martens & Galeotti	p-sitosterol, ursolic acid	Hairy root culture	Sedative, hallucinogenic, skeletal muscle relaxant, analgesic, memory enhancing, anticonvulsant, neuroprotective, anti-Parkinsonian	Savona et al. 2003
Taxus brevifolia Nutt.	Taxol	Biotic and Abiotic Elicitation; Overexpression of DBAT and TXS) in *Taxus marei*	Anticancer	Pavarini et al. 2012; Ho et al. 2005
Vinca rosea L.	Vincristine Vinblastine	Abiotic/biotic stress factor on cell suspension culture	Anti-cancer	Abdal-Rehman et al. 2010
Withania somnifera (L.) Dunal	Withanolide	Root and shoot culture	Anti-inflammatory antistress, antioxidant, immunomodulatory antiangiogenesis anticancer	Sabir et al. 2013

medicinal plants. The successful applications of several other strategies like micro-propagation, *in situ* and *ex situ* conservation, genetic transformation, bioengineering, and polyploidy induction needed for the improvement of medicinal plants are directly or indirectly dependent on plant tissue culture (Grzegorczyk-Karolak et al. 2018).

Also, the effect of diverse experimental parameters and elicitors on the production of secondary metabolites in medicinal plants and elucidation of hormone metabolism, signalling, and transport can be done by culturing the plants under stable *in vitro* conditions (Kumari et al. 2018). Tissue culture not only paves the way for the conservation of plants of medicinal value as well as enhancing their bioactive compounds content but also facilitates the synthesis of engineered molecules (Epinosa-Leal et al. 2018). Thus, it provides opportunities to produce genetically engineered plants capable of producing novel secondary metabolites important for the cosmetic, pharmaceutical, and food industries.

The production of secondary metabolites via tissue culture generally takes place in two steps: (1) biomass accumulation; (2) synthesis and accumulation of secondary metabolites. The production steps can either be carried out in undifferentiated callus and cell suspension culture or differentiated organ (root, shoot, or somatic embryo) culture. Generally, the differentiated organ culture is required for the production of tissue or gland-specific secondary metabolites. For example, the root of *Panax ginseng* produces saponin and therefore root culture under *in vitro* conditions is required for large-scale production of saponin.

Hairy root cultures have emerged as a promising strategy for the *in vitro* production of various important metabolites. The major advantages of this method over conventional *in vitro* cultures are higher genetic and biochemical stability, a higher degree of cell differentiation and rapid growth, and higher biosynthetic capacity (Roychowdhury et al. 2017). Furthermore, the accumulation of secondary metabolites usually occurs in aerial parts of plants. Hairy roots are induced by infecting explants/cultures by the gram-negative soil bacterium *Agrobacterium rhizogenes*. Hairy roots have been a highly exploited biotechnological tool for the production of plants' secondary metabolites. For example, in *Raphanus sativus* higher concentration of phenolic, flavonoid, and quercetin (Balasubramanian et al. 2018), in *Sphagneticola calendulacea* higher concentration of phenolic acid, wedelolactone, and flavonoid (Kundu et al. 2018), and in *Scopolia lurida* higher alkaloid (hyoscyamine, scopolamine, and anisodamine) production (Lan et al. 2018) were achieved using hairy roots. The application of this method in industrial systems is difficult and therefore it cannot be used for the production of secondary metabolites at the commercial level.

Plant cell culture has been widely used for the mass production of several important secondary metabolites. Several benefits of using this culture system over tissue or organ cultures are (1) the provision of a stable system for continuous process operation, (2) uniformity in quality and yield of metabolites, (3) novel metabolites not synthesized by native plants can be produced. Several high-value pharmacologically important compounds, such as ginsenosides, taxol, artemisinin, ajmalicine, and resveratrol, have been produced through exploiting this culture system.

Furthermore, the optimization of several media parameters including culture media type, phytohormones, and biostimulants can enhance the *in vitro* production of valuable secondary compounds of plants (Matkowski, 2008). Compared to field-grown plants, the total phenolic and rosmarinic acid content and ROS activity was found to be higher in *Dracocephalum moldavica in vitro* culture grown on MS medium containing 2.4-D (0.5 mg/L) and BAP (0.2 mg/L) (Weremczuk-Jezyna et al. 2017). In addition to culture media components, environmental factors can also have an impact on the production of metabolites when added to culture systems.

The biotic/abiotic elicitors enhance the production rate of secondary metabolites when applied to culture medium containing cells at different developmental and physiological stages (Rao and Ravishankar 2002). The most common abiotic elicitors are environmental stress conditions – for example, heavy metals, salinity, UV radiation, high pH, and inorganic salts like jasmonic acid and salicylic acid. Zhao et al. (2016) reported increased production of secondary metabolites when *Robinia pseudoacacia* seedlings were treated with higher temperatures, carbon dioxide, and heavy metals. The exposure of UV-B radiation to medicinal plants *W. somnifera* and *Chrysanthemum* has resulted in a significant increase in withaferin A and chlorogenic acid contents, respectively. Also, several reports suggest a higher production of phenolic compounds in plants grown under higher temperatures and reduced water availability stress (Glynn et al. 2004; Alonsa-Amelot et al. 2007).

10.4.2 METABOLIC ENGINEERING

The engineering of metabolic pathways in important medicinal plants offers a new perspective to overproduce secondary metabolites that involve complex developmental regulation and different cell, tissue, and organelles for their biosynthesis (Sharma et al. 2018). Metabolite engineering involves understanding enzymatic reactions related to biosynthetic pathways at gene, transcriptome, and proteome levels. This approach can involve either inhibition or overexpression of the genes encoding rate-limiting enzymes in the biosynthetic processes (Matveeva and Sokornova 2016). However, a lack of comprehensive understanding of the mechanism of the synthesis of several secondary metabolites restricts the exploitation of a competitive approach for enhanced production. Therefore, to fulfil the increasing demand for important metabolites, extensive research is needed for the identification of regulatory and rate-limiting steps in biosynthetic pathways.

10.4.3 GENOME EDITING

The conventional genetic engineering approaches are inefficient to bring about huge changes in the genome of plants, due to copy number variability and random integration of DNA constructs at single or multiple genetic loci on chromosomes (Naqvi et al. 2010). However, the advent of various technologies like site-specific recombinase, 'tunable' transcription factors, genome editing tools, and synthetic promoters has enabled more precise and rapid improvement of plants (Liu et al. 2013). Sequence-specific nucleases such as ZFNs, TALENs, and CRISPR/Cas9 have been widely used to create precise site-specific double-stranded breaks in targeted points of the gene of interest. These engineered nucleases consist of two domains: (1) non-specific DNA cleavage domain, (2) customizable and sequence-specific DNA binding domain (Gaj et al. 2013). Transformation methods, such as the biolistic and *Agrobacterium*-mediated methods, are used for stable and efficient transformation of a plant with gene construct containing the desired gene with sequence-specific nucleases (Ma et al. 2016). Li et al. (2017) have CRISPR/Cas9-mediated knockout of SmCPS1 in the *Salvia miltiorrhiza* hairy root culture. The SmCPS1 gene encodes a chief enzyme involved in the tanshinone biosynthetic pathway. Recently the CRISPR/Cas9 system has been used in several medicinal plants to bring about the desired change in its chemical profile. Thus, plant cells containing only the essential components necessary for the production of specific secondary metabolites can be achieved through the CRISPR/Cas9 system (Noman et al. 2016).

10.4.4 GENETIC TRANSFORMATION

Several techniques such as particle bombardment, liposome fusion, electroporation, laser microbeam, and *Agrobacterium*-mediated transformation have been widely used for the genetic manipulation of plants. The choice of method for genetic manipulation depends on the plant species and its regeneration capacity in tissue culture as well as on the targeted genome (chloroplast, mitochondria, or nuclear genome) for insertion of the gene of interest. Out of all these techniques, the *Agrobacterium*-mediated transformation method has emerged as an efficient transgenic technique to manipulate several pharmaceutically important medicinal plants, including *Echinacea* (Wang and To 2004), *Digitalis* (Pérez-Alonso et al. 2014), *Artemisia* (Chen et al. 2000), *Taxus* (Han et al. 1994), *Scrophularia* (Park et al. 2003), and *Thalictrum* (Samanani et al. 2002). The genes related to the content and composition of bioactive compounds are targeted for transformation, but for the overall development of plants, the basic agronomic characters associated with growth, development, and stability, uniformity, biotic and abiotic resistance have also to be improved. Phytosteroid diosgenin concentration in the transformed *Trigonella foenum-gracium* root was two times higher than in the non-transformed roots (Merkli et al. 1997). Also, the level of bioactive compound solasodine becomes 4.2 times higher in the hairy roots of transgenic *Solanum aviculare* (Argôlo et al. 2000).

10.5 CONCLUSION

Plants have evolved various complex defence mechanisms against different stress conditions which promote the synthesis of numerous secondary metabolites in them. These metabolites serve as a source of medicinal substance for maintaining the good health of the world's increasing population. Recently, biotechnological approaches have become popular for the mass production of numerous pharmacologically important metabolites. These approaches not only boost the production of the metabolites in medicinal plants but also facilitate the engineering of novel bioactive compounds not produced by native plants. Among various techniques, plant cell and tissue culture techniques facilitate the viable, continuous, economical, and sustainable production of a wide range of secondary metabolites. Also, various culture media parameters and environmental factors can act as elicitors to enhance the metabolite content in the culture system. However, in some cases, the lack of complete knowledge of several biosynthetic pathways of bioactive molecules limits the application of this approach to scale up the production of low-yielding metabolites. Modern molecular biology techniques have emerged as a promising approach for the desired alteration in the biochemical profile of plants as well as novel metabolite synthesis via engineering biosynthesis of desired molecule pathways in the plant cells.

REFERENCES

Abdel-Rahman, T.M., T.Y. Kapiel, D.M. Ibrahiem, and E.A. Ali. 2010. Elicitation of alkaloids by biotic and abiotic stress factors in *Catharanthus roseus*. *Egyptian Journal of Botany* 54: 207–224.

Alonso-Amelot, M.E., A. Oliveros-Bastidas, and M.P. Calcagno-Pisarelli. 2007. Phenolics and condensed tannins of high altitude *Pteridium arachnoideum* in relation to sunlight exposure, elevation, and rain regime. *Biochemical Systematics and Ecology* 35: 1–10.

Argôlo, A.C.C., B.V. Charlwood, and M. Pletsch. 2000. The regulation of solasodine production by *Agrobacterium rhizogenes*-transformed roots of *Solanum aviculare*. *Planta medica* 66: 448–451.

Asada, Y., W. Li, and T. Yoshikawa. 1998. Isoprenylated flavonoids from hairy root cultures of Glycyrrhiza glabra. *Phytochemistry* 47: 389–392.

Balasubramanian, M., M. Anbumegala, R. Surendran, M. Arun, and G. Shanmugam. 2018. Elite hairy roots of *Raphanus sativus* (L.) as a source of antioxidants and flavonoids. 3 *Biotech* 8: 1–15.

Bartwal, A., R. Mall, P. Lohani, S.K. Guru, and S. Arora. 2013. Role of secondary metabolites and brassinosteroids in plant defense against environmental stresses. *Journal of Plant Growth Regulation* 32: 216–232.

Borges, C.V., I.O. Minatel, H.A. Gomez-Gomez, G.P.P. Lima. 2017. Medicinal plants: Influence of environmental factors on the content of secondary metabolites. In *Medicinal Plants and Environmental Challenges*, 259–277. Springer, Cham.

British Broadcasting Corporation (BBC) Research. 2017. Plant-derived Drugs: Global Markets. Available from: www.bccresearch.com/market-research/biotechnology/botanical-and plant-derived-drugs-global-markets-bio022h.html [Accessed: 2018-02-08]

Broun, P., Y. Liu, E. Queen, Y. Schwarz, M.L. Abenes, and M. Leibman. 2006. Importance of transcription factors in the regulation of plant secondary metabolism and their relevance to the control of terpenoid accumulation. *Phytochemistry Reviews* 5: 27–38.

Caretto, S., V. Linsalata, G. Colella, G. Mita, and V. Lattanzio. 2015. Carbon fluxes between primary metabolism and phenolic pathway in plant tissues under stress. *International Journal of Molecular Sciences* 16: 26378–26394.

Chandran, H., M. Meena, T. Barupal, and K. Sharma. 2020. Plant tissue culture as a perpetual source for production of industrially important bioactive compounds. *Biotechnology Reports* e00450.

Chen, D.H., H.C. Ye, and G.F. Li. 2000. Expression of a chimeric farnesyl diphosphate synthase gene in *Artemisia annua* L. transgenic plants via *Agrobacterium tumefaciens*-mediated transformation. *Plant Science* 155: 179–185.

Cox-Georgian, D., N. Ramadoss, C. Dona, and C. Basu. 2019. Therapeutic and medicinal uses of terpenes. In *Medicinal Plants*, 333–359. Springer, Cham.

Dias, D.A., S. Urban, and U. Roessner. 2012. A historical overview of natural products in drug discovery. *Metabolites* 2: 303–336.

Espinosa-Leal, C.A., C.A. Puente-Garza, and S. García-Lara. 2018. In vitro plant tissue culture: Means for production of biological active compounds. *Planta* 248: 1–18.

Furze, J.M., M.J. Rhodes, A.J. Parr, R.J. Robins, I.M. Whitehead, and D.R. Threlfall. 1991. Abiotic factors elicit sesquiterpenoid phytoalexin production but not alkaloid production in transformed root cultures of Datura stramonium. *Plant Cell Reports* 10: 111–114.

Gaj, T., C.A. Gersbach, C.F. Barbas. 2013. ZFN, TALEN, and CRISPR/Cas-based methods for genome engineering. *Trends in Biotechnology* 31: 397–405.

Geng, S., Z. Cui, X. Huang, Y. Chen, D. Xu, and P. Xiong. 2011. Variations in essential oil yield and composition during *Cinnamomum cassia* bark growth. *Industrial Crops and Products* 33: 248–252.

Ghorbanpour, M., J. Hadian, S. Nikabadi, and A. Varma. 2017. Importance of medicinal and aromatic plants in human life. In *Medicinal Plants and Environmental Challenges*, 1–23. Springer, Cham.

Glynn, C., A.C. Rönnberg-Wästljung, R. Julkunen-Tiitto, and M. Weih. 2004. Willow genotype, but not drought treatment, affects foliar phenolic concentrations and leaf-beetle resistance. *Entomologia Experimentalis et Applicata* 113: 1–14.

Gonçalves, S., and A. Romano. 2018. Production of plant secondary metabolites by using biotechnological tools. In *Secondary Metabolites, Sources and Applications,* 81–99. IntechOpen.

Grzegorczyk-Karolak, I., A. Wiktorek-Smagur, and K. Hnatuszko-Konka. 2018. An untapped resource in the spotlight of medicinal biotechnology: The genus Scutellaria. *Current Pharmaceutical Biotechnology* 19: 358–371.

Han, K.H., P. Fleming, K. Walker, M. Loper, W.S. Chilton, U. Mocek, M.P. Gordon, and H.G. Floss. 1994. Genetic transformation of mature Taxus: An approach to genetically control the in vitro production of the anticancer drug, taxol. *Plant Science* 95: 187–196.

He, X., W. Huang, W. Chen, T. Dong, C. Liu, Z. Chen, S. Xu, and Y. Ruan. 2009. Changes of main secondary metabolites in leaves of *Ginkgo biloba* in response to ozone fumigation. *Journal of Environmental Sciences* 21: 199–203.

Holopainen, J.K., and J. Gershenzon. 2010. Multiple stress factors and the emission of plant VOCs. *Trends in Plant Science* 15: 176–184.

Kazemi, A., and M. Ghorbanpour. 2017. Introduction to environmental challenges in all over the world. In *Medicinal Plants and Environmental Challenges*, 25–48. Springer, Cham.

Kliebenstein, D.J. 2013. Making new molecules—Evolution of structures for novel metabolites in plants. *Current Opinion in Plant Biology* 16: 112–117.

Kumari, A., P. Baskaran, L. Plačková, H. Omámiková, J. Nisler, K. Doležal, and J. Van Staden. 2018. Plant growth regulator interactions in physiological processes for controlling plant regeneration and in vitro development of *Tulbaghia simmleri*. *Journal of Plant Physiology* 223: 65–71.

Kundu, S., U. Salma, M.N. Ali, A.K. Hazra, and N. Mandal. 2018. Development of transgenic hairy roots and augmentation of secondary metabolites by precursor feeding in *Sphagneticola calendulacea* (L.) Pruski. *Industrial Crops and Products* 121: 206–215.

Lan, X., J. Zeng, K. Liu, F. Zhang, G. Bai, M. Chen, Z. Liao, and L. Huang. 2018. Comparison of two hyoscyamine 6β-hydroxylases in engineering scopolamine biosynthesis in root cultures of *Scopolia lurida*. *Biochemical and Biophysical Research Communications* 497: 25–31.

Li, B., G. Cui, G. Shen, Z. Zhan, L. Huang, J. Chen, and X. Qi. 2017. Targeted mutagenesis in the medicinal plant *Salvia miltiorrhiza*. *Scientific Reports* 7: 1–9.

Liu, W., J.S. Yuan, J.C.N. Stewart. 2013. Advanced genetic tools for plant biotechnology. *Nature Reviews Genetics* 14: 781–793.

Loreto, F., and J.P. Schnitzler. 2010. Abiotic stresses and induced biogenic volatile organic compounds. *Trends in Plant Science* 15: 154–166.

Ma, X., Q. Zhu, Y. Chen, and Y.G. Liu. 2016. CRISPR/Cas9 platforms for genome editing in plants: developments and applications. *Molecular Plant* 9: 961–974.

Mallol, A., R.M. Cusidó, J. Palazón, M. Bonfill, C. Morales, and M.T. Piñol. 2001. Ginsenoside production in different phenotypes of *Panax ginseng* transformed roots. *Phytochemistry* 57: 365–371.

Masci, V.L., S. Bernardini, L. Modesti, E. Ovidi, and A. Tiezzi. 2019. Medicinal plants as a source of alkaloids. In *Medically Important Plant Biomes: Source of Secondary Metabolites*, 85–113. Springer: Singapore.

Matkowski, A. 2008. Plant in vitro culture for the production of antioxidants – A review. *Biotechnology Advances* 26: 548–560.

Matveeva, T.V., and S.V. Sokornova. 2016. Transformation of plants for improvement of yields of secondary metabolites. In *Bioprocessing of Plant In Vitro Systems,* eds A. Pavlov, and T. Bley, 1–42. Springer: Germany.

Merkli, A., P. Christen, and I. Kapetanidis. 1997. Production of diosgenin by hairy root cultures of *Trigonella foenum-graecum* L. *Plant Cell Reports* 16: 632–636.

Naqvi, S., G. Farré, G. Sanahuja, T. Capell, C. Zhu, and P. Christou. 2010. When more is better: Multigene engineering in plants. *Trends in Plant Science* 15: 48–56.

Narula, A., S. Kumar, K.C. Bansal, and P.S. Srivastava. 2004. Biotechnological approaches towards improvement of medicinal plants. In *Plant Biotechnology and Molecular Markers*, 78–116. Springer, Dordrecht.

Nasrollahi, V., A. Mirzaie-Asl, K. Piri, S. Nazeri, and R. Mehrabi. 2014. The effect of drought stress on the expression of key genes involved in the biosynthesis of triterpenoid saponins in liquorice (*Glycyrrhiza glabra*). *Phytochemistry* 103: 32–37.

Ncube, B., J.F. Finnie, and J. Van Staden. 2012. Quality from the field: The impact of environmental factors as quality determinants in medicinal plants. *South African Journal of Botany* 82: 11–20.

Ni, X., S. Wen, W. Wang, X. Wang, H. Xu, and G. Kai. 2011. Enhancement of camptothecin production in *Camptotheca acuminata* hairy roots by overexpressing ORCA3 gene. *Journal of Applied Pharmaceutical Science* 1: 85–88.

Niinemets, Ü. 2010. Mild versus severe stress and BVOCs: Thresholds, priming and consequences. *Trends in Plant Science* 15: 145–153.

Noman, A., M. Aqeel, and S. He. 2016. CRISPR-Cas9: Tool for qualitative and quantitative plant genome editing. *Frontiers in Plant Science* 7: 1740.

Osbourn, A.E., X. Qi, B. Townsend, and B. Qin. 2003. Dissecting plant secondary metabolism – Constitutive chemical defenses in cereals. *New Phytologist* 159: 101–108.

Park, S.U., and P.J. Facchini. 2000. *Agrobacterium*-mediated transformation of opium poppy, *Papaver somniferum*, via shoot organogenesis. *Journal of Plant Physiology* 157: 207–214.

Park, S.U., Y.A. Chae, and P.J. Facchini. 2003. Genetic transformation of the figwort, *Scrophularia buergeriana* Miq., an Oriental medicinal plant. *Plant Cell Reports*, 21: 1194–1198.

Pavarini, D.P., S.P. Pavarini, M. Niehues, and N.P. Lopes. 2012. Exogenous influences on plant secondary metabolite levels. *Animal Feed Science and Technology* 176: 5–16.

Pérez-Alonso, N., B. Chong-Pérez, A. Capote, A. Pérez, Y. Izquierdo, G. Angenon, and E. Jiménez. 2014. *Agrobacterium tumefaciens*-mediated genetic transformation of *Digitalis purpurea* L. *Plant Biotechnology Reports* 8: 387–397.

Pundarikakshudu, K., and N.S. Kanaki. 2019. Analysis and Regulation of Traditional Indian Medicines (TIM). *Journal of AOAC International* 102: 977.

Rao, S.R., and G.A. Ravishankar. 2002. Plant cell cultures: Chemical factories of secondary metabolites. *Biotechnology Advances* 20: 101–153.

Rasool Hassan, B.A. 2012. Medicinal plants (importance and uses). *Pharmaceutica Analytica Acta*, 3: 2153–2435.

Rastegari, A.A., A.N. Yadav, and N. Yadav. 2019. Genetic manipulation of secondary metabolites producers. In *New and Future Developments in Microbial Biotechnology and Bioengineering*, 13–29. Elsevier.

Roychowdhury, D., M. Halder, and S. Jha. 2017. Transformation in medicinal plants: Genetic stability in long-term culture. In *Transgenesis and Secondary Metabolism*, 323–345. Springer, Cham, Switzerland.

Sabir, F., S. Mishra, R.S. Sangwan, J.S. Jadaun, and N.S. Sangwan. 2013. Qualitative and quantitative variations in withanolides and expression of some pathway genes during different stages of morphogenesis in *Withania somnifera* Dunal. *Protoplasma* 250: 539–549.

Samanani, N., S.U. Park, and P.J. Facchini. 2002. In vitro regeneration and genetic transformation of the berberine-producing plant, *Thalictrum flavum* ssp. glaucum. *Physiologia Plantarum* 116: 79–86.

Samuelsson, G. 2004. *Drugs of Natural Origin: A Textbook of Pharmacognosy*. 5th ed.: Swedish Pharmaceutical Press, Stockholm.

Savona, M., C. Mascarello, B. Ruffoni, A. Bisio, G. Romussi, P. Profumo, M. Warchol, and A. Bach. 2003. Salvia cinnabarina Martens et Galeotti. Optimisation of the extraction of a new compound-tissue culture and hairy root transformation. *Agricoltura Mediterranea* (Italy) 133(1): 28–35.

Seo, J.W., J. Wook, J.H. Jeong, C.G. Shin, S.C. Lo, S.S, Han, K.W. Yu, E. Harada, J.Y. Han, Y.E. Choi. 2005. Overexpression of squalene synthase in *Eleutherococcus senticosus* increases phytosterol and triterpene accumulation. *Phytochemistry* 66: 869–877.

Sharma, A. 2018. Gene expression analysis in medicinal plants under abiotic stress conditions. In *Plant Metabolites and Regulation Under Environmental Stress*, 407–414. Academic Press.

Sharma, A., P. Verma, A. Mathur, and A.K. Mathur. 2018. Genetic engineering approach using early Vinca alkaloid biosynthesis genes led to increased tryptamine and terpenoid indole alkaloids biosynthesis in differentiating cultures of *Catharanthus roseus*. *Protoplasma* 255: 425–435.

Shitan, N. 2016. Secondary metabolites in plants: Transport and self-tolerance mechanisms. *Bioscience, Biotechnology and Biochemistry* 80: 1283–1293.

Shukla, P., S.D. Jain, A. Agrawal, and A.K. Gupta. 2019. Indian herbal plants used as antipyretic: A review. *International Journal of Pharmacy & Life Sciences* 10: 6406–6409.

Szathmáry, E., F. Jordán, and C. Pál. 2001. Can genes explain biological complexity? *Science* 292: 1315–1316.

Tada, H., Y. Murakami, T. Omoto, K. Shimomura, and K. Ishimaru. 1996. Rosmarinic acid and related phenolics in hairy root cultures of *Ocimum basilicum*. *Phytochemistry* 42: 431–434.

Thengane, S.R., D.K. Kulkarni, V.A. Shrikhande, S.P. Joshi, K.B. Sonawane, and K.V. Krishnamurthy. 2003. Influence of medium composition on callus induction and camptothecin (s) accumulation in *Nothapodytes foetida*. *Plant Cell, Tissue and Organ Culture* 72: 247–251.

Trujillo-Villanueva, K., J. Rubio-Piña, M. Monforte-González, and F. Vázquez-Flota. 2010. *Fusarium oxysporum* homogenates and jasmonate induce limited sanguinarine accumulation in *Argemone mexicana* cell cultures. *Biotechnology Letters* 32: 1005–1009.

Wang, H.M., and K.Y. To. 2004. *Agrobacterium*-mediated transformation in the high-value medicinal plant *Echinacea purpurea*. *Plant Science* 166: 1087–1096.

Wang, T.Y., Q. Li, and K.S. Bi. 2018. Bioactive flavonoids in medicinal plants: Structure, activity and biological fate. *Asian Journal of Pharmaceutical Sciences* 13: 12–23.

Wawer, I. 2008. Solid-state measurements of drugs and drug formulations. In *NMR Spectroscopy in Pharmaceutical Analysis*, 201–231. Elsevier.

Weremczuk-Jeżyna, I., I. Grzegorczyk-Karolak, B. Frydrych, K. Hnatuszko-Konka, A. Gerszberg, and H. Wysokińska. 2017. Rosmarinic acid accumulation and antioxidant potential of *Dracocephalum moldavica* L. cell suspension culture. *Notulae Botanicae Horti Agrobotanici Cluj-Napoca* 45: 215–219.

Wichers, H.J., R. Wijnsma, J.F. Visser, T.M. Malingré, and H.J. Huizing. 1985. Production of L-DOPA by cell suspension cultures of *Mucuna pruriens*. *Plant Cell, Tissue and Organ Culture* 4: 75–82.

World Health Organization (WHO). 2019. WHO Global Report on Traditional and Complementary Medicine 2019. World Health Organization.

Yan, X., S. Wu, Y. Wang, X. Shang, and S. Dai. 2004. Soil nutrient factors related to salidroside production of *Rhodiola sachalinensis* distributed in Chang Bai Mountain. *Environmental and Experimental Botany* 52: 267–276.

Yang, L., K.S. Wen, X. Ruan, Y.X. Zhao, F. Wei, and Q. Wang. 2018. Response of plant secondary metabolites to environmental factors. *Molecules* 23: 762.

Zhang, X.X., C.J. Li, and Z.B. Nan. 2011. Effects of salt and drought stress on alkaloid production in endophyte-infected drunken horse grass (*Achnatherum inebrians*). *Biochemical Systematics and Ecology* 39: 471–476.

Zhao, Y.H., X. Jia, W.K. Wang, T. Liu, S.P. Huang, and M.Y. Yang. 2016. Growth under elevated air temperature alters secondary metabolites in *Robinia pseudoacacia* L. seedlings in Cd-and Pb-contaminated soils. *Science of the Total Environment* 565: 586–594.

11 Understanding the Metabolomics of Medicinal Plants under Environmental Pollution
Challenges and Opportunities

Prachi Sao, Rahat Parveen, Aryan Khattri,
Shubhra Sharma, Neha Tiwari,
Sachidanand Singh

CONTENTS

11.1 INTRODUCTION

Climate change is continuous, and how it impacts animals and plants is important to consider. The industrialization of the developing world has polluted the environment the most. Also, the use of crude oil, charcoal, and other chemical products to speed up work and transport vehicles and produce packaging materials contributes to all forms of pollution. The soil and water have also been contaminated by agricultural

DOI: 10.1201/9781003178866-11

and residential use of chemicals. Even building dams, industry, residential space, agri space has contributed to vast deforestation that has disrupted natural balance. Environmental contamination is so extreme that various existing physicochemical remediation methods such as Fenton processes, soil vapour extraction, chemical oxidation, Heterogeneous photocatalysis, vitrification, soil washing, electrokinetic, among many others, are expensive and unreachable in remote areas (Castro et al., 2016; Cheng et al., 2016; Rosas et al., 2013; Ortiz et al., 2007; Maurya et al., 2019). Also, this remediation generates by-products that further harm the soil and dramatically alter the way the soil can be used for plant growth (Meneses et al., 2010).

Quality soil yields healthy plants and products that play a vital role in vegetation. Several methods are used to combat the risk of soil erosion and ill plants (Cunningham et al., 1995). *Madseniana*, discovered at Cornell University, breaks down organic materials. Most species of *Paraburkholderia madseniana sp. nov.* are known for their ability to break down aromatic chemicals, and they can develop root nodules to store atmospheric nitrogen (Wilhelm et al., 2020). Worldwide, the detrimental effects of air pollution on the ecosystem are evident (Saxena et al., 2013; Seyyednejad et al., 2011). Smog-free towers are a competent invention to reduce air pollution. The tower sucks in pollution and expels clean air using electricity equivalent to the amount of a water boiler. The tower has faced a backlash from China after the country considered it incompetent because of not achieving desirable results (Laxmipriya et al., 2018). Another invention, the fog catcher, works in the direction of harvesting fog, working on a non-complex mechanism. The mist is caught in vertical nets trickling into a system where it is filtered and mixed with groundwater, ultimately providing clean water to households. The fog catcher has been developed by Qadir et al., but this method is not efficient on a large scale (Qadir et al., 2018).

With this in mind, the use of plants' metabolic activity for phytoremediation is more sustainable, efficient, and economical without threatening the environment. As known, plants are the best example of the mechanism of adaptation and protection to survive in extremely adverse weather changes and pollution (Salt et al., 1998). As, in stress conditions, they adopt sophisticated and efficient strategies at the metabolic level which involve various modifications such as biochemical changes, signalling pathways, molecular cross-talk, redox reaction, change in hormone concentration, gene transcription and, thus, plant metabolomics is a valuable tool for the identification of a multitude of different chemical compounds present in plant cells, with each molecule accounting for several metabolites involved in reducing the environmental stressor. Several rapid waste management technologies for air, water, and soil pollution have been developed, but the reduction of pollution is still a challenge (Rausher, 2001). This method of using plants metabolism to clean up pollutants is known as phytoremediation. It is more economical without generating any harmful by-products, an alternative to mechanical and chemical methods of cleaning pollution which requires further disposal of by-products. Some plants can accumulate pollutants in their leaves or stem in large quantities, especially trace elements such as zinc, lead, copper etc. These plants can absorb heavy metals from contaminated soil and water and are further used as fortified food for animals to improve micronutrients deficiencies (Cunningham and Ow, 1996). By analysing plant metabolic responses in complex environments, researchers can detect the genetic elements that govern plant stress response in adverse conditions.

There are many wild plants such as metallophytes, hyperaccumulators and pseudometallophytes that absorb toxic heavy metals at an unusually high level and metabolize them to a useful product (Gupta et al., 2013). Most of these plants do not have very high biomass production and cause no economic value for processing on a large scale. Plants that can metabolize pollutants are limited; however, those plants whose biomass is high – for instance, aromatic plants, medicinal plants, and edible plants such as maize and wheat – are used for metabolic research for their metabolic activity to clean the environment. The risk associated with edible plants is excess heavy metals might be introduced into our food chain (Vamerali et al., 2010; Saif et al., 2017) Metabolic profiling (metabolomics/metabonomics) is the measurement in biological systems of the complement of low-molecular-weight metabolites and their intermediates that reflects the dynamic response to genetic modification and physiological, pathophysiological, and/or developmental stimuli. The measurement and interpretation of the endogenous metabolite profile from a biological sample (typically urine, serum, or biological tissue extract) have provided many opportunities to investigate the changes induced by external stimuli (e.g., drug treatment) or enhance our knowledge of inherent biological variation within subpopulations. This article will focus on the basic principles of metabolic profiling and how the tools (nuclear magnetic resonance [NMR], liquid chromatography-mass spectrometry [LC-MS]) can be applied in toxicology and pathology. Metabolic profiling can complement conventional methodologies and other 'omics' technologies in investigating preclinical drug development issues. Case studies will illustrate the value of metabolic profiling in improving our understanding of phospholipidosis and peroxisome proliferation. A key message will be that metabolic profiling offers huge potential to highlight biomarkers and mechanisms in support of toxicology and pathology investigations in preclinical drug development (Clarke and Haselden, 2008). Therefore it is important to research metabolites that benefits the species of non-edible crops, the metabolic activity of plants such as aromatic and medicinal plants. Contaminated air, water, and soil can be fixed by studying the metabolic processes of medicinal plants. It is a favoured option for effective phytoremediation of the pollutant as it has the added advantage of biosynthesis of secondary metabolites which accelerate phytoremediation. Metabolomic research on secondary metabolites enables us to better understand how these metabolites are geographically and temporally constrained and how they are influenced by changing abiotic and biotic environments.

Several medicinal plants can metabolize pollutants during the phytoremediation process, such as kenaf *(Hempnha cannabina),* peppermint *(Mentha* sps.), hemp *(Cannabis sativa* L.), tulsi *(Ocimumtenuiflorum) (Ocimumbasilicum* L.), and vetiver *(Chrysopogonzizanioides)* wich can clean both organic and inorganic pollutants. Numerous medicinal plants like *Ricinus communis, Azadirachta indica, Syzygiumcumini, Phyllanthus emblica, Cannabis sativa, Moringa oleifera,* and *Madhuca longifolia* are high-biomass producing plants, and their biomass may be directly used for energy purposes (Jisha et al., 2017; Salt et al., 1998; Weyens et al., 2015; Masinire et al., 2020; Banerjee et al., 2016).

11.2　MEDICINAL PLANTS AND SECONDARY METABOLITES

Medicinal plants have evolved over the years and for many centuries were an essential aspect of human well-being as they are equipped with certain healing properties. Medicinal plants produce secondary metabolites that are not part of the normal process of growth; instead, they play a pivotal role in attracting pollinators, protecting them from environmental hazards and diseases. These secondary metabolites are produced under specific environmental conditions and need to support plants survival in adverse conditions (Verpoorte et al., 2007).

There are three primary subgroups of secondary metabolites: terpenoids, phenolic compounds, and nitrogen compounds. Secondary metabolites are responsible for distinctive and extremely species-specific tastes, colours, phytochemicals, and pharmaceutical compounds. Since plants are immobile, they produce biologically active compounds that are toxic to other species and also have antiviral, antifungal, antibacterial properties to protect themselves from herbaceous animals and pathogens (Hussein and El-Anssary, 2018; Pietra, 2002). This wisdom of plants is exploited by humans and other animals to keep themselves healthy; hence, a comprehensive analysis of different secondary metabolites in plants has significantly increased in metabolomics and the technology to extract high-quality drug substances involves deep research in the field of secondary metabolites processing pathways (Blank et al., 2020; Lazar, 2003; Pichersky and Gang, 2000). Secondary plant metabolites play an important role in the production of the various enzymes that degrade the organic pollutant in nature, still, the interrelationship between secondary plant metabolites and enzymatic diversity has not been fully explored (Singer et al., 2003). Medicinal plants whose metabolomic studies can be used as pollution remediators are listed in Table 11.1.

In the medicinal plant, metabolic research is crucial in understanding routes and pathways of biosynthesis responsible for plant metabolite production because the plant produces more than 200,000 low-molecular-weight molecules and each pathway is unique and needs an in-depth study (Okada et al., 2010; Pichersky and

---

TABLE 11.1
Plants that Can Be Used as Pollution Remediators

Plant	Usage	References
Lemna gibba	Removal of metalloids from stream waters.	Torbati and Keshipour (2020)
Neptunia oleracea	Removal of arsenic from water bodies.	Atabaki et al. (2020)
Salvinia minima	Used in accumulation of lead.	Leal-Alvarado et al. (2018)
Pistia stratiotes	Removes heavy metals.	Wickramasinghe andJayawardana (2018)
Salvinia natans	Removes heavy metals.	Leblebici, Kar, and Yalcin (2018)
Azolla pinnata	Removal of iron from freshwater.	Hasani et al. (2021)

Gang, 2000). For example, each flower has its fragrance or pigmentation, each has its sequence of events leading to the final products, these fragrances and pigmentations are produced to attract pollinators and are very specific to the surrounding environment. Metabolite formation needs a particular precursor and biotic or abiotic elicitor (Hussain et al., 2012). These elicitors are highly sensitive to environmental changes such as high temperature, drought, flood, soil properties and pathogens. Thus, change in climate changes the metabolic pathways involved in cell-cell communication and defence mechanisms.

Medicinal plant metabolomics research includes many disciplines, including biology, chemistry, pharmacology, and toxicity, while the whole metabolome is studied by no single technology. It faces difficulties in extracting phytochemicals, purifying them, and standardizing their dose. Initially, distillation was used for extraction, later chromatography, spectrometry, more recently NMR, microarray, target analysis, gene fingerprinting, genetic engineering, metabolite profiling, metabolic fingerprinting and a combination of technologies are used to make research more precise and accurate (Clarke and Haselden, 2008; Karczewski et al., 2017; Liang et al., 2020; Liu, 2010). Knowledge extraction and analysis in a biological sense of data produced by advanced technologies is another challenge in metabolomics. In the detection of bioactive compounds, computational biology currently plays an important role, in the field of metabolomics in particular. After many whole-genome projects, a large amount of genetic and proteomic data has been gathered, a channelized systemic analysis of this data will show specifics of the secondary metabolite that can be used in mass production of the medical plants (Chen et al., 2011). Computer metabolomics involves complex mathematical, statistical analysis and comprehensive knowledge pathways, clustering of algorithms, and comparative genome analysis.

Metabolic profiling, the classification of metabolites according to a certain metabolic pathway is referred to as 'detection and quantification of a restricted number of specified metabolites', measuring metabolites in cells and tissues of unicellular to multicellular organisms highlights their dynamic response to genetic alteration, as well as physiological, pathological, and/or developmental stimuli (e.g., ageing) (Clarke and Haselden, 2008). Extrinsic stimuli have afforded several opportunities for biological sample testing, including measurements and analysis of the endogenous metabolite profile. Chemicals or biotic and abiotic stress in cells can cause a biochemical reaction cycle that includes receptor activation, gene translation, and transcription to protein products. The diverse array of metabolites produced by this biochemical chain of reactions are abundant in the systemic circulation of plant metabolism, where they are readily accessible for biochemical profiling techniques. Nuclear magnetic resonance (NMR) spectroscopy and mass spectrometry are being utilized to detect metabolites (MS).

11.3 MEDICINAL PLANT METABOLOMICS FOR DEGRADATION OF POLLUTANTS

A huge amount of research has been done on the metabolomics of medicinal plants, to understand the production of secondary metabolites and various related pathways.

11.3.1 Metabolomics of Medicinal Plants and Its Use on Heavy Metals

Heavy metals from natural and man-made sources in the soil or water endanger numerous organisms. Agricultural chemicals, industrial waste, and wastewater irrigation, among other things, have increased the concentration of various metals such as zinc, copper, chromium, cadmium, and so on (Singh et al., 2011). Several medical plant-based metabolic approaches are currently being employed to eliminate heavy metals, with reported positive impacts on the ecosystem (Herrera-Estrella and Guevara-Garca, 2009). Plants can be metal excluders, indicators, or both; when metal is abundant in the soil, they can also behave as hyperaccumulators. Phytoextraction, also known as phytoaccumulation, is a metabolic activity in which plants can absorb heavy metals from contaminated soil and water, allowing them to be stored in the roots and shoots (Memon and Schröder, 2009; Ali et al., 2020). Several techniques, including functional genomics, EST sequencing, and gene expression pattern quantification, will uncover important genes involved in the complex processes that occur within cells. Even if complete genome information is not available, deeper information about a cell's function can be obtained using the metabolomic approach (Roosens et al., 2008).

Solanum nigrum is a high-biomass accumulator and a fast-growing annual medical plant. Due to the well-developed detoxification metabolic activity, in which heavy metals bind to suitable ligands such as organic acids, proteins, and peptides in the presence of enzymes that can also function at a high level of toxic metals, it can metabolize a large number of heavy metals in contaminated soil without causing any toxic effects to the environment (Cui et al., 2007). Several metabolic reactions are adjusted to reduce the negative effects of chronic heavy metals. Furthermore, there is an increase in the number of metabolites such as soluble sugars, proteins, proline, glycine betaine, flavonoids, phenolics, and alkaloids. Among metabolic processes, Amino acid metabolism is critical in plants and metal toxicity tolerance, as it modulates ion transport, stomatal conductance, and ROS scavenging (Rai et al., 2004). Abdel and Sallam observed that irrigation with wastewater increased the accumulations of total free amino acids and proline content in maize (Abdel Latef and Sallam, 2015). Homer et al. also discovered amino acid accumulation in three hyperaccumulator species, *Walsuramonophylla*, *Phyllanthus palwanensis*, and *Dechampetalumglenoides*, in response to heavy metal contamination (Homer et al., 1997; McNear et al., 2010).

According to Hassanein et al. (2013), there is also a rise in flavonoid content after wastewater irrigation. Flavonoids are natural metal chelators that interact with heavy metals to form complexes and inhibit heavy metals from producing free radicals (Agbaire and Akporhonor, 2014). Under stress, the structure of flavonoids is modified by glycosylation, prenylation, and methylation, which improves their antioxidant capabilities and inhibits lipid peroxidation. According to Moons et al., during stress conditions, the plant's metabolic activities change significantly toward the accumulation of protective secondary metabolites (Moons et al.,1997).

The most effective method is planting metal chelation (Memon and Schröder, 2009). Notably, among metals, Cd is known to be a strong promoter of phytochelatins (PCs) in a variety of plants (Anjum et al., 2015). Figure 11.1 illustrates the major transporters for Cd sequestration and storage.

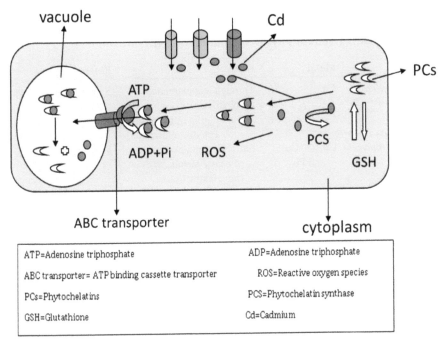

FIGURE 11.1 Major transporters for Cd sequestration and storage are illustrated.

Inside the cell, the concentration of Cd rises, and Cd binds directly to the phytochelatin synthase enzyme, activating it and stimulating phytochelatin production from Glutathione (GSH). The Cd-PCs complexes are then generated and delivered to the vacuole via tonoplast via ABC transporters. The Cd-PCs complex dissociates once it enters vacuoles. PC can be eliminated by the hydrolase enzyme found in vacuoles, or it can be restored to the cytosol. Metallothionein, like PCs, has metal-binding sites. Metallothioneins (MTs) are formed as a result of mRNA translation. An increase in heavy metal content increases the expression of the MTs gene. When MTs attach to cadmium, it detoxifies and maintains cytosolic homeostasis. Furthermore, MTs induce GSH production, which is vital in phytochelatins. Only a few heavy metals are important micronutrients for human health. In a recent study, *Ocimumbasilicum* was found to be suitable for phytoremediation of Cd-contaminated soil, which was enhanced when the plants were given different types of fertilizers (Zahedifar et al., 2015). According to Rai et al. (2004), *Ocimumtenuiflorum* L. can withstand phytotoxicity of Cr by modifying multiple metabolic pathways ever, both of which are harmful to cells when present at excessive levels. Table 11.2 lists some examples of plants whose metabolic processes can aid in heavy metal removal.

11.3.2 METABOLOMICS OF MEDICINAL PLANT FORTIFICATION

Biofortification is a metabolic process that involves enhancing the uptake of crucial mineral nutrients such as Fe, I, Cu, Zn, Mn, Co, Cr, Se, Mo, F, Sn, Si, and V. The

TABLE 11.2
Medicinal Plants Used for Phytoremediation of Heavy Metals

Plant	Metal	Usage	References
Moringa oleifera	Cd	Through leaf extraction (phytoextraction)	Yadav and Srivastava. (2017).
Spartina alterniflora	Cu, Zn and Cr	Accumulation of metal in roots (phytostabilization)	Xu et al. (2018).
Zygosaccharomyces rouxii	Cd, Zn, Cu, and Pb	Extraction and exclusion of heavy metals (phytoextraction)	Liang et al. (2017).
Halimione portulacoides	Zn	Accumulation of metal in tissues (phytostabilization)	Shackira and Puthur (2019)
Commelina communis	Cu	Accumulation of metal in roots (phytostabilization)	Patra et al. (2020).
Portulaca oleracea	Cr	Significant Cr accumulation in harvestable parts (phytoextraction)	Elshamy et al. (2019).
Lavandula vera	Cd, Pb, and Zn	The plant is a good phytoremediator for Pb, Cd, and Zn (phytoextraction)	Pirzadah et al. (2019).
Limnobium laevigatum	Pb, Cr, Ni, and Zn	accumulation of Pb, Cr, Ni, and Zn in roots is higher than in leaves	Arán et al. (2017).
Artemisia annua	As	A. annua could be considered as the top potential candidate for cultivation in As-contaminated soil	Kumari et al. (2018).

uptake of nutrients enriched by metabolic technologies such as genetic engineering and plant breeding has revealed that metabolism of Phytoremediation and Biofortification are closely interlinked. Based on phytoextraction processes these both technologies involve similar aspects such as metal uptake, accumulation and transformation of metabolites from soil (Zhao and McGrath, 2009). Researching the plant metabolites and genetics of hyperaccumulator species revealed that hyperaccumulators concentrate metals in their overground tissues at levels well above those found in the soil or non-accumulating surrounding developing species. Present transcriptome analysis study data has also shown that several gene families and transporter genes are involved in responding to mental stress, by a variety of molecules such as ATPase type P1B, RNAi, metal storage, metal-activated enzymes, channel proteins and metallic enzymes (Liang Zhu Y et al., 1999; Roosens et al., 2008; Zhou and Goldsbrough, 1994).

The metabolism of phytoremediation produces plant materials that are further utilized as sources of foods, animal feeding for fortified meat, and green fertilizers for agricultural products. There is a wide variation in the degree to which metabolic traits respond to environmental pressure ranging from a slight shift in metabolic turnover that changes in C/N/P/K. Due to the enhanced nutrient content of biofortified plants, the nutritional quality of food may contribute to reducing the micronutrients deficiency significantly in areas where all the nutrients are not consumed in the required amount.

Metabolomics of more than 400 species of hyperaccumulators are already known (Memon and Schröder, 2009), it includes several medicinal plants that are also available for fortification, such as selenium and *Adenocaulonhimalaicum*. An example of biofortification is set in the 'world capital of selenium'. Selenium is a nutrient that plays a critical role in metabolism and thyroid function and it helps protect the body from damage caused by oxidative stress. The selenium content in Enshi, China, is reported to be around 20–60 mg/kg DW which is 100 times more than the rest of the world. *Adenocaulonhimalaicum* is a medicinal plant found in Enshi that could potentially be planted worldwide to enrich the content of selenium all over (Yuan et al., 2012). Consumption of Se-rich medicinal plants makes them favourable for reducing the risks of deficiencies and diseases caused by selenium. Several plants, such as purple *Viola calaminaria*, the mustards *Thlaspicalaminare* and *Astragalus* are suitable for the accumulation of selenium and make it available to humans in the form of dry leaves, shoots, and roots. Another plant called *Alyssumcan* stores nickel in dried leaves (Salt et al., 1998). Ni form complexes with carboxylic acids, citrate, malate, and other organic acids accumulated in the vacuoles (McNear et al., 2010). The ability to accumulate organic acid regulate their metal accumulation capability in hyperaccumulators. Organic acid ligands enhance the uptake, translocation, and hyperaccumulation of metals in hyperaccumulators. (Sagner et al, 1998., Boominathan and Doran, 2003). *A.halleri* is a perennial medicinal plant able to both colonize zinc and cadmium polluted sites and accumulate large amounts of these metals in foliar tissues (Meyer and Verbruggen, 2012). Zn forms complexes with malate, it plays an important role in chlorophyll biosynthesis, cell growth regulation, root development, and nitrogen metabolism.

11.3.3 Metabolomics of Medicinal Plants on Polluted Water

Much work is being done to improve the quality of water and, among all the modern-day techniques used, the metabolic approach of using aquatic medicinal plants is most preferable (Abdullah et al., 2020; Raja et al., 2015). The very specific species of macrophytes are key players in the treatment of contaminated water. *Azolla*, *Eichhornia*, *Lemna*, *Potamogeton*, *Spirodela*, and *Wolfiaare* are some plants that have been reported as the highest in their metabolic process of phytoremediators. Being efficient in the metabolic activity of metabolizing the aquatic pollutant, these plants do so through the bioaccumulation of contaminants in their body tissue (Ansari et al., 2020). Since these plants are resistant to the toxicity of contaminants, they play a key role in the cleansing of pollution. *Eichhornia* is highly resistant and can bear the toxicity of major pollutants such as heavy metals, formaldehyde, formic acids, acetic acids even in the place they are present in high concentrations. Some plants like *Lemnaceae* are proven to be highly efficient in reducing the biological oxygen demand (BOD) and chemical oxygen demand (COD). It is a cost-effective bioremediation method that uses aquatic macrophytes, which is proven to be an important technique to improve the quality of contaminated water (Darajeh et al., 2016).

The uptake efficiency of plants could be enhanced by increasing the key metabolites responsible for the accumulation of contaminants. The plants may act as an excluder or accumulator, because of their uptake of metals. The contaminants get accumulated in the aerial tissues of the plants, where they are metabolized and biotransformed into active forms. The ratio of shoot and root in hyperaccumulator is greater than that in non-accumulator plants, the further one being able to flourish in adverse environments with minimal maintenance yielding higher biomass. Heavy metals get accumulated by hyperaccumulators in high quantities as compared to non-accumulators. Aquatic macrophytes in water bodies alter the physiochemical environment of the body, they absorb contaminants and store them in their biomass.

11.3.4 METABOLOMICS OF MEDICINAL PLANTS FOR POLLUTED AIR

Many medicinal plants can metabolize toxins from the air with the help of bacterial metabolism, both indoor and outdoor air. A large number of pollutants are metabolized by the leaf and the stem surface of these plants (Agarwal et al., 2019). They accumulate and transport pollutants to the roots of the rhizosphere, where further metabolization, degradation or sequestration of specific pollutants occurs with the help of certain bacteria such as phyllospheric and endophytic bacteria, therefore understanding the association and associated pathways may reveal genes that may enhance plant phytoremediation properties when modified (Weyens et al., 2015). Domestic exhaust, vehicle exhaust, industrial pollutants, and fossil fuels together produce particulate matter that is toxic to humans, plants, and animals. Some of this particulate matter contains heavy metals, harmful gases, organic substances, PHAs, and various other chemicals (Myers and Maynard, 2005). The plant absorbs this particulate matter and processes it to eliminate toxic effects. One such example is Nox and So_2 which gets into the leaves by going through the stomata and is processed to the organic molecules through the reduction process, plants release enzymes that can perform the task of breaking down pollutants to free sugars, organic acids, lipids, cell-wall components, glycine etc. Therefore, genetic engineering might create magic vegetables by increasing the concentrations of enzymes in plants. Therefore, genetic engineering could give them the ability to overproduce enzymes to clean larger areas. *Chrysanthemum X morifolium* is a medicinal and a great phytoremediator for benzene and toluene (Yang and Liu, 2011).

Azadirachta indica A. Juss, commonly known as neem, is known for its medicinal and phytoremediation properties, it can heal various diseases and also purify the air. It can take a high amount of SO_2 from the contaminated air (Abdullateef et al., 2014). It has been reported that the plants subjected to varied environmental conditions accumulate osmoprotectant molecules, which help preserve cellular homeostasis and the detoxification of ROS (Rontein et al., 2002). The metabolomic of a plant's stress tolerance can explain its dependency on the balance between ROS buildup and the cellular antioxidative defence machinery's ability to detoxify it (Duccer and Ting, 1970).

11.4   ADVANCES IN METABOLOMICS OF MEDICINAL PLANTS

11.4.1   GENETIC ENGINEERING APPLICATION

Genetic engineering can also be called genetic alteration and can be described as changes made by humans in the genetic structure or arrangement of some species. These methods are widely used by scientists to enhance the essential characteristics of spices such as disease tolerance, adverse environmental contamination, and increased yield (Rausher, 2001). Using genetically modified plants for phytoremediation improves the efficacy of the technology. Extensive research on the metabolomics of phytoremediation plants, at the molecular and pathways level, researchers were able to selectively modify genes of the plant to enhance phytoremediation properties.

Some of the plants are designed to release enzymes into the rhizosphere that can do chemical degradation, and the advantage of this strategy is that these plants do not need to ingest the chemicals and absorb them in their processes. These enzymes may bind themselves to the rhizosphere bacteria and convert chemical compounds into the soil itself. Genetically engineered *Arabidopsis* plants released root-specific laccase (LAC1) and demonstrated increased resistance to many allelochemicals and 2,4,6-trichlorophenol (El-Ramady et al., 2020). Due to the accumulation of heavy metals in the soil, few transgenic plants were engineered to increase the tolerance of heavy metals such as tobacco *(Nicotiana tabacum)* plants in which the yeast metallothionein (MT) gene was expressed to increase the tolerance of cadmium (Farooq and Khanday, 2020). The bacterial gene encoding mercuric reductase was overexpressed in the transgenic poplar plant for mercury remediation.

The removal of heavy metals and toxic compounds is very important as they interact with the biota (Karczewski et al., 2017). Various aquatic plants are known for having been able to extract heavy metals from the aquatic environment such as water hyacinth *(Eichhornia crassipes),* water ferns *(Salvinia minima),* duckweed *(Lemna minor, Spirodelaintermedia)*, water lettuce *(Pistiastratoites)*, watercress *(Nasturtium officinale)* (Cui et al., 2020). Due to the rise in traffic, industrialization and urbanization have been major factors. Plants, which are an integral part of the environment, play a major role in removing toxic contaminants found in the air. Various plants are under observation to research their function in environmental cleanings, such as *Pongamia pinnata, Tabernae montana divaricata, Ipomea carnea, Ficus religiosa, Ficus benghalensis*, and *Quisqualis indica* (Ozyigit et al., 2021).

11.4.2   METABOLOMICS OF MEDICINAL PLANTS IN BIOFUEL PRODUCTION

Any form of valorization for the crude biomass from the phytoremediation process would be very necessary because it can produce a comparatively clean and carbon-neutral alternative to fossil fuels that ensure sustainable development of the environment. Using fast-growing hyperaccumulating plants to allow biomass-based energy production is gradually becoming more common, it is favourable for energy needs or as green fertilizer. Studies show that *Hibiscus cannabin* (kenaf) is a medicinal plant and also a renewable choice for the manufacture of environmentally safe and effective bioenergy. Kenaf is an annual and spring plant that is present in tropical and temperate climates (Nizam et al., 2016). It is very well known that kenaf plants possess therapeutic properties and also serve as a good source of industrial fibre for producing eco-friendly bags, mats, furniture, and other items. The oil from kenaf seeds includes antioxidants and polyunsaturated fatty acids that cure multiple health problems, such as cholesterol level, cancer, arthritis, prevent erythrocyte haemolysis nerve disorders, and high blood pressure (Jisha et al., 2017).

Metabolomic research suggests kenaf can extract 10–20 tonnes per acre CO_2 from the air and the solid fuel generated from pretreatment and hydrolysis of kenaf biomass generates heat and electricity. It can absorb heavy metals, such as iron, lead, chromium, cadmium, arsenic, zinc, and oil, from the contaminated soil and water and can grow in low fertile lands (Abioye et al., 2010). Both root and shoots are used for hyperaccumulation. It is proposed that a steam distillation procedure used to extract the essential oil will eliminate the impurities in the oil and the residuary biomass can be used as a source of energy that requires low input, low water for irrigation, and low nutrients and provides a high yield of biomass.

11.4.3 MEDICINAL PLANT METABOLITES AS A POLLUTION BIOMARKER

A common application of environmental pollution testing is determining the chemical make-up of plants around the polluted area. There are various mechanisms by which plants may suck up elements such as via their roots, or through the water cycle. In India, the leaves of the neem tree have been used as a measure of air pollution since the fourteenth century to show the quality of water (Abdullateef et al., 2014).

It is now not unknown that pollution caused by anthropogenic activities is causing plenty of changes in our environment, whether it is the body or surroundings. Similarly, changes can be seen in plants also, in a study on the plant it was found that there is a relation between changing air quality and changing in biochemical features (chlorophyll and ascorbic acid content, etc.) in plant leaves, this concluded that plants can be used as biomarkers of pollution (Thawale et al., 2011).

11.5 CHALLENGES

Medicinal plants are very specific to geographical areas and thus cannot be used in different locations and must be identified by region. As a result, phytoremediation might not be very rapid, requiring time to clean up the environment. Since medicinal plants produce secondary metabolites that are highly reactive, thorough research on the metabolic interaction of secondary metabolites and accumulated pollutants should be required, as human ingestion can result in serious side effects. To study how effective plants could be biomarkers of pollution, it is necessary to carry out a large number of studies on different flora under different climatic conditions, which could aid in the preparation of a biologically sensitive flora map. Because plants that indicate the existence of pollution usually display changes in their metabolism, it is often difficult to report such minor changes. Uncertainties due to lack of sufficient experience in certain areas and to the fact that the field is naturally variable can often lead to an increased error in results or an inflated value. Plants generally display a minimal amount of changes/responses to the stressors present in the environment, and more often only respond to some of them, making them ineffective as substantial biomarkers.

11.6 CONCLUSION

An integrated evaluation of metabolite modifications and a host pathway review is more likely to yield more pertinent knowledge about phytoremediation medicinal plants. Genetically modifying the plants with the associated microbe will exhibit multiple traits and could be an excellent technique to enhance metal detoxification of contaminated sites. The plants used can be economically useful as they can be used for biofuel or energy production and also recovery of metal could be eco-environmental convenient. Though phytoremedic plants are a good technique, still a few areas should be researched for using this technique, like risks related to transgenic plants and understanding the plant-microbe-metal-soil interaction for contaminated sites. It is a green technology that uses plants for remediation and for restoring the environment and this technique can be integrated with other remediation techniques for high efficiency. This article concludes the utilization of biotechnological tools for pollution management with the help of phytoremediation in association with plants and microbes to reduce the risks of pollutants present in the environment. The plants selected must be able to deal with the presence of a high concentration of metals.

REFERENCES

Abdel Latef, A.A., and Sallam, M.M., 2015. Changes in growth and some biochemical parameters of maize plants irrigated with sewage water. *Austin Journal of Plant Biology* 1(1): 1004.

Abdullah, S.R.S., Al-Baldawi, I.A., Almansoory, A.F., Purwanti, I.F., Al-Sbani, N.H., and Sharuddin, S.S.N., 2020. Plant-assisted remediation of hydrocarbons in water and soil: Application, mechanisms, challenges and opportunities. *Chemosphere* 247: 125932. https://doi.org/10.1016/j.chemosphere.2020.125932

Abdullateef, B., Kolo, B.G., Waziri, I., and Idris, M.A. 2014. Assessment of neem tree (*Azadirachta indica*) leaves for pollution status of Maiduguri environment, Borno State, Nigeria. *International Journal of Engineering and Science* 3(9):31–35.

Abioye, P.O., Abdul Aziz, A., and Agamuthu, P. 2010. Enhanced biodegradation of used engine oil in soil amended with organic wastes. *Water, Air, & Soil Pollution* 209: 173–179. https://doi.org/10.1007/s11270-009-0189-3

Agarwal, P., Sarkar, M., Chakraborty, B., and Banerjee, T. 2019. Phytoremediation of Air Pollutants: Prospects and Challenges. In: Pandey, V.C., Bauddh, K. (eds), Phytomanagement of Polluted Sites. Elsevier, pp. 221–241. https://doi.org/10.1016/B978-0-12-813912-7.00007-7

Agbaire, P.O., and Akporhonor, E.E. 2014. The effects of air pollution on plants around the vicinity of the Delta Steel Company, Ovwian-Aladja, Delta State, Nigeria. *IOSR Journal of Environmental Science, Toxicology and Food Technology* 8(7): 61–65.

Ali, S., Abbas, Z., Rizwan, M., Zaheer, I., Yavaş, İ., Ünay, A., Abdel-Daim, M., Bin-Jumah, M., Hasanuzzaman, M., and Kalderis, D. 2020. Application of floating aquatic plants in phytoremediation of heavy metals polluted water: A review. *Sustainability* 12: 1927. https://doi.org/10.3390/su12051927

Anjum, N.A., Singh, H.P., Khan, M.I.R., Masood, A., Per, T.S., Negi, A., et al. (2015). Too much is bad – An appraisal of phytotoxicity of elevated plant-beneficial heavy metal ions. *Environmental Science and Pollution Research* 22: 3361–3382. DOI: 10.1007/s11356-014-3849-9

Ansari, A.A., Naeem, M., Gill, S.S., and AlZuaibr, F.M. 2020. Phytoremediation of contaminated waters: An eco-friendly technology based on aquatic macrophytes application. *Egyptian Journal of Aquatic Research* 46: 371–376. https://doi.org/10.1016/j.ejar.2020.03.002

Arán, D.S., Harguinteguy, C.A., Fernandez-Cirelli, A., and Pignata, M.L. (2017). Phytoextraction of Pb, Cr, Ni, and Zn using the aquatic plant *Limnobiumlaevigatum* and its potential use in the treatment of wastewater. *Environmental Science and Pollution Research* 24(22): 18295–18308.

Atabaki, N., Shaharuddin, N.A., Ahmad, S.A., Nulit, R., and Abiri, R. (2020). Assessment of water mimosa (*Neptunia oleracea* Lour.) morphological, physiological, and removal efficiency for phytoremediation of arsenic-polluted water. *Plants* 9(11): 1500.

Banerjee, R., Goswami, P., Pathak, K., and Mukherjee, A. 2016. Vetiver grass: An environment clean-up tool for heavy metal contaminated iron ore mine-soil. *Ecological Engineering* 90: 25–34. https://doi.org/10.1016/j.ecoleng.2016.01.027

Boominathan, R., and Doran P.M. 2002. Nickel induced oxidative stress in roots of Ni hyperaccumulator *Alyssum bertolonii*. *New Phytologist* 156: 205–215.

Blank, D.E., Alves, G.H., Nascente, P.D.S., Freitag, R.A., and Cleff, M.B. 2020. bioactive compounds and antifungal activities of extracts of *Lamiaceae* species. *Journal of Agricultural Chemistry and Environment* 9: 85–96. https://doi.org/10.4236/jacen.2020.93008

Castro, D.C., Cavalcante, R.P., Jorge, J., Martines, M.A.U., Oliveira, L.C.S., Casagrande, G.A., Machulek Jr., A. 2016. Synthesis and characterization of mesoporous Nb_2O_5 and its application for photocatalytic degradation of the herbicide methylviologen. *Journal of the Brazilian Chemical Society* 27: 303–313. https://doi.org/10.5935/0103-5053.20150244

Chen, S., Xiang, L., Guo, X., and Li, Q. 2011. An introduction to the medicinal plant genome project. *Frontiers in Medicine* 5: 178–84. https://doi.org/10.1007/s11684-011-0131-0

Cheng, M., Zeng, G., Huang, D., Lai, C., Xu, P., Zhang, C., and Liu, Y., 2016. Hydroxyl radicals based advanced oxidation processes (AOPs) for remediation of soils contaminated with organic compounds: A review. *Chemical Engineering Journal* 284: 582–598. https://doi.org/10.1016/j.cej.2015.09.001

Clarke, C.J., and Haselden, J.N., 2008. metabolic profiling as a tool for understanding mechanisms of toxicity. *Toxicologic Pathology* 36: 140–147. https://doi.org/10.1177/0192623307310947

Cui, H., Li, H., Zhang, S., Yi, Q., Zhou, Jing, Fang, G., and Zhou, J. 2020. Bioavailability and mobility of copper and cadmium in polluted soil after phytostabilization using different plants aided by limestone. *Chemosphere* 242: 125252. https://doi.org/10.1016/j.chemosphere.2019.125252

Cui, S., Zhou, Q., and Chao, L., 2007. Potential hyper-accumulation of Pb, Zn, Cu, and Cd in endurant plants distributed in an old smeltery, northeast China. *Environmental Geology* 51: 1043-1048

Cunningham, S., and Ow, D., 1996. Promises and prospects of phytoremediation. *Plant Physiology* 110: 715–719. https://doi.org/10.1104/pp.110.3.715

Cunningham, S.D., Berti, W.R., and Huang, J.W. 1995. Phytoremediation of contaminated soils. *Trends in Biotechnology* 13: 393–397. https://doi.org/10.1016/S0167-7799(00)88987-8

Darajeh, N., Idris, A., Fard Masoumi, H.R., Nourani, A., Truong, P., and Sairi, N.A. 2016. Modeling BOD and COD removal from palm oil mill secondary effluent in floating wetland by *Chrysopogon zizanioides* (L.) using response surface methodology. *Journal of Environmental Management* 181: 343–352. https://doi.org/10.1016/j.jenvman.2016.06.060

Duccer, W.M., and Ting, I.P., 1970. Air pollution oxidants – Their effects on metabolic processes in plants. *Annual Review of Plant Physiology* 21: 215–234. https://doi.org/10.1146/annurev.pp.21.060170.001243

El-Ramady, H., El-Henawy, A., Amer, M., Omara, A.E.-D., Elsakhawy, T., Elbasiouny, H., Elbehiry, F., Abou Elyazid, D., and El-Mahrouk, M. 2020. Agricultural waste and its nano-management: Mini review. *Egyptian Journal of Soil Science* 60(4): 349–364. https://doi.org/10.21608/EJSS.2020.46807.1397

Elshamy, M.M., Heikal, Y.M., and Bonanomi, G. (2019). Phytoremediation efficiency of *Portulaca oleracea* L. naturally growing in some industrial sites, Dakahlia District, Egypt. *Chemosphere* 225: 678–687.

Farooq, S., and Khanday, M.U.D., 2020. Phytoremediation of heavy metals: A strategy for the removal of toxic metals from the environment using plants. *International Journal of Chemical Studies* 8: 318–326. https://doi.org/10.22271/chemi.2020.v8.i3e.9244

Gupta, A.K., Verma, S.K., Khan, K., and Verma, R.K., 2013. Phytoremediation using aromatic plants: A sustainable approach for remediation of heavy metals polluted sites. *Environmental Science and Technology*: 130906155737004. https://doi.org/10.1021/es403469c

Hasani, Q., Pratiwi, N.T.M., Effendi, H., Wardiatno, Y., Raja GukGuk, J.A., Maharani, H. W., and Rahman, M. 2021. Azolla pinnata as phytoremediation agent of iron (Fe) in ex sand mining waters. *Chiang Mai University Journal of Natural Sciences* 20(1): e2021017.

Hassanein, R.A., Hashem, H.A., El-Deep, M.H., and Shouman, A. 2013. Soil contamination with heavy metals and its effect on growth, yield and physiological responses of vegetable crop plants (turnip and lettuce). *Journal of Stress Physiology & Biochemistry* 9(4): 145–162.

Herrera-Estrella, L.R., and Guevara-García, A.A., 2009. Heavy Metal Adaptation. In: *ELS*. *American Cancer Society*. https://doi.org/10.1002/9780470015902.a0001318.pub2

Homer, F.A., Reeves, R.D., and Brooks, R.R., 1997. The possible involvement of amino acids in nickel chelation in some nickel accumulating plants. *Current Topics in Phytochemistry* 14: 31–33.

Hussain, Md.S., Fareed, S., Ansari, S., Rahman, Md.A., Ahmad, I.Z., and Saeed, Mohd. 2012. Current approaches toward production of secondary plant metabolites. *Journal of Pharmacy and Bioallied Sciences* 4: 10–20. https://doi.org/10.4103/0975-7406.92725

Hussein, R.A., and El-Anssary, A.A., 2018. Plants Secondary Metabolites: The Key Drivers of the Pharmacological Actions of Medicinal Plants. In: Herbal Medicine. IntechOpen. https://doi.org/10.5772/intechopen.76139

Jisha, C.K., Bauddh, K., and Shukla, S.K., 2017. Phytoremediation and Bioenergy Production Efficiency of Medicinal and Aromatic Plants In: Bauddh, K., Singh, B., and Korstad, J. (eds). Phytoremediation Potential of Bioenergy Plants. Springer Singapore, Singapore, pp. 287–304. https://doi.org/10.1007/978-981-10-3084-0_11

Karczewski, K., Riss, H.W., and Meyer, E.I. 2017. Comparison of DNA-fingerprinting (T-RFLP) and high-throughput sequencing (HTS) to assess the diversity and composition of microbial communities in groundwater ecosystems. *Limnologica* 67: 45–53. https://doi.org/10.1016/j.limno.2017.10.001

Kumari, A., Pandey, N., and Pandey-Rai, S. 2018. Exogenous salicylic acid-mediated modulation of arsenic stress tolerance with enhanced accumulation of secondary metabolites and improved size of glandular trichomes in *Artemisia annua* L. *Protoplasma* 255(1): 139–152.

Laxmipriya, S., AjayKumar, A., Aravinthan, S., Arunachalam, N. 2018. Reduction of air pollution using smog-free tower: A review paper. *International Research Journal in Advanced Engineering and Technology* 4 (2): 3251–3255.

Lazar, T. 2003. Taiz, L. and Zeiger, E. Plant Physiology. 3rd edn. *Annals of Botany* 91(6), 750–751. https://doi.org/10.1093/aob/mcg079

Leal-Alvarado, D.A., Estrella-Maldonado, H., Sáenz-Carbonell, L., Ramírez-Prado, J.H., Zapata-Pérez, O., and Santamaría, J.M. 2018. Genes coding for transporters showed a rapid and sharp increase in their expression in response to lead, in the aquatic fern (*Salvinia minima* Baker). *Ecotoxicology and Environmental Safety* 147: 1056–1064.

Leblebici, Z., Kar, M., and Yalcin, V. (2018). Comparative study of Cd, Pb, and Ni removal potential by *Salvinia natans* (L.) All. and *Lemna minor* L.: Interactions with growth parameters. *Romanian Biotechnological Letters* 23: 13235–13248.

Liang, L., Cheng, X., Dai, T., Wang, Z., Li, J., Li, X., Lei, B., Liu, P., Hao, J., and Liu, X. 2020. Metabolic fingerprinting for identifying the mode of action of the fungicide SYP-14288 on *Rhizoctonia solani*. *Frontiers in Microbiology* 11: 3066. https://doi.org/10.3389/fmicb.2020.574039

Liang, L., Liu, W., Sun, Y., Huo, X., Li, S., and Zhou, Q. 2017. Phytoremediation of heavy metal contaminated saline soils using halophytes: Current progress and future perspectives. *Environmental Reviews* 25(3): 269–281.

Liang Zhu, Y., Pilon-Smits, E.A.H., Jouanin, L., and Terry, N. 1999. Overexpression of glutathione synthetase in indian mustard enhances cadmium accumulation and tolerance. *Plant Physiology* 119(1): 73–80. https://doi.org/10.1104/pp.119.1.73

Liu, W.J.H. 2010. Introduction to Traditional Herbal Medicines and their Study. In: Liu, W.J.H. (ed.). Traditional Herbal Medicine Research Methods. John Wiley & Sons, Hoboken, NJ, pp. 1–26. https://doi.org/10.1002/9780470921340.ch1

Masinire, F., Adenuga, D., Tichapondwa, S., and Chirwa, E., 2020. remediation of chromium(VI) containing wastewater using *Chrysopogon zizanioides* (Vetiver Grass). *Chemical Engineering Transactions* 79: 385–390. https://doi.org/10.3303/CET2079065

Maurya, P., Malik, D.S., and Sharma, A. 2019. Impacts of Pesticide Application on Aquatic Environments and Fish Diversity. In: Contaminants in Agriculture and Environment: Health Risks and Remediation, pp. 111–128. https://doi.org/10.26832/AESA-2019-CAE-0162-09

McNear, D.H., Chaney, R.L., and Sparks, D.L. 2010. The hyperaccumulator *Alyssum murale* uses complexation with nitrogen and oxygen donor ligands for Ni transport and storage. *Phytochemistry* 71: 188–200. DOI: 10.1016/j.phytochem.2009.10.023

Memon, A.R., and Schröder, P. 2009. Implications of metal accumulation mechanisms to phytoremediation. *Environmental Science and Pollution Research* 16: 162–175. https://doi.org/10.1007/s11356-008-0079-z

Meneses, M., Pasqualino, J.C., Céspedes-Sánchez, R., and Castells, F. 2010. Alternatives for reducing the environmental impact of the main residue from a desalination plant. *Journal of Industrial Ecology* 14: 512–527. https://doi.org/10.1111/j.1530-9290.2010.00225.x

Meyer, C.-L., and Verbruggen, N. 2012. Use of the model species *Arabidopsis halleri* towards phytoextraction of cadmium polluted soils. *New Biotechnology* 30: 9–14.

Moons, A., Prinsen, E., Bauw, G., and Van Montagu, M. 1997. Antagonistic effects of abscisic ccid and jasmonates on salt stress-inducible transcripts in rice roots. *The Plant Cell* 9(12): 2243–2259. DOI: 10.1105/tpc.9.12.2243.

Myers, I., and Maynard, R.L. 2005. Polluted air – Outdoors and indoors. *Occupational Medicine* 55: 432–438. https://doi.org/10.1093/occmed/kqi137

Nizam, M., Wahid-U-Zzaman, M., Rahman, Md.M., and Kim, J.-E. 2016. Phytoremediation potential of Kenaf (*Hibiscus cannabinus* L.), Mesta (*Hibiscus sabdariffa* L.), and Jute (*Corchorus capsularis* L.) in arsenic-contaminated soil. *Korean Journal of Environmental Agriculture* 35: 111–120. https://doi.org/10.5338/KJEA.2016.35.2.15

Okada, T., Mochamad Afendi, F., Altaf-Ul-Amin, Md., Takahashi, H., Nakamura, K., and Kanaya, S. 2010. Metabolomics of medicinal plants: The importance of multivariate analysis of analytical chemistry data. *Current Computer-Aided Drug Design* 6: 179–196. https://doi.org/10.2174/157340910791760055

Ortiz, M., Raluy, R.G., and Serra, L. 2007. Life cycle assessment of water treatment technologies: wastewater and water-reuse in a small town. *Desalination* 204(1–3): 121–131. https://doi.org/10.1016/j.desal.2006.04.026

Ozyigit, I., Can, H., and Dogan, I. 2021. Phytoremediation using genetically engineered plants to remove metals: A review. *Environmental Chemistry Letters* 19: 669–698. https://doi.org/10.1007/s10311-020-01095-6

Pichersky, E., and Gang, D.R. 2000. Genetics and biochemistry of secondary metabolites in plants: an evolutionary perspective. *Trends in Plant Science* 5: 439–445. https://doi.org/10.1016/s1360-1385(00)01741-6

Patra, D.K., Pradhan, C., and Patra, H.K. 2020. Toxic metal decontamination by phytoremediation approach: Concept, challenges, opportunities and future perspectives. *Environmental Technology & Innovation* 18: 100672.

Pietra, F. 2002. Evolution of the secondary metabolite versus evolution of the species. *Pure and Applied Chemistry* 74: 2207–2221. https://doi.org/10.1351/pac200274112207

Pirzadah, T.B., Malik, B., and Dar, F.A. 2019. Phytoremediation potential of aromatic and medicinal plants: A way forward for green economy. *Journal of Stress Physiology & Biochemistry* 15: 62–75.

Qadir, M., Jiménez, G.C., Farnum, R.L., Dodson, L.L., and Smakhtin, V. 2018. Fog water collection: Challenges beyond technology. *Water* 10: 372. https://doi.org/10.3390/w10040372

Rai, V., Vajpayee, P., Singh, S., and Mehrotra, S. 2004. Effect of chromium accumulation on photosynthetic pigments, oxidative stress defense system, nitrate reduction, proline level and eugenol content of *Ocimum tenuiflorum* L. *Plant Science* 167: 1159–1169. https://doi.org/10.1016/j.plantsci.2004.06.016

Raja, S., Cheema, H.M.N., Babar, S., Khan, A.A., Murtaza, G., and Aslam, U. 2015. Socio-economic background of wastewater irrigation and bioaccumulation of heavy metals in crops and vegetables. *Agricultural Water Management* 158: 26–34. https://doi.org/10.1016/j.agwat.2015.04.004

Rausher, M.D. 2001. Co-evolution and plant resistance to natural enemies. *Nature* 411: 857–864. https://doi.org/10.1038/35081193

Rontein, D., Basset, G., and Hanson, A.D. 2002. Metabolic engineering of osmoprotectant accumulation in plants. *Metabolic Engineering* 4(1): 49–56. doi: 10.1006/mben.2001.0208

Roosens, N.H.C.J., Willems, G., and Saumitou-Laprade, P. 2008. Using *Arabidopsis* to explore zinc tolerance and hyperaccumulation. *Trends in Plant Science* 13: 208–215. https://doi.org/10.1016/j.tplants.2008.02.006

Rosas, J.M., Vicente, F., Santos, A., and Romero, A. 2013. Soil remediation using soil washing followed by Fenton oxidation. *Chemical Engineering Journal* 220: 125–132. https://doi.org/10.1016/j.cej.2012.11.137

Sagner, S., Kneer, R., Wanner, G., Cosson, J.P., Deus-Neumann, B., and Zenk, M.H. 1998 Hyperaccumulation, complexation and distribution of nickel in *Sebertia acuminate*. *Phytochemistry* 47: 339–343.

Saif, S., Khan, Mohd., Saghir, Zaidi, A., Rizvi, A., and Shahid, M. 2017. Metal Toxicity to Certain Vegetables and Bioremediation of Metal-Polluted Soils. In: Zaidi, A., Khan, Mohammad and Saghir (eds). Microbial Strategies for Vegetable Production. Springer, Cham, pp. 167–196. https://doi.org/10.1007/978-3-319-54401-4_8

Salt, D.E., Smith, R.D., and Raskin, I. 1998. Phytoremediation. *Annual Review of Plant Physiology and Plant Molecular Biology* 49: 643–668. https://doi.org/10.1146/annurev. arplant.49.1.643

Saxena, M., Saxena, J., Nema, R., Singh, D., and Gupta, A. 2013. Phytochemistry of medicinal plants. *Journal of Pharmacognosy and Phytochemistry* 1(6): 168–182.

Seyyednejad, S.M., Niknejad, M., and Koochak, H. 2011. A review of some different effects of air pollution on plants. *Research Journal of Environmental Sciences* 5: 302–309. https:// doi.org/10.3923/rjes.2011.302.309

Shackira, A.M., and Puthur, J.T. 2019. Phytostabilization of Heavy Metals: Understanding of Principles and Practices. In: Plant–Metal Interactions. Springer, Cham, pp. 263–282.

Singer, A., Crowley, D., and Thompson, I. 2003. Secondary plant metabolites in phytoremediation and biotransformation. *Trends in Biotechnology* 21: 123–130. https:// doi.org/10.1016/S0167-7799(02)00041-0

Singh, R., Gautam, N., Mishra, A., and Gupta, R. 2011. Heavy metals and living systems: An overview. *Indian Journal of Pharmacology* 43(3): 246–253. https://doi.org/10.4103/ 0253-7613.81505

Thawale, P., Babu, S., Wakode, R., Singh, S., Kumar, S., and Juwarkar, A., 2011. Biochemical changes in plant leaves as a biomarker of pollution due to anthropogenic activity. *Environmental Monitoringand Assessment* 177: 527–535. https://doi.org/10.1007/ s10661-010-1653-7

Torbati, S., and Keshipour, S. 2020. Application of *Lemnagibba* L. and a bio-based aerogel for the removal of metal (loid) s from stream waters near three gold deposits in northwestern Iran. *Environmental Technology & Innovation* 20: 101068.

Vamerali, T., Bandiera, M., and Mosca, G. 2010. Field crops for phytoremediation of metal-contaminated land. A review. *Environmental Chemistry Letters* 8: 1–17. https://doi.org/ 10.1007/s10311-009-0268-0

Verpoorte, R., Choi, Y.H., and Kim, H.K. 2007. NMR-based metabolomics at work in phyto-chemistry. *Phytochemistry Reviews* 6: 3–14. https://doi.org/10.1007/s11101-006-9031-3

Weyens, N., Thijs, S., Popek, R., Witters, N., Przybysz, A., Espenshade, J., Gawronska, H., Vangronsveld, J., and Gawronski, S.W. 2015. The role of plant–microbe interactions and their exploitation for phytoremediation of air pollutants. *International Journal of Molecular Sciences* 16: 25576–25604. https://doi.org/10.3390/ijms161025576

Wickramasinghe, S., and Jayawardana, C.K. 2018. Potential of aquatic macrophytes *Eichhornia crassipes*, *Pistia stratiotes* and *Salvinia molesta* in phytoremediation of tex-tile wastewater. *Journal of Water Security* 4 : 1–8.

Wilhelm, R.C., Murphy, S.J.L., Feriancek, N.M., Karasz, D.C., DeRito, C.M., Newman, J.D., and Buckley, D.H. 2020. *Paraburkholderia madseniana* sp. nov., a phenolic acid-degrading bacterium isolated from acidic forest soil. *International Journal of Systematic and Evolutionary Microbiology* 70: 2137–2146. https://doi.org/10.1099/ ijsem.0.004029

Xu, Y., Sun, X., Zhang, Q., Li, X., and Yan, Z. 2018. Iron plaque formation and heavy metal uptake in *Spartina alterniflora* at different tidal levels and waterlogging conditions. *Ecotoxicology and Environmental Safety* 153: 91–100.

Yadav, S., and Srivastava, J. 2017. Cadmium phytoextraction and induced antioxidant gene response in *Moringa oleifera* Lam. *American Journal of Plant Physiology* 12(2): 58–70.

Yang, H., and Liu, Y. 2011. Phytoremediation on Air Pollution. In: The Impact of Air Pollution on Health, Econony, Environment and Agricultural Sources. IntechOpen. https://doi.org/10.5772/19942

Yuan, L., Yin, X., Zhu, Y., Li, F., Huang, Y., Liu, Y., and Lin, Z. 2012. Selenium in Plants and Soils, and Selenosis in Enshi, China: Implications for Selenium Biofortification. In: Yin, X., and Yuan, L. (eds). Phytoremediation and Biofortification. Springer Netherlands, Dordrecht, pp. 7–31. https://doi.org/10.1007/978-94-007-1439-7_2

Zahedifar, M., Moosavi, A.A., Shafigh, M., Zarei, Z., and Karimian, F. 2015. Cadmium accumulation and partitioning in *Ocimum basilicum* as influenced by application of various potassium fertilizers. *Archives of Agronomy and Soil Science* 62: 150721163937009. https://doi.org/10.1080/03650340.2015.1074187

Zhao, F.-J., and McGrath, S.P. 2009. Biofortification and phytoremediation. *Current Opinion in Plant Biology* 12(3): 373–380. DOI: 10.1016/j.pbi.2009.04.005.

Zhou, J., and Goldsbrough, P.B. 1994. Functional homologs of fungal metallothionein genes from Arabidopsis. *Plant Cell* 6: 875–884. https://doi.org/10.1105/tpc.6.6.875

12 Understanding the Proteomics of Medicinal Plants under Environmental Pollution
Challenges and Opportunities

Pooja Singh, V.K. Mishra, Rohit Kashyap, Rahul Rawat

CONTENTS

12.1 INTRODUCTION

Pollution of the environment occurs when pollutants are introduced into the natural environment, causing negative impacts on the ecosystem. It has four key effect points: air, water, soil, and habitat. Airborne, soil, and water pollutants may enter plants mostly via leaves, roots, or the whole surface, respectively. They can cause rapid and direct effects, such as oxidative stresses, while prolonged exposure to pollutants interferes with plant metabolism leading to impaired growth, reduced harvest, economic losses, and disturbed ecosystem functions. There exists a complex

DOI: 10.1201/9781003178866-12

213

interplay between pollutants and their effects on plants, and in turn, plants tend to adapt, recover, and mitigate the effects of the pollution.

Secondary metabolites from medicinal plants are of key therapeutic importance. Several contaminants in the environment have an impact on plant growth, development, and secondary metabolite production. To cope with all of these challenges, medicinal plants, like any other plant, have effective stress avoidance or tolerance strategies that allow them to adapt and defend themselves under stress. These adaptations manifest at the morphological, anatomical, biochemical, and molecular levels. Medicinal plant products are generally considered to be safe because of the notion of their 'natural' origin equated to being 'safe' –free of toxic effects and adverse reactions. However, the consumption of herbs could produce prominent toxic effects either due to inherent toxicity or to contaminants (heavy metals, microorganisms, pesticides, toxic organic solvents, radioactivity, etc.) (Youns et al. 2010; Neergheen-Bhujun 2013; Bernstein et al. 2021). High toxicity is the most common reason why new agents drop out of drug development in the pharmaceutical industry. Approaches to apply 'omics'-based technologies to medicinal plant research may help us to understand the mechanism of plant responses to environmental stress, and also detect toxicity related to phytotherapy.

The novel 'omics' technologies enable researchers to identify the genetics underlying medicinal plant responses to adaptation mechanisms providing impetus to investigate the complex interplay between medicinal plants, their metabolism, secondary metabolite production, and the effect of polluting environment. The genome can be defined as the complete set of genes inside a cell. Genomics is, therefore, the study of the genetic make-up of organisms. Transcriptomics is the complete set of transcripts in a cell, and their abundance, for a specific developmental stage or physiological condition (Wang et al. 2009). Proteins play an important role in biological processes by providing structural support as well as physiological functions (Figure 12.1). The complete set of proteins in a cell refers to as proteome (Park 2004). Metabolomics is the latest technique, is defined as the quantitative complement of low-molecular-weight metabolites present in a cell under a given set of physiological conditions (Kell et al. 2005). In view of these, the changes at the cellular or subcellular level due to impacting influence of environmental stimulus can be more precisely understood through omics technologies (Figure 12.2).

12.2 PROTEOMIC METHODOLOGIES USED TO STUDY ADVERSE POLLUTION STRESS-INDUCED DYNAMIC CHANGES IN PROTEOME OF MEDICINAL PLANTS

12.2.1 Gel-Based Proteomics Technologies

Protein extract is a mixture of a large number of individual proteins. One of the most frequently applied approaches is to provide protein separation by means of gel electrophoresis, which is based on the migration of molecules in an electric field in a gel medium. One- and two-dimensional polyacrylamide gel electrophoresis are common separation methods employed in the analysis of proteins. However, 1-DE has a major disadvantage in resolving protein mixtures overlap of protein bands due to very similar molecular weight. Two-dimensional electrophoresis (2-DE) combines

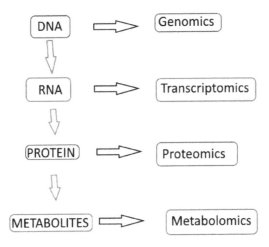

FIGURE 12.1 Genomics, transcriptomics, proteomics, and metabolomics.

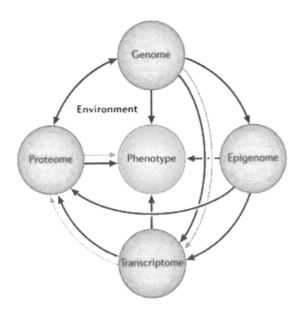

FIGURE 12.2 Omics and environment.

isoelectric focusing (IEF), the first dimension, which separates proteins according to their isoelectric point, and SDS-PAGE, the second dimension, which separates them further according to their molecular mass (MW). Each spot on the resulting 2-D gel corresponds to a single protein in the sample. The next steps in gel-based proteomics are gel image analysis and finally protein identification by mass spectrometry (MS) techniques. Methodological advances in 2-DE have led to the development of 2-D differential in-gel electrophoresis (2-D DIGE) (Unlü et al. 1997). Different pools of proteins are labelled with different fluorescent dyes (up to three, Cy2, Cy3, or Cy5). The samples are then mixed and resolved on the same 2-D gel. Proteins that are common to the samples appear as 'spots' with a fixed ratio of fluorescent signals, whereas proteins that differ between the samples have different fluorescence ratios (Minden 2012). This technique can reveal the presence of several hundred proteins in a highly reproducible manner. For a high-resolution separation of enzymatically active protein complexes, two-dimensional blue native/sodium dodecyl sulfate-poly-acrylamide gel electrophoresis (2D BN/SDS-PAGE) is a method of choice (Lasserre and Ménard 2012).

Stained gels are scanned on a 'visible' or 'fluorescent' scanner. The image can then be imported to specific software to be analysed and compared. Software, such as Image Master 2D Platinum (GE Healthcare), Progenesis (Nonlinear Dynamics), PDQuest (Bio-Rad), and Samespots, can be used to detect spots and to compare the spot intensity (Unlü et al. 1997; Rosengren et al. 2003; Nebrich et al. 2005; Wheelock and Buckpitt 2005; Martin 2006; Wheelock and Goto 2006; Clark and Gutstein 2008; Kang et al. 2009).

12.3 MASS SPECTROMETRY (MS) FOR PROTEIN IDENTIFICATION

MS is an analytical technique used for the identification and quantification of protein analytes in a gaseous state based upon the mass to charge ratio (m/z) in a vacuum environment. During the past two decades, mass spectrometry has become established as the primary method for protein identification from complex mixtures of biological origin. In proteomic analysis, the goal is to characterize all cellular proteins. MS in conjunction with the development of comprehensive protein databases and advances in computational methods has been successful in the high-throughput characteriza-tion and identification of proteins.

Matrix-assisted laser desorption ionization (MALDI) and MS and electrospray ionization (ESI) are two technologies that are commonly used for protein ionization. Time-of-flight (TOF) and quadrupole mass analysers have been developed for use in a mass spectrometer (Yates 1998a; Mann et al. 2001).

The protein spots obtained in 2-DE are excised from 2-D gel and then the target protein is digested with appropriate protease into peptide fragments. Then a set of masses of these peptides is constructed by MS such as MALDI-TOF or ESI-TOF, called 'peptide mass fingerprinting' (PMF) which is characteristic for the target pro-tein and is used to search peptide masses generated by theoretical fragmentation of protein sequences of databases (Yates 1998b). Matches from at least three to six peptides derived from the same protein are required to positively identify a protein (Pappin et al. 1993; Yates 1998b).

Tandem MS (MS/MS) has been used to obtain the amino acid sequence of peptides. Peptide ions generated from an ESI source are separated based on the m/z ratio and further dissociated by collision with an inert gas (Mann et al. 2001). The resultant tandem spectra of amino acid composition can be searched against pro-tein, expressed sequence tags (ESTs), and genome databases to identify the protein.

Various programs used for tandem spectra are Sequest (Eng et al. 1994; Clauser et al. 1999), PROWL (Qin et al. 1997), Protein Prospector (Clauser et al. 1999), and MASCOT (Perkins et al. 1999).

Mass spectrometry with MALDI-TOF/TOF and Tandem-MS-MS and being widely used equipment is the central among current proteomics (Aslam et al. 2017) including proteomics of plant responses to environmental stresses (Ahsan et al. 2009; Kosová et al. 2011; Pérez-Clemente et al. 2013). However, utilization of proteomics facilities, including the software for equipment and databases and the requirement for skilled personnel, substantially increases the costs, thereby limiting their wider use, especially in the developing world.

12.4 SECOND-GENERATION GEL-FREE MS-BASED PROTEOMIC TECHNOLOGIES

Two-dimensional gel electrophoresis is now a mature and well-established technique; however, it suffers from some ongoing concerns regarding quantitative reproducibility and limitations on the ability to study certain classes of proteins (Abdallah et al. 2012). Alternatives to 2-DE gel-based quantification of proteins are gel-free MS-based proteomics. Currently, label-based proteomic approaches, such as isotope-coded affinity tags (ICAT), isotope-coded protein labelling (ICPL) isobaric tags for relative and absolute quantification (iTRAQ), tandem mass tag (TMT), stable isotopic labelling with amino acids in cell culture (SILAC), and label-free LC/MS represent attractive alternatives. The workflow in proteomics is depicted in Figure 12.3.

12.5 PROTEOMICS OF MEDICINAL PLANTS

Expression profiling is an important tool to investigate how a plant responds to environmental changes. In recent years, with rapid developments in protein separation and protein identification techniques based on mass spectrometry and bioinformatics, proteomics has become important to assess the changes in protein types and their expression levels under different stresses (Hazen et al. 2003). The DNA sequences of a genome integrate vital information of the origin, development, and epigenomic regulation of a plant, Recently, the sequenced genome of grapes (Velasco et al. 2007; Jaillon et al. 2007), *Phalaenopsis equestris* (Cai et al. 2015), *Brassica napus* L. (Chalhoub et al. 2014), *Capsicum annuum* L. (Qin et al. 2014; Kim et al. 2014), *Momordica charantia* L. (Urasaki et al. 2017), *Coffea canephora* (Denoeud et al. 2014), *Salvia miltiorrhiza* (Ma et al. 2012), *Ziziphus jujuba* Mill. (Li et al. 2014), *Glycyrrhiza uralensis* Fisch. Ex DC. (Mochida et al. 2017), *Dendrobium officinale* (Yan et al. 2015), *Azadirachta indica* A. Juss. (Shivaraj et al. 2015) and *Catharanthus roseus* L. chloroplast and genome (Ku et al. 2013; Kellner 2015) and the chloroplast of *Pogostemon cablin* (Blanco) Benth. (He et al. 2016) have been sequenced. Integrating genome sequencing with transcriptomics and proteomics data will enable deeper information of gene, protein/ enzyme and function under various stimuli, including stresses.

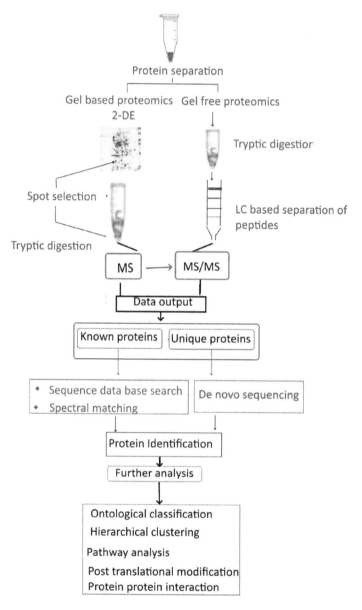

FIGURE 12.3 Workflow in proteomics.

12.6 PROTEOMICS TO EVALUATE EFFECT OF POLLUTION ON MEDICINAL PLANTS/PRODUCTS

With industrialization and urbanization, medicinal plants are persistently facing environmental stresses, such as light intensity, UV exposure, heavy metal stress, low and high temperature, and air pollution exposure. To cope with environmental stresses, plants have evolved sophisticated defence mechanisms and stress recovery responses (Figure 12.4).

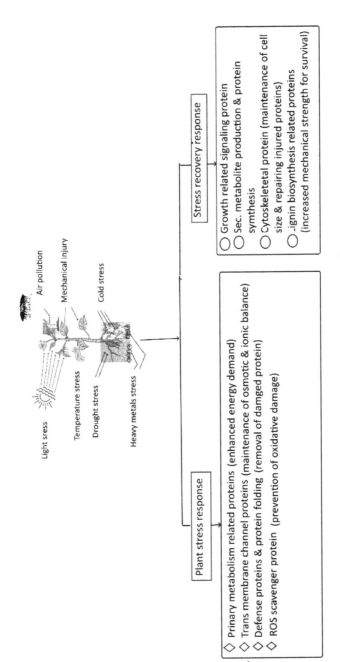

FIGURE 12.4 Influence of environmental pollution on medicinal plants – stress and recovery responses.

12.7 INADEQUATE OR EXCESSIVE LIGHT

Solar radiation is a key environmental signal in the regulation of plant secondary metabolism. Photosynthetically active radiation (PAR) plays an important part in climate change and ecological processes (Hu et al. 2007). Quantity of photosynthetically active radiation (PAR), irradiance, has crucial effects on primary photosynthetic processes (Murchie and Horton 1997). Excessive irradiance revealed upregulation of fibrillins; antioxidative enzymes (Cu/Zn-SOD, Fe-SOD); chaperones (cpHSP70-1 and 2); PSI assembly (YCF37 protein); biosynthesis of carotenoids, tocopherols, chinons, linoleic acid, and jasmonic acid in *Arabidopsis* (Giacomelli et al. 2006; Ytterberg et al. 2006; Kosová et al. 2011). Metabolic responses to light and ultraviolet (UV) radiation exposure are known to depend on the ratio of spectral ranges (UV-B/PAR). Ultraviolet-B and photosynthetically active radiation interactively affect the yield and pattern of monoterpenes in leaves of peppermint (*Mentha × piperita* L.) (Behn et al. 2010). A study showed that the highest essential oil yield was achieved during flowering at high PAR and approximate ambient UV-B radiation). However, low PAR and the absence of UV-B radiation led to reduced menthol and increased menthone contents and thereby to a substantial decrease in oil quality (Behn et al. 2010).

12.8 UV IRRADIATION

In recent years, greater concerns have been expressed over UV-B radiation together with other climate change factors influencing terrestrial organisms and ecosystems (Bornman et al. 2015). Ultraviolet (UV) radiation of the solar spectrum has been conventionally divided into UV-A (320-400 nm), UV-B (280-320 nm), and UV-C (less than 280 nm) of which UV-A and longer wavelengths of UV-B are of biological significance. Depletion of the stratospheric ozone layer has caused higher levels of UV-B radiation to reach the Earth. Plants have adaptive mechanisms to cope with the changes caused by UV-B radiation. Several reports focus on the effect of ultraviolet (UV) irradiation on the proteome and metabolome (Table). UV-A treatment of *Taxus chinensis* activated a photosynthetic reaction, and enhanced glycolysis, four glycolysis-related key enzymes increased, provided precursors for secondary metabolism. The 1-deoxy-D-xylulose-5-phosphate reductoisomerase and 4-hydroxy-3-methylbut-2-enyl diphosphate reductase were identified significantly increased during UV-A radiation, which resulted in paclitaxel enhancement (Zheng et al. 2016). UV-B irradiation can also cause the reduction of allergic compounds such as ginkgolic acids in Ginkgo biloba (Zheng et al. 2016). UV-B irradiation with dark treatment of Clematis terniflora, a source of antitumor drugs, showed an enhanced flow of S-adenosylmethionine synthetase and cysteine synthase, 1,1-diphenyl-2-picrylhydrazyl scavenging activity, (+) dihydrolipoyl dehydrogenase/glutamate dehydrogenase and the content of γ-aminobutyric acid (B. Yang et al. 2016). Clematis terniflora responses to UV-B irradiation demonstrated the accumulation of indole alkaloids (Gao et al. 2016), upregulation of flavonoid biosynthesis (B. Yang et al. 2016), and increased aminobutyric acid levels (B. Yang et al. 2016). Metabolic content in medicinal plants was improved by changes in abundance in 10-hydroxygeraniol oxidoreductase in

Catharanthus roseus (Zhu et al. 2015). Proteomics analysis of UV-irradiated *Lonicera japonica* Thunb. showed increased expression of bioactive metabolites enhancement of 1-deoxy-d-xylulose 5-phosphate reductoisomerase and 5-enolpyruvylshikimate-phosphate synthase (Zhang et al. 2013) which was crucial to supply more precursors for the secondary metabolites including caffeoylquinic acids and iridoids. Integrated analysis of proteomics and transcriptomic data revealed the involvement of plant hormone signal transduction and phosphatidylinositol signalling in UV-B irradiation effect of *Mirabilis himalaica* (Gu et al. 2018) which might be the key metabolic strategy of UV-B radiation to improve the biosynthesis of rotenoid in *M. himalaica*. Transcriptome analysis via genome-wide microarray in grape berry skin after UV-C irradiation revealed 238 genes up-regulated more than 5-fold, enrichment analysis of Gene Ontology terms showed that genes encoding stilbene synthase were enriched in the up-regulated genes. Metabolome analysis using liquid chromatography-quadrupole time-of-flight mass spectrometry detected 2,012 metabolite peaks. Principal component analysis using the peaks showed that only one metabolite peak, identified as resveratrol, was highly induced by UV-C light. Metabolic pathway map of grape in the Kyoto Encyclopedia of Genes and Genomes (KEGG) database and in the KaPPA-View 4 KEGG system projected the transcriptome and metabolome data on a metabolic pathway map showed specific induction of the resveratrol synthetic pathway by UV-C light (Suzuki et al. 2015).

12.9   HEAVY METAL POLLUTION

Pollution of soils by heavy metals is an ever-growing problem throughout the world. Heavy metal contamination may affect the production of secondary metabolites (SMs) in medicinal plants (Husen 2021a, 2021b). A review of the literature suggests that there are both positive and negative alterations of SMs: heavy metal pollution has decreased secondary production in some plants, such as *Hypericum perforatum* L. (Murch et al. 2003) and *Catharanthus roseus* (Pandey et al. 2007) while other reports claim an increase in secondary metabolite production, such as in *Artemisia annua* L. (Rai et al. 2011), *Bacopa monnieri* L. (Sinha and Saxena 2006) and *Mentha pulegium* L. (Lajayer et al. 2017). Generally, enhanced secondary metabolite production results from increased synthesis of precursors (Zheng and Wu 2004). Alteration in secondary metabolism may be a strategy of the plant to survive and grow in adverse conditions (including growth in the presence of phytotoxic metals (Cobbett and Goldsbrough 2002).

Recent reviews on proteomics analysis of plants' responses to heavy metals have provided a deeper understanding of protein function and the interactions among the pathways involved in detoxification of toxic metals in plant cells (Ahsan et al. 2009; Liang et al. 2013; Cvjetko et al. 2014; Hasan et al. 2017). Recent reports of plant proteome analysis in heavy metal stress is presented in Table 12.2. Plants operate various mechanisms in which proteins are involved. Proteins differentially regulated in plants in response to metal toxicity, as identified by proteomic analyses are of the class antioxidative defence proteins, chaperones, proteins involved in signalling molecules and secondary metabolites, CO_2 assimilation and photosynthesis and PR proteins (Ahsan et al. 2009).

TABLE 12.1
Proteomic Responses of Some Selected Medicinal Plants to UV Radiation

Pollution	Medicinal plant	Secondary metabolite production (SMs)	Proteomic analysis	References
UV-B	*Catharanthus roseus*	(+) alkaloid biosynthesis	(+)10-hydroxygeraniol oxidoreductase	(Zhu et al. 2015)
UV-B	*Catharanthus roseus*	+ SM	+Phosphoproteins related to calcium-calmodulin, calcium-dependent kinase, and heat shock proteins increased; proteins related to glycolysis and the reactive oxygen species scavenging system were changed	(Zhuo et al. 2019)
Combined ultraviolet and darkness regulation	*Mahonia bealei*	+alkaloid and flavonoid contents	+ calmodulin-calcium signalling and ATP-binding cassette transporter involved in transport of berberine	(Zhu et al. 2021)
UV	*Lonicera japonica*	+caffeoylquinic acids and iridoids	(+) protein level for caffeoylquinic acids and iridoids components; (+) 1-deoxy-ᴅ-xylulose 5-phosphate reductoisomerase and 5-enolpyruvylshikimate-phosphate synthase; proteins SMs (caffeoylquinic acids and iridoids) biosynthetic-related proteins, photosynthesis, carbohydrate and energy metabolism, stress, DNA, transport-related proteins, lipid metabolism, amino acid metabolism, cell wall	(Zhang et al. 2013)
UV-B irradiation with dark treatment	*Clematis terniflora =*	Indole alkaloid	(+) S-adenosylmethionine synthetase and cysteine synthase as well as 1,1-diphenyl-2-picrylhydrazyl scavenging activity; (+) dihydrolipoyl dehydrogenase/glutamate dehydrogenase and the content of γ-aminobutyric acid	(B. Yang et al. 2016)
UV	*Ginkgo biloba*	(+) Flavonoids; (-) ginkgolic acids (GAs)	(+) antioxidants and stress-responsive proteins; (-) photosynthesis	(Zheng et al. 2015)
UV A	*Taxus chinensis*	anticancer – paclitaxel		

TABLE 12.2
Plant Proteome Analysis in Heavy Metal Stress

Proteins	Plant Species	Metals	References
	Heat shock protein		
HSPs70	*Populus*	Cd	Lomaglio et al. 2015
	Arabidopsis thaliana		Neumann et al. 1994
HSPs 60	*Solanum lycopersicum*	Cd	Rodríguez-Celma et al. 2010
	Arabidopsis thaliana	Cd	Sarry et al. 2006
HSPs 90	*Arabidopsis thaliana*	As	Haralampidis et al. 2002
	Oryza sativa	Cu	Song et al. 2014
HSPs 100	*Oryza sativa*	As	Agarwal et al. 2003
sHSPs	*Lycopersicon peruvianum*	Cd	Neumann et al. 1994
	Populus alba	Cu, Zn	Lingua et al. 2012
	Antioxidative defence proteins		
APXs	*Brassica juncea*	Ni	Wang et al. 2012
SOD, CAT, APXs	*Ocimum basilicum*	Ni, Cu and Zn	Georgiadou et al. 2018
APX, GR, SOD	*Oryza sativa*	Cd	L. Yang, et al. 2016
	Oryza sativa	As	Ahsan et al. 2008
	Signalling molecules		
Salicylic acid	Pea, maize, barley, rice	Hg, Cd	Shao et al. 2010; Krantev et al. 2008; Agami and Mohamed 2013
Jasmonic acids	*Arabidopsis, Brassica napus*, Wheat	Pb, As, Cu	Zhu et al. 2016; Farooq et al. 2018, Li et al. 2013
calcium–calmodulin	*Arabidopsis*	Cd, Ni, Pb	Chen et al. 2019; Snedden and Fromm 2001
Abscisic acid	Pea, wheat	Hg	Shao et al. 2010; Kang et al. 2015
Polyamines	bean	Cd	Hossain et al. 2010; Hossain et al. 2012
Auxin	*Arabidopsis*	Cd	Hu et al. 2013; Singh et al. 2021

Proteomic analysis of copper stress on the root *Cannabis sativa* induced the suppression of two proteins (thioredoxin-dependent peroxidase and 60S ribosomal protein L12), the down-regulation of seven proteins enolase, cyclophilin, ABC transporter substrate-binding protein, glycine-rich RNA binding protein (GRP), putative peroxidase, and elicitor inducible protein, while five proteins were up-regulated (aldo/keto reductase, putative auxin-induced protein, 40S ribosomal protein S20, mitochondrial formate dehydrogenase and actin (Bona et al. 2007). Proteomic analysis for Cd hyperaccumulation and tolerance of B. juncea using 2-D DIGE and iTRAQ showed the involvement of peptide methionine sulfoxide reductase, 2-nitropropane dioxygenase O-acetylserine sulfhydrylase, glutathione-S-transferase and glutathione-conjugate membrane transporter (Alvarez et al. 2009). Proteomic analysis of *Brassica juncea* exposed to Ni stress revealed that differential expression of 61 protein spots of which 37 protein spots were ambiguously identified by matrix-assisted laser desorption ionization/time-of-flight mass spectrometry (MALDI-TOF MS). The majority of these identified proteins were found to be involved in sulphur metabolism, protection against oxidative stress and induced expression of photosynthesis and ATP generation-related proteins (Wang et al. 2012) indicating the involvement of heavy metal sequestration and antioxidant system and the tolerance and accumulation of Ni is an energy-demanding process.

Differential responses to arsenic (As) stress adaptation have been reported two B. napus cultivars through pulse amplitude modulated fluorometer and isobaric tags-based proteomics (iTRAQ) showing altered activities and expression level of antioxidants enzymes and reduced level of photosynthetic activity (Farooq et al. 2021). Proteomic analysis of chromium stress and sulphur deficiency responses in leaves of two canola (*Brassica napus* L.) cultivars differing in Cr(VI) tolerance revealed that 58 protein spot (2-DE), of these, 39 protein spots were identified by MALDI-TOF/TOF mass spectrometry. The study showed differentially regulated proteins predominantly had functions in photosynthesis, energy metabolism, stress defence, protein folding and stabilization, signal transduction, redox regulation and sulphur metabolism (Yıldız and Terzi 2016). The protein profile of saffron in response to Cd stress showed that 15 proteins were up-regulated in the leaves which were primarily related to metabolism, signal transduction, stress and defence response and energy. Influence of heavy metals (Ni, Cu, and Zn) in basil (*Ocimum basilicum* L.) showed induction in activity of nitro-oxidative stress response enzymes (SOD, CAT, APX, NR). Cu stress increased concentration profiling, an allergenic protein, decrease in the concentration of total proteins and antioxidant capacity. And also, severe Cu stress resulted in the accumulation of specific proteins related to transpiration and photosynthetic processes. However, Ni stress in basil plants appears to be less damaging and with lower allergenic potential compared with Cu and Zn stress, while Cu-stressed basil plants experience the most detrimental effects and display the highest allergen production (Georgiadou et al. 2018). Proteomic effects of Se stress in pepper (*Capsicum annuum* L.) seedlings were analysed through tandem mass tag labelling, high-performance liquid chromatography fractionation, and mass spectrometry showed a total of 4,693 proteins were identified, of 3,938 of which yielded quantitative information. Among them, the expression of 172 proteins was up-regulated, and the expression of 28 proteins was down-regulated. Systematic bioinformatics analysis revealed that differentially expressed proteins were most strongly associated with the terms 'metabolic process', 'posttranslational modification, protein turnover, chaperones', and 'protein-processing in endoplasmic reticulum', according to Gene Ontology, eukaryotic orthologous groups classification, and Kyoto Encyclopedia of Genes and Genomes enrichment analysis, respectively. Furthermore, several heat shock proteins were identified as differentially expressed proteins (Zhang et al. 2019). The study revealed that the responses of pepper to Se stress involve various pathways.

12.10   CONCLUSIONS AND FUTURE PROSPECTS

To conclude, proteomic research of medicinal plants offers an in-depth understanding of cellular and subcellular and whole-plant responses to the adverse influences of environmental pollution. It is generally envisioned that sequencing of medicinal plants, more proteomic studies on medicinal plants are likely to be translated into improving medicinal plants against ever-changing environmental factors. However, it appears that a great deal of work remains to be done in the enormous and nebulous field of environmental pollution-induced changes of medicinal plants. From the literature searches, it is evident that so far only a few research groups have been involved in the proteomics of environmental pollution on medicinal plants with the majority of attention focused on the safety of medicinal plant uses as drugs. However, some progress which has been made in proteomic research in a few medicinal plants to some pollution-related stresses, i.e. insufficient or excessive light, UV irradiation, and heavy metal pollution, etc., and medicinal plants, is covered here. In our opinion, medicinal plant proteomics towards improvement of medicinal plants' productivity and environmental stress resistance should be better focused on cellular or subcellular proteomics and analysis of post-translational modifications of proteins to gain knowledge about the essential role of cellular compartmentalization and the mechanisms underlying protein/gene targeting and trafficking. Therefore, the medicinal adverse pollution stress response should be analysed at a cellular or subcellular level, integrated with studies on whole plants, organs, or tissues, to discriminate the specific responses of different cell types to abiotic stress. In the post-genomic era, integrated transcriptomic, proteomic, and metabolomic approaches will enable the system biology of adverse pollution stress responses of medicinal plants to be analysed. More information on other pollution stress-induced proteomic changes in many more medicinal plants is likely to accumulate in the near future.

REFERENCES

Abdallah, C., E. Dumas-Gaudot, J. Renaut, & K. Sergeant. 2012. Gel-based and gel-free quantitative proteomics approaches at a glance. *International Journal of Plant Genomics* 2012: 494572. https://doi.org/10.1155/2012/494572

Agami, A.R., & G.F. Mohamed. (2013). Exogenous treatment with indole-3-acetic acid and salicylic acid alleviates cadmium toxicity in wheat seedlings. *Ecotoxicology and Environmental Safety* 94:164–171.

Agarwal, M., C. Sahi, S. Katiyar-Agarwal, S. Agarwal, T. Young, D.R. Gallie, V.M. Sharma, K. Ganesan, & A. Grover. 2003. Molecular characterization of rice hsp101: complementation of yeast hsp104 mutation by disaggregation of protein granules and differential expression in *indica* and *japonica* rice types. *Plant Molecular Biology* 51: 543–553.

Ahsan, N., D.G., Lee, I., Alam, P.J., Kim, J.J., Lee, Y.O., Ahn, S.S., Kwak, I.J., Lee, J.D., Bahk, K.Y. Kang, & J. Renaut. 2008. Comparative proteomic study of arsenic-induced differentially expressed proteins in rice roots reveals glutathione plays a central role during As stress. *Proteomics* 8: 3561–3576.

Ahsan, N., J. Renaut, & S. Komatsu. 2009. Recent developments in the application of proteomics to the analysis of plant responses to heavy metals. *Proteomics* 9: 2602–2621.

Alvarez, S., B.M. Berla, J. Sheffield, R.E. Cahoon, J.M. Jez, & L.M. Hicks. 2009. Comprehensive analysis of the *Brassica juncea* root proteome in response to cadmium exposure by complementary proteomic approaches. *Proteomics* 9: 2419–2431.

Aslam, B., M. Basit, M.A. Nisar, M. Khurshid, & M.H. Rasool. 2017. Proteomics: technologies and their applications. *Journal of Chromatographic Science* 55: 182–196.

Behn, H., A. Albert, F. Marx, G. Noga, & A. Ulbrich. 2010. Ultraviolet-B and photosynthetically active radiation interactively affect yield and pattern of monoterpenes in leaves of Peppermint (*Mentha* × *piperita* L.). *Journal of Agricultural and Food Chemistry* 58: 7361–7367.

Bernstein, N., M Akram, Z. Yaniv-Bachrach, & M. Daniyal. 2021. Is it safe to consume traditional medicinal plants during pregnancy? *Phytotherapy Research* 35: 1908–1924.

Bona, E., F. Marsano, M. Cavaletto, & G. Berta. 2007. Proteomic characterization of copper stress response in *Cannabis sativa* roots. *Proteomics* 7: 1121–1130.

Bornman, J.F., Barnes, P.W., S.A. Robinson, C.L. Ballaré, S.D. Flint, & M.M. Caldwell. 2015. Solar ultraviolet radiation and ozone depletion-driven climate change: effects on terrestrial ecosystems. *Photochemical and Photobiological Sciences* 14: 88–107.

Cai, J.; X. Liu, K. Vanneste, S. Proost, W.C. Tsai, K.W. Liu, L.J Chen, Y He, Q. Xu, C. Bian, & Z. Zheng, 2015. The genome sequence of the orchid *Phalaenopsis equestris*. *Nature Genetics* 47: 65–72.

Chalhoub, B., F. Denoeud, S. Liu, I.A. Parkin, H. Tang, X. Wang, J. Chiquet, H. Belcram, C. Tong, B. Samans, & M. Corréa. 2014. Early allopolyploid evolution in the post-Neolithic *Brassica napus* oilseed genome. *Science* 345: 950–953.

Chen, J., R.X. Duan, W.J. Hu, N.N. Zhang, X.Y. Lin, J.H. Zhang, & H.L. Zheng. 2019. Unravelling calcium-alleviated aluminium toxicity in *Arabidopsis thaliana:* insights into regulatory mechanisms using proteomics. *Journal of Proteomics* 199: 15–30.

Clark, B.N., & H.B. Gutstein. 2008. The myth of automated, high-throughput two-dimensional gel analysis. *Proteomics* 8:1197–1203.

Clauser, K.R., P. Baker, & A.L. Burlingame. 1999. Role of accurate mass measurement (+/- 10 ppm) in protein identification strategies employing MS or MS/MS and database searching. *Analytical Chemistry* 71: 2871–2882.

Cobbett, C., & P. Goldsbrough. 2002. Phytochelatins and metallothioneins: roles in heavy metal detoxification and homeostasis. *Annual Review of Plant Biology* 53(1):159–182.

Cvjetko, P., M. Zovko, & B. Balen. 2014. Proteomics of heavy metal toxicity in plants. *Archives of Industrial Hygiene and Toxicology* 65: 1–18.

Denoeud, F., L. Carretero-Paulet, A. Dereeper, G. Droc, R. Guyot, M. Pietrella, C. Zheng, A. Alberti, F. Anthony, G. Aprea, J.M. Aury, et al. 2014. The coffee genome provides insight into the convergent evolution of caffeine biosynthesis. *Science* 345: 1181–1184.

Eng, J.K., A.L. McCormack, & J.R.Yates. 1994. An approach to correlate tandem mass spectral data of peptides with amino acid sequences in a protein database. *Journal of the American Society of Mass Spectrometry* 5: 976–989.

Farooq, M.A., K. Zhang, F. Islam, J. Wang, H.U. Athar, A. Nawaz, Z.U. Zafar, J. Xu, & W. Zhou. 2018. Physiological and iTRAQ-based quantitative proteomics analysis of methyl jasmonate–induced tolerance in *Brassica napus* under arsenic stress. *Proteomics* 18: 1700290.

Farooq, M.A., Z. Hong, F. Islam, Y. Noor, F. Hannan, Y. Zhang, A. Ayyaz, T.M. Mwamba, W. Zhou, & W. Song, 2021. Comprehensive proteomic analysis of arsenic induced toxicity reveals the mechanism of multilevel coordination of efficient defense and energy metabolism in two *Brassica napus* cultivars. *Ecotoxicology and Environmental Safety* 208: 111744.

Gao, C., B. Yang, D., Zhang, M. Chen, & J. Tian. 2016. Enhanced metabolic process to indole alkaloids in *Clematis terniflora* DC. after exposure to high level of UV-B irradiation followed by the dark. *BMC Plant Biology* 16: 1–15.

Georgiadou, E.C., E. Kowalska, K. Patla, K. Kulbat, B. Smolińska, J. Leszczyńska, & V. Fotopoulos. 2018. Influence of heavy metals (Ni, Cu, and Zn) on nitro-oxidative stress responses, proteome regulation and allergen production in basil (*Ocimum basilicum* L.) plants. *Frontiers in Plant Science* 9: 862.

Giacomelli, L., A. Rudella, & K.J. van Wijk. 2006. High light response of the Thylakoid proteome in *Arabidopsis* wild type and the ascorbate-deficient mutant vtc2-2. A comparative proteomics study. *Plant Physiology* 141: 685–701.

Gu, L., W. Zheng, M. Li, H. Quan, J. Wang, F. Wang, W. Huang, Y. Wu, X. Lan, & Z. Zhang. 2018. Integrated analysis of transcriptomic and proteomics data reveals the induction effects of rotenoid biosynthesis of *Mirabilis himalaica* caused by UV-B radiation. *International Journal of Molecular Sciences* 19: 3324.

Haralampidis, K., D, Milioni, S. Rigas, & P. Hatzopoulos. 2002. Combinatorial interaction of Cis elements specifies the expression of the *Arabidopsis* AtHsp 90-1gene. *Plant Physiology* 129: 1138–1149.

Hasan, Md.K., Y. Cheng, M.K. Kanwar, X.-Y. Chu, G.J. Ahammed, & Z.Y. Qi. 2017. Responses of plant proteins to heavy metal stress–A review. *Frontiers in Plant Science* 8: 1492.

Hazen, S.P., Y. Wu, & J.A. Kreps. 2003. Gene expression profiling of plant responses to abiotic stress. *Functional and Integrative Genomics* 3: 105–111.

He, Y., H. Xiao, C. Deng, L. Xiong, J. Yang, & C. Peng. 2016. The complete chloroplast genome sequences of the medicinal plant *Pogostemon cablin*. *International Journal of Molecular Science* 17: 820–83.

Hossain, M.A., M. Hasanuzzaman, & M. Fujita. 2010. Up-regulation of antioxidant and glyoxalase systems by exogenous glycine betaine and proline in mung bean confer tolerance to cadmium stress. *Physiology and Molecular Biology of Plants* 16: 259–272.

Hossain, M.A., P. Piyatida, J.A.T. Silva, & M. Fujita, 2012. Molecular mechanism of heavy metal toxicity and tolerance in plants: central role of glutathione in detoxification of reactive oxygen species and methylglyoxal and in heavy metal chelation. *Journal of Botany* 2012: 1–37.

Hu, B., Y. Wang, & G. Liu. 2007. Spatiotemporal characteristics of photosynthetically active radiation in China. *Journal of Geophysical Research: Atmospheres* 112 (D14).

Hu, Y.F., G., Zhou, X.F. Na, L. Yang, W.B. Nan, X. Liu, Y.Q. Zhang, J.L. Li, & Y.R. Bi. 2013. Cadmium interferes with maintenance of auxin homeostasis in *Arabidopsis* seedlings. *Journal of Plant Physiology* 170: 965–975.

Husen, A. 2021a. Harsh Environment and Plant Resilience (Molecular and Functional Aspects). Springer Nature Switzerland. DOI: 10.1007/978-3-030-65912-7

Husen, A. 2021b. Plant Performance Under Environmental Stress (Hormones, Biostimulants and Sustainable Plant Growth Management). Springer Nature Switzerland. DOI: 10.1007/978-3-030-78521-5

Jaillon, O., J.M. Aury, B. Noel, A. Policriti, C. Clepet, A. Casagrande, N. Choisne, S. Aubourg, N. Vitulo, C. Jubin, & A. Vezzi, 2007. French-Italian public consortium for grapevine genome C: The grapevine genome sequence suggests ancestral hexaploidization in major angiosperm phyla. *Nature* 449: 463–467.

Kang, G., G. Li, L. Wang, L. Wei, Y. Yang, P. Wang, Y. Wang, W. Feng, C. Wang, & T. Guo. 2015. Hg-responsive proteins identified in wheat seedlings using iTRAQ analysis and the role of ABA in Hg stress. *Journal of Proteome Research* 14: 249–267.

Kang, Y., T. Techanukul, A. Mantalaris, & J.M. Nagy. 2009. Comparison of three commercially available DIGE analysis software packages: minimal user intervention in gel-based proteomics. *Journal of Proteome Research* 8: 1077–1084.

Kell, D.B., M. Brown, H.M. Davey, W.B. Dunn, I. Spasic, & S.G. Oliver. 2005. Metabolic footprinting and systems biology: the medium is the message. *Nature Reviews Microbiology* 3: 557–565.

Kellner, A. 2015. Genome Sequence of *Catharanthus roseus*. Medicinal Plants Genomic Resource. (http://medicinalplantgenomics.msu.edu/).

Kim, S., M. Park, S.I. Yeom, Y.M. Kim, J.M. Lee, H.A. Lee, E. Seo, J. Choi, K. Cheong, K.T. Kim, & K. Jung. 2014. Genome sequence of the hot pepper provides insights into the evolution of pungency in *Capsicum* species. *Nature Genetics.* 46: 270–278.

Kosová, K., P. Vítámvás, I.T. Prášil, & J. Renaut. 2011. Plant proteome changes under abiotic stress – Contribution of proteomics studies to understanding plant stress response. *Journal of Proteomics* 74: 1301–1322.

Krantev, A., R. Yordanova, T. Janda, G. Szalai, & L. Popov. 2008. Treatment with salicylic acid decreases the effect of cadmium on photosynthesis in maize plants. *Journal of Plant Physiology* 165: 920–931.

Ku, C., W.C. Chung, L.L. Chen, & C.H. Kuo. 2013. The complete plastid genome sequence of Madagascar Periwinkle *Catharanthus roseus* (L.) G. Don: Plastid genome evolution, molecular marker identification, and phylogenetic implications in Asterids. *PLoS ONE* 8: e68518.

Lajayer, H.A., G. Savaghebi, J. Hadian, M. Hatami, & M. Pezhmanmehr. 2017. Comparison of copper and zinc effects on growth, micro- and macronutrients status and essential oil constituents in pennyroyal (*Mentha pulegium* L.). *Brazilian Journal of Botany* 40: 379–388.

Lasserre, J.P., & A. Ménard. 2012. Two-dimensional blue native/SDS gel electrophoresis of multiprotein complexes. *Methods in Molecular Biology* 869: 317–337.

Li, G., X. Peng, H., Xuan, L., Wei, Y., Yang, T. Guo, & G. Kang. 2013. Proteomic analysis of leaves and roots of common wheat (*Triticum aestivum* L.) under copper-stress conditions. *Journal of Proteome Research* 12: 4846–4861.

Li, Y., C. Xu, & X. Lin. 2014. De novo assembly and characterization of the fruit transcriptome of Chinese Jujuba (*Ziziphus jujuba* Mill) using 454 pyrosequencing and the development of novel trinucleotide SSR markers. *PLoS ONE* 9: e106438.

Liang, X., L. JianFeng, S. ShengQing, W. Yuan, C. ErMei, G. Ming, & J. ZePing, 2013. A review on the progress of proteomic study on plant responses to heavy metal stress. *Acta Prataculturae Sinica* 22: 300–311.

Lingua, G., E., Bona, V. Todeschini, C. Cattaneo, F. Marsano, G. Berta, & M. Cavaletto. 2012. Effects of heavy metals and arbuscular mycorrhiza on the leaf proteome of a selected poplar clone: a time course analysis. *PloS ONE* 7: e38662.

Lomaglio, T., M. Rocco, D. Trupiano, E. De Zio, A. Grosso, M. Marra, S. Delfine, D. Chiantante, D. Morabito, & G.S Scippa. 2015. Effect of short-term cadmium stress *on Populus nigra* L. detached leaves. *Journal of Plant Physiology* 182: 40–48.

Ma, Y., L. Yuan, B. Wu, X. Li, S. Chen, & S. Lu. 2012. Genome-wide identification and characterization of novel genes involved in terpenoid biosynthesis in *Salvia miltiorrhiza*. *Journal of Experimental Bot*any 63: 2809–2823.

Mann, M., R.C. Hendrickson, & A. Pandey. 2001. Analysis of proteins and proteomes by mass spectrometry. *Annual Review of Biochem*istry 70: 437–473.

Martin, H.M. 2006. Software analysis of two-dimensional gels in proteomic experiments. *Current Bioinformatics* 1: 255–262.

Minden, J.S. 2012. Two-Dimensional Difference Gel Electrophoresis (2D DIGE). In: Conn, P.M. (ed.). Laboratory Methods in Cell Biology. Elsevier, pp. 111–141.

Mochida, K., T. Sakurai, H. Seki, T. Yoshida, K. Takahagi, S. Sawai, H. Uchiyama, T. Muranaka, & K. Saito. 2017. Draft genome assembly and annotation of *Glycyrrhiza uralensis*, a medicinal legume. *Plant Journal* 89: 181–194.

Murch, S.J., K. Haq, H.P.V. Rupasinghe, & P.K. Saxena. 2003. Nickel contamination affects growth and secondary metabolite composition of St. John's wort (*Hypericum perforatum* L.). *Environmental and Experimental Botany* 49: 251–257.

Murchie, E.H., & P. Horton. 1997. Acclimation of photosynthesis to irradiance and spectral quality in British plant species: chlorophyll content, photosynthetic capacity and habitat preference. *Plant, Cell and Environment* 20: 438–448.

Nebrich, G., H. Liegmann, M. Wacker, M. Herrmann, D. Sagi, A. Landowsky, & J. Klose. 2005. Proteomer, a novel software application for management of proteomic 2DE-gel data-II application. *Molecular & Cellular Proteomics* 4(8): S296–S296.

Neergheen-Bhujun, V.S. 2013. Underestimating the toxicological challenges associated with the use of herbal medicinal products in developing countries. *BioMed Research International* 2013: 804086. https://doi.org/10.1155/2013/804086

Neumann, D., O. Lichtenberger, D, Günther, K. Tschiersch, & L. Nover. 1994. Heat-shock proteins induce heavy-metal tolerance in higher plants. *Planta* 194: 360–367.

Pandey, S. Gupta, K., & A.K. Mukherjee. 2007. Impact of cadmium and lead on *Catharanthus roseus* – A phytoremediation study. *Journal of Environmental Biology* 2007. PMID: 18380091.

Pappin, D.J., P. Hojrup, & A.J., Bleasby. 1993. Rapid identification of proteins by peptide-mass fingerprinting. *Current Biology* 3: 327–332.

Park, O.K. 2004. Proteomic studies in plants. *Journal of Biochemistry and Molecular Biology* 37: 133–138.

Pérez-Clemente, R.M., V. Vives, S.I. Zandalinas, M.F. López-Climent, V. Muñoz, & A. Gómez-Cadenas. 2013. Biotechnological Approaches to Study Plant Responses to Stress. *BioMed Research International* 2013: e654120.

Perkins, D.N., D.J. Pappin, D.M. Creasy, & J.S. Cottrell. 1999. Probability-based protein identification by searching sequence databases using mass spectrometry data. *Electrophoresis* 20: 3551–3567.

Qin, C., C. Yu, Y. Shen, X. Fang, L. Chen, J. Min, J. Cheng, S. Zhao, M. Xu, Y. Luo, and Y. Yang. 2014. Whole-genome sequencing of cultivated and wild peppers provides insights into capsicum domestication and specialization. *Proceedings of the National Academy of Sciences* 111: 5135–5140.

Qin, J., D. Fenyö, Y. Zhao, W.W. Hall, D.M. Chao, C.J. Wilson, R.A. Young, & B.T. Chait. 1997. A strategy for rapid, high-confidence protein identification. *Analytical Chemistry* 69: 3995–4001.

Rai, R., S. Pandey, & S.P. Rai. 2011. Arsenic-induced changes in morphological, physiological, and biochemical attributes and artemisinin biosynthesis in *Artemisia annua*, an antimalarial plant. *Ecotoxicology* 20: 1900–1913.

Rodríguez-Celma, J., R. Rellán-Álvarez, A. Abadía, J. Abadía & A.F. López-Millán (2010). Changes induced by two levels of cadmium toxicity in the 2-DE protein profile of tomato roots. *Journal of Proteomics* 73: 1694–1706.

Rosengren, A.T., J.M. Salmi, T. Aittokallio, J. Westerholm, R. Lahesmaa, T.A. Nyman, & O.S. Nevalainen. 2003. Comparison of PDQuest and Progenesis software packages in the analysis of two-dimensional electrophoresis gels. *Proteomics* 3:1936–1946.

Sarry, J.E., L. Kuhn, C. Ducruix, A. Lafaye, C. Junot, V. Hugouvieux, A. Jourdain, O. Bastien, J.B. Fievet, D. Vailhen, B. Amekraz, et al. 2006. The early responses of *Arabidopsis thaliana* cells to cadmium exposure explored by protein and metabolite profiling analyses. *Proteomics* 6: 2180–2198.

Shivaraj, Y., S. Govind, S. Jogaiah, & D. Sannaningaiah. 2015. Functional analysis of medicinal plants using system biology approaches. *International Journal of Pharmacy and Pharmaceutical Science* 7: 41–43.

Singh, H., J.A., Bhat, V.P. Singh, F.J. Corpas, & S.R. Yadav. 2021. Auxin metabolic network regulates the plant response to metalloids stress. *Journal of Hazardous Materials* 405: 124250.

Snedden, W., & A. Fromm. 2001. Calmodulin as a versatile calcium signal transducer in plants. *New Phytology* 151:35–66.

Song, W.-Y., D.G. Mendoza-Cózatl, Y. Lee, J.I. Schroeder, S.N. Ahn, H.S. Lee, T. Wicker, E. Martinoia. 2014. Phytochelatin–metal(loid) transport into vacuoles shows different substrate preferences in barley and *Arabidopsis*. *Cell and Plant Environment* 37: 1192–1201.

Sinha, S., & R. Saxena. 2006. Effect of iron on lipid peroxidation, and enzymatic and non-enzymatic antioxidants and bacoside-A content in medicinal plant *Bacopa monnieri* L. *Chemosphere* 62: 1340–1350.

Shao, H.B., L.Y. Chu, F.T. Ni, D.G. Guo, H. Li, & W.X. Li. 2010. Perspective on phytoremediation for improving heavy metal contaminated soils. In: Ashraf, M., Ozturk, M., & Ahmad, M. (eds). Plant adaptation and Phytoremediation. Springer, Dordrecht, pp. 227–244.

Suzuki M., R. Nakabayashi, Y. Ogata, N. Sakurai, T. Tokimatsu, S. Goto, M. Suzuki, M. Jasinski, E. Martinoia, S. Otagaki, S. Matsumoto, K. Saito, & K. Shiratake. 2015. Multiomics in grape berry skin revealed specific induction of the stilbene synthetic pathway by ultraviolet-C irradiation. *Plant Physiology* 168: 47–59.

Urasaki, N., H. Takagi, S. Natsume, A, Uemura,.,N Taniai, N.,Miyagi, M. Fukushima, S. Suzuki, K. Tarora, M. Tamaki, & M.Sakamoto. 2017. Draft genome sequence of bitter gourd (*Momordica charantia*), a vegetable and medicinal plant in tropical and subtropical regions. *DNA Research* 24: 51–58.

Unlü, M., M.E. Morgan, & J.S. Minden. 1997. Difference gel electrophoresis: a single gel method for detecting changes in protein extracts. *Electrophoresis* 18: 2071–2077.

Velasco, R., A. Zharkikh, M. Troggio, D.A. Cartwright, A. Cestaro, D. Pruss, M. Pindo, L.M. Fitzgerald, S. Vezzulli, J. Reid, & G. Malacarne. 2007. A high quality draft consensus sequence of the genome of a heterozygous grapevine variety. *PLoS ONE*: 2, e132.

Wang, Y., L.Y. Hu, Zhu, & X.X. Li. 2012. Response to nickel in the proteome of the metal accumulator plant *Brassica juncea*. *Journal of Plant Interactions* 7: 230–237.

Wang, Z., M. Gerstein, & M. Snyder. 2009. RNA-Seq: a revolutionary tool for transcriptomics. *Nature Reviews Genetics* 10: 57–63.

Wheelock, A.M., & A.R. Buckpitt. 2005. Software-induced variance in two-dimensional gel electrophoresis image analysis. *Electrophoresis* 26, 4508–4520.

Wheelock, A.M., & S. Goto. 2006. Effects of post-electrophoretic analysis on variance in gel-based proteomics. *Expert Review of Proteomics* 3: 129–142.

Yan, L., X. Wang, H. Liu, Y. Tian, J. Lian, R. Yang, S. Hao, X. Wang, & S. Yang. 2015. The genome of *Dendrobium officinale* illuminates the biology of the important traditional Chinese orchid herb. *Molecular Plant* 8: 922–934

Yang, B., X. Wang, C. Gao, M. Chen, Q. Guan, J. Tian, & S. Komatsu. 2016. Proteomic and metabolomic analyses of leaf from *Clematis terniflora* DC. exposed to high-level ultraviolet-B irradiation with dark treatment. *Journal of Proteome Research* 15:2643–2657.

Yang, L., J. Ji, K.R. Harris-Shultz, H. Wang, E.F. Abd-Allah, Y. Luo, & Hu. 2016. The dynamic changes of the plasma membrane proteins and the protective roles of nitric oxide in rice subjected to heavy metal cadmium stress. *Frontiers in Plant Science* 7: 190.

Yates, J.R. 1998a. Mass spectrometry and the age of the proteome. *Journal of Mass Spectrometry* 33: 1–19.

Yates, J.R. 1998b. Database searching using mass spectrometry data. *Electrophoresis* 19: 893–900.

Yıldız, M., & H. Terzi. 2016. Proteomic analysis of chromium stress and sulfur deficiency responses in leaves of two canola (*Brassica napus* L.) cultivars differing in Cr(VI) tolerance. *Ecotoxicology and Environmental Safety* 124: 255–266.

Youns, M., J.D. Hoheisel, & T. Efferth. 2010. Toxicogenomics for the prediction of toxicity related to herbs from traditional Chinese medicine. *Planta Medica* 76: 2019–2025.

Ytterberg, A.J., J.-B. Peltier, & K.J. van Wijk. 2006. Protein profiling of plastoglobules in chloroplasts and chromoplasts. A surprising site for differential accumulation of metabolic enzymes. *Plant Physiology* 140: 984–997.

Zhang, C., B. Xu, W. Geng, Y. Shen, D. Xuan, Q. Lai, C. Shen, C. Jin, & C. Yu. 2019. Comparative proteomic analysis of pepper (*Capsicum annuum* L.) seedlings under selenium stress. *PeerJ* 7: e8020.

Zhang, L., X. Li, W. Zheng, Z. Fu, W. Li, L. Ma, K. Li, L. Sun & J. Tian. 2013. Proteomics analysis of UV-irradiated *Lonicera japonica* Thunb. with bioactive metabolites enhancement. *Proteomics* 13: 3508–3522.

Zheng, W., S. Komatsu, W. Zhu, L. Zhang, X. Li, L. Cui, & J. Tian. 2016. Response and defense mechanisms of *Taxus chinensis* leaves under UV-A radiation are revealed using comparative proteomics and metabolomics analyses. *Plant Cell Physiology* 57: 1839–1853.

Zheng, W., X. Li, L. Zhang, Y. Zhang, X. Lu, & J. Tian. 2015. Improved metabolites of pharmaceutical ingredient grade *Ginkgo biloba* and the correlated proteomics analysis. *Proteomics* 15: 1868–1883.

Zheng, Z., and M. Wu. 2004. Cadmium treatment enhances the production of alkaloid secondary metabolites in *Catharanthus roseus*. *Plant Science* 166(2): 507–514.

Zhu, F.Y., W.L. Chan, M.X. Chen, R.P. Kong, C. Cai, Q. Wang, J.H. Zhang, & C. Lo. 2016. SWATH-MS quantitative proteomic investigation reveals a role of jasmonic acid during lead response in *Arabidopsis*. *Journal of Proteome Research* 15: 3528–3539.

Zhu, W., H. Han, A. Liu, Q. Guan, J. Kang, L. David, C. Dufresne, S. Chen, & J. Tian. 2021. Combined ultraviolet and darkness regulation of medicinal metabolites in *Mahonia bealei* revealed by proteomics and metabolomics. *Journal of Proteomics* 233: 104081.

Zhu, W., B. Yang, S. Komatsu, X. Lu, X. Li, & J. Tian. 2015. Binary stress induces an increase in indole alkaloid biosynthesis in *Catharanthus roseus*. *Frontiers in Plant Sciences* 6: 582.

ZhuoHeng, Z., L. ShengZhi, Z. Wei, O. YuTing, H. Yamaguchi, K. Hitachi, K. Tsuchida, T. JingKui, & S. Komatsu. 2019. Phosphoproteomics reveals the biosynthesis of secondary metabolites in *Catharanthus roseus* under ultraviolet-B radiation. *Journal of Proteome Research* 18: 3328–3341.

13 Genome-Editing Strategy for Medicinal Plants Growing under Adverse Environmental Pollution
Challenges and Opportunities

Arvind Arya

CONTENTS

13.1   INTRODUCTION

With the constant changes in the environment, it is imperative to increase awareness of these changes among humans. Around the globe, people are confronted with many small to large environmental problems. Some issues are severe and make humanity vulnerable to disasters. An urgent focus is needed to address the current environmental problems. One of the major current environmental problems is pollution. Unfavourable or undesirable changes in the surroundings including air, water, and soil are termed pollution. There are many ways to define pollution; one way is that

DOI: 10.1201/9781003178866-13

pollution occurs when nature is unable to clear an element brought to it in an unnatural way (Figure 13.1).

Pollution is a very serious issue as life on this planet is in absolute need of the natural elements. The main cause of pollution is the indiscriminate activities of humans, or it is due to natural ecosystems, such as air pollution due to storms and forest fires. The major difference between human-made and natural pollution is that human-made pollution is a greater threat and more harmful than self-replenishing natural pollutants (Figure 13.2).

There are many technologies available to alleviate pollution. Due to the cost and their demerits in cleaning up specific pollutants, technologies are not providing a complete solution and are not being adopted. Despite so many advanced treatment methods and technologies, the fight against pollution is still not consistent. Thus, such methods are largely being replaced by natural means of pollution control. New technologies are being developed to clean a wide range of pollutants from the air, water, and soil. One of the main methods of pollution management is the use of plants. In the recent past, the scientific community has realized the importance of plants in fighting the pollution problem (Saxena et al. 2020; Awa and Hadibarata 2020).

Research on plant-mediated pollution control has gained further momentum following the participation of more advanced methods of genome editing, transgenesis, and nanotechnology. Plants are very sensitive to pollutants and can present distinct symptoms against specific pollutants (Weinstein et al. 1990). Plants are being used to get quantitative information on the quality of the environment, i.e., soil, water, and air (Madejón et al. 2006). Plants are also being used to monitor the presence of radionuclides (^{137}Cs, ^{40}K, ^{210}Pb, and ^{7}Be) through accumulation (Dushenkov et al. 1997; Terry and Banuelos 1999; Todorović et al. 2013). Plants are popularly known as the sinks of pollution. They can not only monitor pollution but have the potential of mitigation (Joshi et al. 2020). The surface morphology of plants has a direct correlation with the absorption of dust particles (Sharma et al. 2005; Ram et al. 2014).

FIGURE 13.1 Physicochemical methods available for cleaning of radionuclide pollutants.

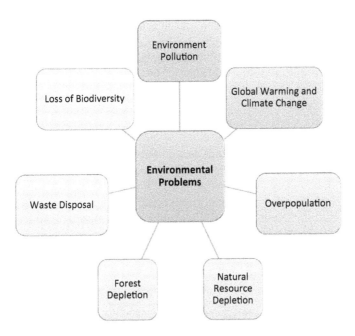

FIGURE 13.2   Major environmental problems.

For the air pollutants, plant leaves serve as a major organ that helps in sticking particulate and gaseous pollutants and absorption of soluble particles. The size and morphology viz. The surface texture and shape of the leaves also affect their dust particle-capturing capacity (Das and Prasad 2012; Liu et al. 2012). Plants also help remove water pollution as they assist in preventing soil erosion and stabilize the sediments. Plants shelter shorelines from waves and wind currents and thus protect the soil from erosion. By stabilizing the sediments and preventing erosion they also help in water purification.

Radionuclides are another hazardous pollutant that comes from nuclear power plants or the explosion of atomic bombs. The cleaning of the affected area requires studies on the chemistry of the environment of that place along with the rate of radioactive decay and character of deposition. There is a shift towards the biological method of cleaning radionuclide pollutants over the other physicochemical methods. Plants are being used for the treatment of several radionuclide pollutants (^{3}H, U, Pu, ^{137}Cs, and ^{90}Sr).

Plants possess different metabolic systems and thus the use of selective herbicides is benefiting in the remediation of contaminated soil. The association of microorganisms with plants also provides efficacy of degradation of toxic compounds by altering the metabolic system.

Medicinal plants and endemic plants are very important for the livelihood of humans. More than 60 per cent of the world's population depend on plant-derived medicines (Boukhatem and Setzer 2020). Medicinal plants along with other plants are affected a lot by climate change. Climate change is one of the most serious challenges to humankind. According to the Intergovernmental Panel on Climate Change (IPCC), there will be unpredictable changes in the climate with a likely increase in temperature from 1.4°C to 5.8°C. Some of the major key factors affecting endemic populations are habitat destruction, a decline in natural pollinators (due to excessive insecticide use), over-harvesting, and climate change (Lahn 2021).

Medicinal plants have been used for centuries for the well-being of humans. They have many purposes from aesthetic to medicinal. They are known for their pharmaceutical, cosmetics, and nutritional value. Medicinal plant knowledge is recorded in many ancient works of literature of different civilizations worldwide. Around 25 per cent of contemporary medicines are derived from medicinal plants (Shukla 2020). The overall trade in medicinal plants and products alone is expected to be around 5 trillion USD by 2050 (Jadhav et al. 2020). It is estimated that worldwide more than 70,000 plants are being used in medicines (Mehta et al. 2020). Medicinal plants are vital for livelihoods, as many communities earn their living by selling medicinal non-wood products. There is an increased demand for medicinal plants in developed as well as developing countries. Medicinal plants yield essential oils through distillation, along with the market value of their plant parts. These medicinal essential oils are in high demand by the cosmetic, pharmaceutical, and food industries.

Due to the loss of biodiversity and erosion of traditional knowledge, medicinal plants are facing a major threat. One way to preserve the relationship between plants and people is to utilize traditional practices that help increase the biodiversity of these plants. In this chapter, it is shown how medicinal plants are being reinvigorated for their properties to fight and control the pollution problem in view of recent plant modification methods.

13.2   SELECTION OF PLANT SPECIES TO OVERCOME THE POLLUTION PROBLEM IN THE URBAN ENVIRONMENT

Roads are a major contributor to pollution in urban areas (Angelevska et al. 2021). Planting along the side of the road can help mitigate the pollution problem to a substantial degree (Altaf et al. 2021). However, the availability of space for planting trees on the roadside and central verge is a big issue. Despite planting on the central verge having many benefits, it is often neglected. Numerous studies have suggested that trees and shrubs resistant to drought/frost demonstrate tolerance toward pollution (Álvarez et al. 2020; Han et al. 2020; Jordanovska et al. 2020). The selection of such trees is further facilitated by researchers identifying the important characteristics of plants to mitigate pollution (Figure 13.3).

In urban areas, planting is now being focused on as it is the best way to mitigate particulate air pollution. Plants such as *Tamarindus indicus, Mangifera indica, Polyalthea longifolia, Derris indica*, and *Thespepsia populnea* are highly effective at collecting particulate pollutants (Kundu et al. 2017). The dust-collecting properties are attributed to the size and shape of the leaves of these plants. The importance of trees in the urban environment is being noted for their aesthetic and sustainable value. The need is to identify suitable, disease-resistant, evergreen plants to maintain the green landscape. Many other trees, viz. *Azadirachta indica, Bombax ceiba, Cassia fistula, Delonix regia, Ficus religiosa, Jacaranda mimosifolia,*

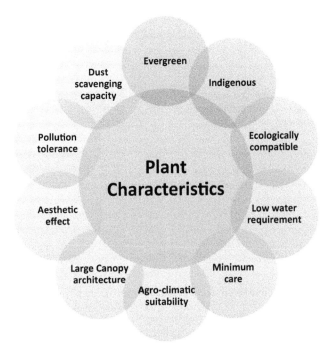

FIGURE 13.3 Characteristics of plants used for pollution control.

Lagerstroemia indica, *Plumeria rubra*, and *Syzygium cumini*, are also very popular and effective at controlling air pollution in urban areas (Sen 2020; Velmurugan et al. 2020). However, it is also imperative to study the relationship of these plants with humans and the environment. Some of the plants are responsible for causing allergies and fever and thus studies related to this type of adverse effect are prerequisite (Simard and Benoit 2010; Damaiyani et al. 2018). Alternatively, the plants with a short flowering period and less pollen productivity should be given preference (Chakre 2006).

13.3   PHYTOREMEDIATION

In situ removal of contaminants from the environment (soil, water, air, etc.) by plants is called phytoremediation. With the advances of science and technology, engineered plants are being used for the removal, degradation, and containment of contaminants (Sarma et al. 2021). There are two major categories of pollutants: organic and inorganic. Further, organic pollutants are divided into PCBs (polychlorinated biphenyls), PAHs (polycyclic aromatic hydrocarbons), halogenated hydrocarbons, and chlorinated solvents (Borji et al. 2020). Being less reactive and non-accumulative, organic pollutants are less toxic to plants (Tripathi et al. 2019). Inorganic contaminants are heavy metals (mercury, lead, cadmium), non-metallic compounds (arsenic), and radionuclides (uranium, caesium, chromium, etc.). Heavy metals are also important for the growth and development of plants but, at the same time, they

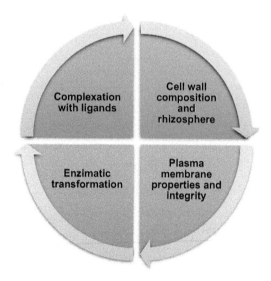

FIGURE 13.4   Cellular and molecular mechanisms involved in phytoremediation.

are toxic at higher concentrations. They cause damage to the cells and tissues by inducing oxidative damage, generate free radicles, interfere with the structure of proteins, and replace other essential nutrients (Cherian and Oliveira 2005; Iqbal 2015; Ansari et al. 2021) (Figure 13.4). Some pollutants usually enter the plant body through the natural system of nutrient absorption. Plants use various means to protect themselves from these pollutants. Some plants called hyperaccumulators can accumulate a huge number of toxic metals and other radionuclides in their tissues (Peer et al. 2006). Information on different plant species and heavy metal contaminants is compiled in Table 13.1. However, the absorption of any metal by plants depends on its bioavailability in the soil or water. Further, pH or other changes in the soil chemistry may increase the bioavailability of these metals. The use of mycorrhizae was also found to have a profound effect on the absorption of metals by plants (Shetty et al. 1994).

From the economic perspective, researchers have focused more on hyperaccumulator plants since these plants may accumulate more than 3 per cent of heavy metals without any sign of damage to the plants (Baker and Brooks 1989; Chaney et al. 1997).

Phytoremediation is one of the energy-efficient and aesthetic ways of cleaning the environment. There are several advantages of phytoremediation, one of which is its low cost (25 to 100 USD per tonne of soil; 0.60 to 6.00 USD per 1,000 gallons of polluted water). Phytoremediation costs only half of what the other methods offer. The other advantage of phytoremediation is permanent remediation. However, some of the properties of plants, such as the length of roots of the plants and their ability to tolerate the contaminants, are areas of concern. Sometimes the hyperaccumulator plants present great concern since they can become the source of accumulated contaminants entering the food chain through grazing animals.

TABLE 13.1
Hyperaccumulator Plant Species, their Families and Accumulated Heavy Metal/Metalloids

Family	Species	Heavy Metal Contaminants	Key References
Azollaceae or Salviniaceae	*Azolla pinnata*	Cd	(Rai 2008)
Asteraceae	*Bidens pilosa*	Cd	(Sun et al. 2009)
	Sonchus asper	Pb and Zn	(Yanqun et al. 2005)
	Helianthus annuus	Cd, Cr and Ni	(Turgut et al. 2004)
Brassicaceae	*Alyssum bertolonii*	Ni	(Li et al. 2003)
	Alyssum murale	Ni	(Bani et al. 2010)
	Arabidopsis thaliana	Cu, Mn, Pb and Zn	(Lasat 2002)
	Arabidopsis halleri	Cd and Zn	(Cosio et al. 2004)
	Brassica junceae	Ni and Cr	(Saraswat and Rai 2009)
	Brassica oleracea	Cd	(Kumar et al. 2005)
	Cardaminopsis halleri	Cd and Zn	(Sun et al. 2007)
	Rorippa globosa	Cd	(Wei et al. 2008)
	Thlaspi caerulescens	Cd, Ni and Zn	(Lombi et al. 2001)
Crassulaceae	*Sedum alfredii*	Cd, Pb and Zn	(Sun et al. 2007)
Euphorbiaceae	*Euphorbia cheiradenia*	Pb	(Chehregani Rad and Malayeri. 2007)
Lamiaceae	*Clerodendrum infortunatum*	Cu	(Rajakaruna and Bohm 2002)
	Haumaniastrum katangense	Cu	(Duvign et al. 2010)
Fabaceae	*Astragalus racemosus* *Astragalus bisulcatus*	Se	(Vallini et al. 2005)
Pteridaceae	*Pteris vittata*	As, Cr and Se	(Baldwin and Butcher 2007)
Solanaceae	*Solanum nigrum*	Cd	(Sun et al. 2008)
	S. photeinocarpum	Cd	(Zhang et al. 2011)
Violaceae	*Viola baoshanensis*	Cd	(Wu et al. 2010)

13.4 GENOME EDITING-ASSISTED PLANT-BASED REMEDIATION

In phytoremediation, plants typically break down or degrade the organic pollutants or remove the metal contaminants from soil and water. The whole process of removing metal contaminants and organic pollutants using plants varies. This low-cost technology has great potential to mitigate the impact of harmful pollutants and reinstall the natural state of the ecosystem. Based on growth rate, biomass, and bio-concentration levels, etc., several plants have been identified as possible candidates for phytoremediation. Scientists have further enhanced the abilities of such plants by using genetic engineering and genome editing. Modification in plants could increase the capabilities of phytoremediation by 50 to 500 times.

Phytoextraction, phytoaccumulation, and phytovolatilization are some of the key determinants of genetic modification in plants (Figure 13.5). A very specific

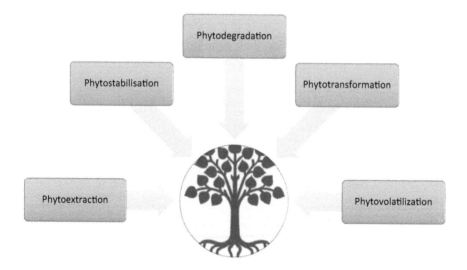

FIGURE 13.5   Different methods of phytoremediation.

approach towards genome modification, viz. CRISPR Cas9, can be used to target the key mechanisms involved in phytoremediation. Genome-sequencing data of plants becomes pivotal in increasing the remediation properties of different plant species. Many phytoremediators have been sequenced, viz. *Noccaea caerulescens* (formerly known as *Thlaspi caerulescens*), *Arabidopsis halleri*, *Pteris vittata*, *Brassica juncea*, and *Populus trichocarpa*. Some of the energy crops were also engineered to increase their tolerance to pollutants. The *Populus* species (one of the hyperaccumulators) has also been modified using the CRISPR Cas9 method of genome editing. With the advent of sequencing methods and bioinformatics tools, it is now possible to apply the codon-optimized version of CRISPR Cas9 for several monocots and dicot plant species.

Enhanced expression of enzymes such as metallothionein, metal reductions etc., and the pathways of pollution transport (vacuolar compartmentation of heavy metals or the transport of pollutants to the rhizosphere) are the primary targets of phytoremediation. The low efficiency of Cas9 gRNA-mediated cloning for enhancing phytoremediation has been carried out in some selected plants only (Belhaj et al. 2015). Scientists suggested the use of an all-in-one plasmid approach by amalgamating diverse gRNAs with Cas9 in a single T-DNA to improve the editing (Mikami et al. 2015; Bortesi and Fischer 2015). Due to the complex nature of plant genomes, it is very difficult to manage site-specific mutagenesis in plants and therefore Cas9 gRNA could be wondrous in facilitating the targeting of multiple sequences and traits simultaneously. CRISPR-aided genome engineering has great potential for exploiting plant genomes for enhancing phytoremediation.

Knowledge of tolerance is helping researchers to introduce the genetic characteristics in desired plant species with increased metal tolerance (Figure 13.6). It is also being suggested to incinerate the plant biomass with accumulated heavy

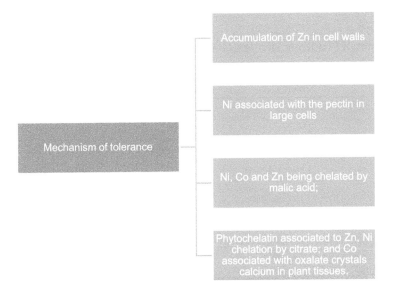

FIGURE 13.6 Mechanism of tolerance.

metal for their recycling. Further, this biomineration could be used for the mining of economically important metals.

13.5 PHYTOEXTRACTION (PHYTOACCUMULATION)

Phytoextraction is the adoption of contaminants from the root system followed by its accumulation in the aerial parts of the plants (Table 13.2). The aerial parts of the plants with accumulated heavy metal contaminants are harvested and processed (Chaney et al. 1997). Plants used for phytoextraction should have the following characteristics (Kanwar et al. 2020):

(i)   Tolerance against toxic levels of metals
(ii)   High biomass yield
(iii) Accumulation of metal contaminants in leaves or easy-to-harvest parts.

The heavy metal-containing biomass is treated and reduced in volume and weight by several treatments, followed by its disposal as hazardous waste or sometimes also for re-extraction of trace elements (Lombi et al. 2001). The types of plant species for phytoextraction fall into different categories, such as hyperaccumulators, plants with high biomass production, and genetically engineered plants.

High biomass-producing plants and natural hyperaccumulator plants can be selected and employed for decontaminating polluted sites. However, some features such as low biomass of hyperaccumulator plants or low phytoextraction efficiency limit the use of such plant species. The bioengineered plants have an edge over the

TABLE 13.2
Well-Known Plants for Phytoextraction

Plants	Species	Heavy metal contaminants	Key references
Indian mustard	*Brassica juncea*	As, B, Cd, Cr(VI), Cu, Ni, Pb, Se, Sr and Zn	(Raskin and Ensley 1999; Salido et al. 2003)
Alpine Pennycress	*Thlaspi caerulescens*	Cd and Zn	(McGrath and Zhao 2003; Cosio et al. 2004)
Alyssum	*Alyssum wulfenianum*	Ni	(Reeves and Brooks 1983)
Canola	*Brassica napus*	Se	(Bañuelos et al. 1998)
Kenaf	*Hibiscus cannabinus* L. cv. Indian		
Tall Fescue	*Festuca arundinacea* Schreb cv. Alta		
Poplar	*Populus* sp.	As and Cd	(Pierzynski et al. 1994)
Sunflower	*Helianthus annuus*	Cs and Sr	(Adler 1996)
Sudangrass	*Sorghum vulgare* L.	Pb, Zn, Hg and Ni	(Salido et al. 2003)
Alfalfa	*Medicago sativa*		
Maize	*Zea mays*		

natural plant species in terms of high biomass, heavy metal tolerance, and high accumulation (Bhargava et al. 2012).

The transgenesis approach allows root-to-shoot translocation followed by vacuolization of heavy metals in the above-ground plant parts that are easy to harvest and thus does not affect plant growth (Bhargava et al. 2012; Mosa et al. 2016). Such genetic modification is also proposed to increase the trace elements (biofortification) to support the nutrition of plants when grown on non-contaminated sites (Singh et al. 2017; Wu et al. 2018).

Genetic engineering-mediated alteration in the process of membrane transportation was found suitable for phytoextraction. Li et al. (1996) used a protein transporter (yeast cadmium factor gene from *Saccharomyces cerevisiae*) for transgenesis. This YCF1 gene was also transferred to *B. juncea* and *Populus* sp. and was found to increase the absorption of Cd and Pd (Bhuiyan et al. 2011; Shim et al. 2013). A vast range of heavy metal transporters are reviewed by Suman et al. (2018).

Another strategy is to increase the heavy metal-binding site in plant cells without compromising the toxicity resistance. By increasing the production of enzymes such as γ-glutamylcysteine synthetase (γ-ECS) and cysteine synthetase (GS) the overproduction of peptidic chelators can be managed (Koprivova et al. 2002; Bittsánszky et al. 2005).

13.6 RHIZOFILTRATION

Heavy metal-polluted soil and water is a major problem that can be resolved by using plants. Rhizofiltration is a plant-based method to remove the contamination. Many methods are available for the removal of such pollutants from water, such as ion exchange, chemical and microbiological precipitation, etc. These methods have drawbacks of low efficiency, cost, and selective treatment. In some other approaches, living and non-living bacteria and algae have been used for the treatment and recovery of heavy metals from the aqueous environment. However, a few methods in which the cultured cells of some plants and fungi were used for the removal of heavy metals were discouraged due to their high cost. Aquatic plants were also researched for their efficacy to treat the heavy metal pollution from the water table. The limitation of this method is the slow growth of the plants and slow-growing roots. Further, the higher water content of aquatic plants makes their drying difficult. Recently, researchers have also tested terrestrial plants as they have a long and well-developed root system which provides a high surface area, and these plants can be easily dried in the open air. The plant roots are being used for the removal of heavy metal pollutants from the water. This method is known as rhizofiltration. Many different heavy metal pollutants, such as Cu^{2+}, Cd^{2+}, Cr^{6+}, Ni^{2+}, Pb^{2+}, and Zn^{2+}, have been removed using hydroponically grown terrestrial plants (Table 13.3). In Chernobyl, radioactive contaminants were also cleared from the groundwater by the use of sunflowers (Dushenkov 2003). Between 23 and 30 per cent of uranium was reported to be taken up by the *Brassica juncea* hairy roots cultures developed through *Agrobacterium rhizogenes*-mediated transformation (Eapen et al. 2003). Many plants, such as sunflower, rye, mustard, and tobacco, are being employed for rhizofiltration. Due to the presence of a fibrous

TABLE 13.3
Some Common Plants Used for Rhizofiltration

Plants	Heavy metal contaminants	Key references
Azolla caroliniana	As	(Favas et al. 2012)
Eichhornia crassipes	As, Cs, Co, Hg and Mn	(Karkhanis et al. 2005; Chattopadhyay et al. 2012)
Lemna minor and *Lemna gibba*	Ag, Au and Ni	(Khellaf and Zerdaoui 2010; Pratas et al. 2012; Sasmaz and Obek 2012)
Typha angustifolia	As, Cd, Cr, Cu, Fe, Mn, Ni, Pb and Zn	(Chandra and Yadav 2010)
Indian mustard	Cu, Cd, Cr, Ni, Pb and Zn	(Dushenkov et al. 1995; Prasad and De Oliveira Freitas 2003)
Sunflower	Cu, Cd, Cr, 137Cs, Ni, Pb, 90Sr, U and Zn	(Prasad and De Oliveira Freitas 2003; Jadia and Fulekar 2008)
Brassica juncea	Cd and Pb	(Dushenkov et al. 1997)
Medicago sativa	Cr, Cu, Ni, Pb and Zn	(Gardea-Torresdey et al. 1998)

TABLE 13.4
Some Common Plants Used for Phytostabilization

Plants	Heavy metal contaminants	Key references
Sorghum sp.	Cd, Cu Ni, Pb and Zn	(Jadia and Fulekar 2009)
Solanum nigrum	Ni	(Ferraz et al. 2012)
Eucalyptus urophylla and *Eucalyptus saligna*	Zn	(Magalhães et al. 2011)
Vigna unguiculata	Pb and Zn	(Kshirsagar and Aery 2007)

root system, terrestrial plants are widely preferred for rhizofiltration. Such plants are natural hyperaccumulators with almost 100 times more accumulating capacity than non-accumulator plant species.

13.7 PHYTOSTABILIZATION

Phytostabilization refers to the decrease in the mobility of heavy metals in soil by decreasing the wind-mediated blown dust, and by reducing the solubility of metals in soil and/or leaching. Plants have the potential to reduce the accumulation of contaminants by their roots. Soil with a high amount of heavy metals becomes toxic and therefore the plant used for phytostabilization should have high root biomass with the capability to immobilize contaminants, and should be tolerant to metal toxicity. Phytostabilization is efficient in lowering the level of heavy metals and radionuclides. The method reduces the leaching of metals by changing their solubility (Table 13. 4). Phytostabilization is applicable only at sites where there is low contamination. Pajuelo et al. (2016) reported metal phytostabilization by the legume-rhizobium symbiosis. Using the genetic-engineering approach, both the symbiotic partners were made metal-tolerant. A similar study was also performed by Pérez-Palacios et al. (2017). They reported the genetic engineering of a double symbiotic system for Cu phytostabilization. The metallothionein gene mt4a from *Arabidopsis thaliana* was transferred to *Medicago truncatula* and the copper-resistant gene copAB from *Pseudomonas fluorescens* was transferred to the *Ensifer medicae* strain.

13.8 PHYTODEGRADATION (PHYTOTRANSFORMATION)

The breakdown of contaminants by plants is termed phytodegradation or phytotransformation. Plants take up the contaminants and break them down with the help of enzymes through the metabolic process (Table 13.5). Enzymes catalyse and accelerate the degradation of organic pollutants and convert them into simpler forms which further aid in plant growth. Trinitrotoluene (TNT), an explosive, has contaminated many sites due to the limited success of remediation techniques. However, many plant species were found able to break down TNT. The phytodegradation of TNT has major limitations as it affects the growth of the plants. One of the soil bacteria (*Entereo cloaca*) was found to utilize this explosive as its nitrogen source. Two enzymes (PETN

TABLE 13.5
Some Common Plants Used for Phytodegradation

Plants	Contaminants	Key references
Hybrid poplars (*Populus trichocarpa* × *Populus deltoides*)	Phytodegradation of trichloroethylene	(Newman et al. 1997)
Leucaena leucocephala	Phytodegradation of ethylene dibromide	(Doty et al. 2003)
Catharanthus roseus	Phytodegradation of RDX and HMX	(Bhadra et al. 2001)
Datura innoxia	TNT, HMX and Heavy Metals	(Elyse 1997; Lucero et al. 1999)
Lycopersicon peruvianum	TNT and Heavy Metals	(Lucero et al. 1999)
Blumea malcolmii	Malachite Green dye	(Kagalkar et al. 2011)
Erythrina crista-galli	petroleum	(De Farias et al. 2009)
Chlorella pyrenoidosa	pentachlorophenol	(Headley et al. 2008)

reductase and nitroreductase) were responsible for degrading the TNT into a harmless product. The gene responsible for producing these two enzymes were transferred into tobacco plants. This led to the formation of a transgenic tobacco plant to degrade TNT more efficiently without compromising its growth.

13.9 RHIZODEGRADATION

The degradation of contaminants in the rhizosphere by means of microorganisms is known as rhizodegradation. Several microorganisms, such as fungi, bacteria, and yeast, feed on these contaminants. In this process, harmful pollutants are broken down into less or non-toxic, harmless products by these microorganisms. Further, plants aid in this biodegradation by providing additional nutrients to the microorganisms (Table 13.6).

Transgenic plants have been developed to improve the plant microbial interaction. The natural ability of plants to secrete natural substances is being enhanced to stimulate the microbial activity.

13.10 PHYTOVOLATILIZATION

Plants are also able to absorb contaminants from the soil and release them into the atmosphere. This process of making an organic contaminant volatile by plants is termed phytovolatilization (Table 13.7). In the process of volatilization, contaminants are partitioned into the air spaces, followed by their release into the air. Many volatile organic compounds may be volatilized either directly (from stem or leaves) or indirectly (from the soil due to plant root activity). Although phytovolatilization is beneficial as it results in photochemical decay in the atmosphere, in urban areas it is viewed as a risk for degrading the air quality.

TABLE 13.6
Some Common Plants Used for Rhizodegradation

Soil contaminants	Plant species	Key references
Polychlorinated Biphenyls	Orange (*Citrus sp.*) Apple (*Pyrus sp.*) Mulberry (*Morus sp.*)	(Donnelly et al. 1994) (Gilbert and Crowley 1997) (Kvesitadze et al. 2006)
Perchlorate	*Salix nigra*	(Yifru and Nzengung 2008)
Atrazine	Orchard Grass (*Dactylis glomerata*) Smooth Bromegrass (*Bromus inermis*), Tall Fescue (*Festuca arundinacea*) Illinois Bundle Flower (*Desmanthus illinoensis*) Perennial Ryegrass (*Lolium perenne*) Switchgrass (*Panicum virgatum*) Eastern Gamagrass (*Tripsacum dactyloides*)	(Lin et al. 2011)
Phenanthrene and Pyrene	*Kandelia candel*	(Lu et al. 2011)
Petroleum Hydrocarbons	*Sesbania cannabina*	(Maqbool et al. 2012)

TABLE 13.7
Some Common Plants Used for Phytovolatilization

Plants	Contaminants	Key references
Rice, Rabbit foot grass, Azolla, Pickleweed	Se	(Bañuelos et al. 1997; Hansen et al. 1998; De Souza et al. 2000; Lin et al. 2000)
Parrot's feather (*Myriophyllum brasiliense*), iris-leaved rush (*Juncus xiphioides*) cattail (*Typha latifolia*) club-rush (*Scirpus robustus*)	Se	(Pilon-Smits et al. 1999)
Brassica juncea	Se	(Raskin et al. 1997; Bañuelos et al. 1998)

Many inorganic compounds, such as Se, As, and Hg, can be volatilized from plants. Direct phytovolatilization differs from transpiration as most of the compounds are moderately hydrophobic and thus found transformed by the plants. However, in indirect phytovolatilization, the flux of volatile contaminants is increased. Phytovolatilization is equally important for traditional as well as other compounds present in the plant roots and subsurface areas. Mercury is one such chemical that has been remedied using phytovolatilization and converted into less toxic elementary mercury.

13.11   PHYTOREMEDIATION POTENTIAL OF MEDICINAL PLANTS

Medicinal plants can accumulate a high concentration of toxic metals present in the soil (Masarovičová et al. 2010). However, the use of aromatic and medicinal plants is not well researched as prominent phytoremediation crops (Vamerali et al. 2010). The least significance has been given to aromatic and medicinal plants, while the use of non-aromatic crops for phytoremediation is not a viable option due to the accumulation of toxic metals in their edible parts followed by their entry into the food chain (Tables 13.8 and 13.9). Therefore, the major thrust is being given to plants producing non-edible aromatic oils as there are efficient checkpoints to prevent the metal translocation to essential oil (Scora and Chang 1997; Bernstein et al. 2009). The health of contaminated soil can be rejuvenated by the use of aromatic and medical plants (Lal et al. 2013; Pirzadah et al. 2019).

TABLE 13.8
Remediation of Possible Heavy Metals by Medicinal Plants Using Different Phytotechnology

Phyto-technologies	Plants	Heavy metals	Key references
Phytostabilizer	Vetiver	Cd, Pb, Cu, As, Hg, Cr, Ni, Sn, Zn	(Minh et al. 2011, Chen et al. 2012)
	Lemongrass	Cd, Hg, Pb, Cu, Ni, Cr	(Lal et al. 2013; Gautam et al. 2017)
	Palmarosa	Cr, Cd, Pb, Ni, Fe, Zn, Cu	(Pandey et al. 2015; 2019, Pirzadah et al. 2019)
	Citronella	Cd, Cr	(Boruah et al. 2000)
	Ocimum	Cr, Cd, Ni, Pb, As, Zn	(Prasad et al. 2011, Jisha et al. 2017, Dinu et al. 2020)
	Mint	Cr, Pb, Ni, Cd, Cu, Zn,	(Pandey et al. 2019; Dinu et al. 2021)
Hyper accumulator for certain heavy metals	Salvia	Cd, Pb, Cr	(Pandey et al. 2019)
Potential bio-monitor, Phytostabilizer and a hyperaccumulator	Rosemary	Ni. Hg, Pb, Cu, Zn, Cd, Ni, Fe, As, Sb	(Jisha et al. 2017; Berg and Borges 2020)
Facultative metallophytes	Chamomile	Cd, Zn, Cu, Pb, Ni	(Grejtovský and Pirč 2000; Chizzola et al. 2003; M Kummerová et al. 2008; Voyslavov et al. 2013)
Hyperaccumulator	Geranium	Cd, Ni, Pb, Zn	(Saxena et al. 1999; Mahdieh et al. 2013)
	Lavender	Cd, Pb, Cu, Zn, Fe	(Divrikli et al. 2006; Mashhoor Roodi et al. 2012; Angelova et al. 2015)

TABLE 13.9
Heavy Metal Contaminants and their Removal by Medicinal Plants

Heavy metal/ Contaminants	Plant species	Key references
Al	Mentha species	(Nabi et al. 2020)
As	Cymbopogon citratus	(Jisha et al. 2017)
	Ocimum tenuiflorum	(Siddiqui et al. 2013)
Cd	Catharanthus roseus	(Ahmad and Misra 2014)
	Cymbopogon citratus	(Jisha et al. 2017)
	Euphorbia hirta	(Hamzah et al. 2016)
	Hypericum perforatum	(Schneider and Marquard 1996)
	Hypericum sp.	(Tirillini et al. 2006)
	Lavandula vera	(Angelova et al. 2015)
	lemongrass	(Gautam et al. 2017)
	M. chamomilla	(Voyslavov et al. 2013)
	Matricaria chamomilla	(Stancheva et al. 2014)
	Mentha species	(Nabi et al. 2020)
	Ocimum basilicum	(Prasad et al. 2011; Dinu et al. 2020; Youssef 2021)
	Rosmarinus officinalis	(Jisha et al. 2017; Dinu et al. 2020)
	Vetiveria zizanioides	(Sinha et al. 2013)
Cr	Aloe vera	(Prasad et al. 2011)
	Catharanthus roseus	(Ahmad and Misra 2014)
	Cymbopogon flexuosus	(Gautam et al. 2017; Patra et al. 2018)
	Cymbopogon winterianus	(Boruah et al. 2000; Sulastri and Tampubolon 2019; Singh et al. 2020)
	Mentha species	(Zurayk et al. 2002; Prasad et al. 2010; Nabi et al. 2020)
	Ocimum basilicum	(Dinu et al. 2020)
	Ocimum tenuiflorum	(Sundaramoorthy et al. 2010; Chauhan and Vimala 2014; Fahimirad and Hatami 2017)
Cu	Cymbopogon citratus	(Jisha et al. 2017)
	Mentha arvensis	(Manikandan et al. 2015)
	Vetiveria zizanioides	(Chen et al. 2012)
Hg	Chrysopogon zizanioides	(Danh et al. 2009; Lomonte et al. 2014)
Ni	Catharanthus roseus	(Dar et al. 2020)
	Cymbopogon citratus	(Kumar and Maiti 2015; Gautam et al. 2017)
	Mentha species	(Zurayk et al. 2002; Nabi et al. 2020)
	Ocimum basilicum	(Dinu et al. 2020; Youssef 2021)
	Vetiveria zizanioides	(Chen et al. 2012)
Pb	Catharanthus roseus	(Ahmad and Misra 2014)
	Cymbopogon citratus	(Mans et al. 2004; Sinha et al. 2013)
	Lavandula vera	(Angelova et al. 2015)
	lemongrass	(Gautam et al. 2017)
	M. chamomilla	(Voyslavov et al. 2013)
	Matricaria chamomilla	(Stancheva et al. 2014)
	Mentha crispa	(Sá et al. 2015)
	Ocimum basilicum	(Dinu et al. 2020; Youssef 2021)

(continued)

TABLE 13.9 (Continued)
Heavy Metal Contaminants and their Removal by Medicinal Plants

Heavy metal/ Contaminants	Plant species	Key references
	Pelargonium sps.	(Abdullah and Sarem 2010)
	Rosmarinus officinalis	(Gaida et al. 2013; Ardalan et al. 2014)
	Vetiveria zizanioides	(Chen et al. 2012)
Sn	*Vetiveria zizanioides*	(Chen et al. 2012; Sulastri and Tampubolon 2019)
Zn	*Cymbopogon citratus*	(Kumar and Maiti 2015; Gautam et al. 2017)
	Lavandula vera	(Pandey et al. 2018)
	lemongrass	(Gautam et al. 2017; Patra et al. 2018)
	Matricaria chamomilla	(Stancheva et al. 2014)
	Mentha arvensis	(Zaid et al. 2020)
	Ocimum basilicum	(Siddiqui et al. 2013; Youssef 2021)
	Vetiveria zizanioides	(Chen et al. 2012; Sinha et al. 2013)
Radioactive waste	*Euphorbia hirta*	(Barthwal et al. 2008; Chakravarty et al. 2017)

Hypericum perforatum and *Matricaria recutita* were reported to take up a high level of Cd present in the soil (Marquard and Schneider 1998). Chamomile was researched by Chizzola (1998) and reported as a Cd accumulator. Chizzola and Mitteregger (2005) also found a similar Cd accumulation result with *M. recutita*. Specific metabolites, viz., hypericin, apigenin, and quercetin, produced by medicinal plants were assumed to help in Cd sequestration (Masarovičová et al. 2004). Metal contaminants and mineral nutrient interaction are also important in governing metal sequestration by plants (Fox and Guerinot 1998). With the increasing concentration of nutrients, metal contaminants (Cu) accumulation was found to decrease in *Brassica pekinesis* when cultivated in a hydroponic system (Xiong et al. 2002). Similar inter-action studies were also performed in a hydroponic system (*H. perforatum, M. recutita,* and *Salvia officinalis*) and soil (*M. recutita*) by Kráľová and Masarovičová (2004) and Grejtovský and Pirč (2000) respectively.

13.12 CONCLUSION

Plant-based control of pollution is a cost-effective and sustainable method compared to other chemical-based methods and technology. Plants are capable of taking up contaminants (phytoextraction), transforming them (phytodegradation), emitting (phytovolatilization), and/or stabilizing them (phytostabilization). Further, the poten-tial of plants is being enhanced through genetic engineering, phytohormones, and nanoparticles and by enhancing plant–microbial interaction. The search for more effi-cient and useful plants and new genes, along with more efficient and advanced technolo-gies, has widened our understanding of plant-mediated remediation of contaminants. In future, this research will yield hyperaccumulator plants to fight pollution.

REFERENCES

Abdullah, W.S., and S.M. Sarem. 2010. The potential of chrysanthemum and pelargonium for phytoextraction of lead-contaminated soils. *Jordan Journal of Civil Engineering* 4: 409–416.

Adler, T. 1996. Botanical Cleanup Crews. *Science News* 150: 42.

Ahmad, R., and N. Misra. 2014. Evaluation of phytoremediation potential of *Catharanthus roseus* with respect to chromium contamination. *American Journal of Plant Sciences* 5: 2378–2388.

Altaf, R., S. Altaf, M. Hussain, R.U. Shah, R. Ullah, M.I. Ullah, A. Rauf, M.J. Ansari, S.A. Alharbi, S. Alfarraj, and R. Datta. 2021. Heavy metal accumulation by roadside vegetation and implications for pollution control. *PLoS ONE* 16: e0249147.

Álvarez, S.P., E.F.H. Ardisana, and R.P. Leal. 2020. Plant Biotechnology for Agricultural Sustainability. In Resources Use Efficiency in Agriculture, pp. 389–425. Singapore: Springer Singapore.

Angelevska, B., V. Atanasova, and I. Andreevski. 2021. Urban air quality guidance based on measures categorization in road transport. *Civil Engineering Journal (Iran)* 7: 253–267.

Angelova, V.R., D.F. Grekov, V.K. Kisyov, and K.I. Ivanov. 2015. Potential of lavender (*Lavandula vera* L.) for phytoremediation of soils contaminated with heavy metals. *International Journal of Biological, Food, Veterinary and Agricultural Engineering* 9: 465–472.

Ansari, M.K.A., A. Ahmad, S. Umar, M. Iqbal, M.H. Zia, A. Husen, and G. Owens. 2021. Suitability of Indian mustard genotypes for phytoremediation of mercury-contaminated sites. *South African Journal of Botany* 142: 12–18.

Ardalan, F., M. Vakili, and S. Kourepaz. 2014. Lead phytoremediation of *Rosmarinus officinalis* and its effect on the plant growth. *International Journal of Biosciences* 4: 75–79.

Awa, S.H., and T. Hadibarata. 2020. Removal of heavy metals in contaminated soil by phytoremediation mechanism: A review. *Water, Air, and Soil Pollution* 231: 47.

Baker, A.J.M., and R.R. Brooks. 1989. Terrestrial higher plants which hyperaccumulate metallic elements – A review of their distribution, ecology and phytochemistry. *Biorecovery* 1: 81–126.

Baldwin, P.R., and D.J. Butcher. 2007. Phytoremediation of arsenic by two hyperaccumulators in a hydroponic environment. *Microchemical Journal* 85: 297–300.

Bani, A., D. Pavlova, G. Echevarria, A. Mullaj, R.D. Reeves, J.L. Morel, and S. Sulçe. 2010. Nickel hyperaccumulation by the species of Alyssum and Thlaspi (Brassicaceae) from the ultramafic soils of the Balkans. *Botanica Serbica* 34: 3–14.

Bañuelos, G.S., H.A. Ajwa, N. Terry, and A. Zayed. 1997. Phytoremediation of selenium laden soils: A new technology. *Journal of Soil and Water Conservation* 52: 426–430.

Bañuelos, G.S., H.A. Ajwa, L. Wu, and S. Zambrzuski. 1998. Selenium accumulation by *Brassica napus* grown in Se-Laden soil from different depths of Kesterson Reservoir. *Soil and Sediment Contamination* 7: 481–496.

Barthwal, J., S. Nair, and P. Kakkar. 2008. Heavy metal accumulation in medicinal plants collected from environmentally different sites. *Biomedical and Environmental Sciences* 21: 319–324.

Belhaj, K., A. Chaparro-Garcia, S. Kamoun, N.J. Patron, and V. Nekrasov. 2015. Editing plant genomes with CRISPR/Cas9. *Current Opinion in Biotechnology* 32: 76–84.

Berg, E.C., and A.C. Borges. 2020. Use of plants in the remediation of arsenic-contaminated waters. *Water Environment Research* 92: 1669–1676.

Bernstein, N., D. Chaimovitch, and N. Dudai. 2009. Effect of irrigation with secondary treated effluent on essential oil, antioxidant activity, and phenolic compounds in oregano and rosemary. *Agronomy Journal* 101: 1–10.

Bhadra, R., D.G. Wayment, R.K. Williams, S.N. Barman, M.B. Stone, J.B. Hughes, and J.V. Shanks. 2001. Studies on plant-mediated fate of the explosives RDX and HMX. *Chemosphere* 44: 1259–1264.

Bhargava, A., F.F. Carmona, M. Bhargava, and S. Srivastava. 2012. Approaches for enhanced phytoextraction of heavy metals. *Journal of Environmental Management* 105: 103–120.

Bhuiyan, M.S.U., S.R. Min, W.J. Jeong, S. Sultana, K.S. Choi, W.Y. Song, Y. Lee, Y.P. Lim, and J.R. Liu. 2011. Overexpression of a yeast cadmium factor 1 (YCF1) enhances heavy metal tolerance and accumulation in *Brassica juncea*. *Plant Cell, Tissue and Organ Culture* 105: 85–91.

Bittsánszky, A., T. Kömives, G. Gullner, G. Gyulai, J. Kiss, L. Heszky, L. Radimszky, and H. Rennenberg. 2005. Ability of transgenic poplars with elevated glutathione content to tolerate zinc(2+) stress. *Environment International* 31: 251–254.

Borji, H., G.M. Ayoub, M. Al-Hindi, L. Malaeb, and H.Z. Hamdan. 2020. Nanotechnology to remove polychlorinated biphenyls and polycyclic aromatic hydrocarbons from water: A review. *Environmental Chemistry Letters* 18: 729–746.

Bortesi, L., and R. Fischer. 2015. The CRISPR/Cas9 system for plant genome editing and beyond. *Biotechnology Advances* 33: 41–52.

Boruah, H.P.D., A.K. Handique, and G.C. Borah. 2000. Response of Java citronella (*Cymbopogon winterianus* Jowitt) to toxic heavy metal cadmium. *Indian Journal of Experimental Biology* 38: 1267–1269.

Boukhatem, M.N., and W.N. Setzer. 2020. Aromatic herbs, medicinal plant-derived essential oils, and phytochemical extracts as potential therapies for coronaviruses: Future perspectives. *Plants* 9: 1–23.

Chakravarty, P., K. Bauddh, and M. Kumar. 2017. Phytoremediation: A Multidimensional and Ecologically Viable Practice for the Cleanup of Environmental Contaminants. In Phytoremediation Potential of Bioenergy Plants, pp. 1–46. Springer Singapore.

Chakre, O.J. 2006. Choice of eco-friendly trees in urban environment to mitigate airborne particulate pollution. *Journal of Human Ecology* 20: 135–138.

Chandra, R. and S. Yadav. 2010. Potential of *Typha angustifolia* for phytoremediation of heavy metals from aqueous solution of phenol and melanoidin. *Ecological Engineering* 36: 1277–1284.

Chaney, R.L., M. Malik, Y.M. Li, S.L. Brown, E.P. Brewer, J.S. Angle, and A.J.M. Baker. 1997. Phytoremediation of soil metals. *Current Opinion in Biotechnology* 8: 279–284.

Chattopadhyay, S., R.L. Fimmen, B.J. Yates, V. Lal, and P. Randall. 2012. Phytoremediation of mercury- and methyl mercury-contaminated sediments by water hyacinth (*Eichhornia crassipes*). *International Journal of Phytoremediation* 14: 142–161.

Chauhan, P., and Y. Vimala. 2014. Biochemical screening of sugar mill effluent-contained heavy metal accumulation and tolerance in *Ocimum tenuiflorum* L. *Progressive Agriculture* 14: 69–75.

Chehregani Rad, A., and B. Malayeri. 2007. Removal of heavy metals by native accumulator plants. *International Journal Of Agriculture & Biology* 9: 462–465.

Chen, K.F., T.Y. Yeh, and C.F. Lin. 2012. Phytoextraction of Cu, Zn, and Pb enhanced by chelators with Vetiver (*Vetiveria zizanioides*): Hydroponic and pot experiments. *International Scholarly Research Notices* 2012: 729693.

Cherian, S., and M.M. Oliveira. 2005. Transgenic plants in phytoremediation: Recent advances and new possibilities. *Environmental Science and Technology* 39: 9377–9390.

Chizzola, R. 1998. Aufnahme und verteilung von cadmium in sonnenblume, kamille und johanniskraut. *Z Arznei- und Gewürzpfl* 3: 91–95.

Chizzola, R., H. Michitsch, and C. Franz. 2003. Monitoring of metallic micronutrients and heavy metals in herbs, spices and medicinal plants from Austria. *European Food Research and Technology* 216: 407–411.

Chizzola, R., and U.S. Mitteregger. 2005. Cadmium and zinc interactions in trace element accumulation in chamomile. *Journal of Plant Nutrition* 28: 1383–1396.

Cosio, C., E. Martinoia, and C. Keller. 2004. Hyperaccumulation of Cadmium and Zinc in *Thlaspi caerulescens* and *Arabidopsis halleri* at the leaf cellular level. *Plant Physiology* 134: 716–725.

Damaiyani, J., Y.A. Purwestri, and I. Sumardi. 2018. The orbicules and allergenic protein of African tulip tree (*Spathodea campanulata* P.Beauv.): A roadside ornamental plant in Malang, Indonesia. *Berkala Penelitian Hayati* 24: 43–46.

Danh, L.T., P. Truong, R. Mammucari, T. Tran, and N. Foster. 2009. Vetiver grass, *Vetiveria zizanioides*: A choice plant for phytoremediation of heavy metals and organic wastes. *International Journal of Phytoremediation* 11: 664–691.

Dar, F.A., T.B. Pirzadah, and B. Malik. 2020. Accumulation of Heavy Metals in Medicinal and Aromatic Plants. In Plant Micronutrients, pp. 113–127. Springer International Publishing.

Das, S., and P. Prasad. 2012. Particulate matter capturing ability of some plant species: Implication for phytoremediation of particulate pollution around Rourkela Steel Plant, Rourkela, *India Nature Environment and Pollution Technology* 11: 657–665.

Dinu, C., S. Gheorghe, A.G. Tenea, C. Stoica, G.G. Vasile, R.L. Popescu, E.A. Serban, and L.F. Pascu. 2021. Toxic metals (As, Cd, ni, Pb) impact in the most common medicinal plant (*Mentha piperita*). *International Journal of Environmental Research and Public Health* 18: 3904.

Dinu, C., G.G. Vasile, M. Buleandra, D.E. Popa, S. Gheorghe, and E.M. Ungureanu. 2020. Translocation and accumulation of heavy metals in *Ocimum basilicum* L. plants grown in a mining-contaminated soil. *Journal of Soils and Sediments* 20: 2141–2154.

Divrikli, U., N. Horzum, M. Soylak, and L. Elci. 2006. Trace heavy metal contents of some spices and herbal plants from western Anatolia, Turkey. *International Journal of Food Science and Technology* 41: 712–716.

Donnelly, P.K., R.S. Hegde, and J.S. Fletcher. 1994. Growth of PCB-degrading bacteria on compounds from photosynthetic plants. *Chemosphere* 28: 981–988.

Doty, S.L., T. Q. Shang, A.M. Wilson, A.L. Moore, L.A. Newman, S.E. Strand, and M.P. Gordon. 2003. Metabolism of the soil and groundwater contaminants, ethylene dibromide and trichloroethylene, by the tropical leguminous tree, *Leuceana leucocephala*. *Water Research* 37: 441–449.

Dushenkov, S. 2003. Trends in phytoremediation of radionuclides. *Plant and Soil* 249: 167–175.

Dushenkov, S., Y. Kapulnik, M. Blaylock, B. Sorochisky, I. Raskin, and B. Ensley. 1997. Phytoremediation: A novel approach to an old problem. *Studies in Environmental Science* 66: 563–572.

Dushenkov, V., P.B.A. Nanda Kumar, H. Motto, and I. Raskin. 1995. Rhizofiltration: The use of plants To remove heavy metals from aqueous streams. *Environmental Science and Technology* 29: 1239–1245.

Duvign, S.M.P.A., M. Faucon, M. Ngongo, P. Meerts, and N. Verbruggen. 2010. Copper tolerance in the cuprophyte *Haumaniastrum katangense*. *Plant and Soil* 328: 235–244.

Eapen, S., K.N. Suseelan, S. Tivarekar, S.A. Kotwal, and R. Mitra. 2003. Potential for rhizofiltration of uranium using hairy root cultures of *Brassica juncea* and *Chenopodium amaranticolor*. *Environmental Research* 91 : 127–133.

Elyse, M. 1997. Biotransformation of high explosives by the solanaceous plant *Datura innoxia*. New Mexico State University.

Fahimirad, S., and M. Hatami. 2017. Heavy Metal-Mediated Changes in Growth and Phytochemicals of Edible and Medicinal Plants. In Medicinal Plants and Environmental Challenges, pp. 189–214. Springer International Publishing.

De Farias, V., L.T. Maranho, E.C. De Vasconcelos, M.A. Da Silva Carvalho Filho, L.G. Lacerda, J.A.M. Azevedo, A. Pandey, and C.R. Soccol. 2009. Phytodegradation potential of *Erythrina crista-galli* L., Fabaceae, in petroleum-contaminated soil. *Applied Biochemistry and Biotechnology* 157: 10–22.

Favas, P.J.C., J. Pratas, and M.N.V. Prasad. 2012. Accumulation of arsenic by aquatic plants in large-scale field conditions: Opportunities for phytoremediation and bioindication. *Science of the Total Environment* 433: 390–397.

Ferraz, P., F. Fidalgo, A. Almeida, and J. Teixeira. 2012. Phytostabilization of nickel by the zinc and cadmium hyperaccumulator *Solanum nigrum* L. Are metallothioneins involved? *Plant Physiology and Biochemistry* 57: 254–260.

Fox, T.C., and M. Lou Guerinot. 1998. Molecular biology of cation transport in plants. *Annual Review of Plant Biology* 49: 669–696.

Gaida, M.M., N.R. Landoulsi, M.N. Rejeb, and S. Smiti. 2013. Growth and photosynthesis responses of *Rosmarinus officinalis* L. to heavy metals at Bougrine mine. *African Journal of Biotechnology* 12: 150–161.

Gardea-Torresdey, J.L., J.H. Gonzalez, K.J. Tiemann, O. Rodriguez, and G. Gamez. 1998. Phytofiltration of hazardous cadmium, chromium, lead and zinc ions by biomass of *Medicago sativa* (Alfalfa). *Journal of Hazardous Materials* 57: 29–39.

Gautam, M., D. Pandey, and M. Agrawal. 2017. Phytoremediation of metals using lemongrass (*Cymbopogon citratus* (D.C.) Stapf.) grown under different levels of red mud in soil amended with biowastes. *International Journal of Phytoremediation* 19 : 555–562.

Gilbert, E.S., and D.E. Crowley. 1997. Plant compounds that induce polychlorinated biphenyl biodegradation by *Arthrobacter sp.* Strain B1B. *Applied and Environmental Microbiology* 63: 1933–1938.

Grejtovský, A., and R. Pirč. 2000. Effect of high cadmium concentrations in soil on growth, uptake of nutrients and some heavy metals of *Chamomilla recutita* (L.) Rauschert. *Journal of Applied Botany* 74: 172–174.

Iqbal, M., A. Ahmad, M.K.A. Ansari, M.I. Qureshi, I.M. Aref, P.R. Khan, S.S. Hegazy, H. El-Atta, A. Husen, and K.R. Hakeem. 2015. Improving the phytoextraction capacity of plants to scavenge metal(loid)-contaminated sites. *Environmental Reviews* 23: 44–65.

Hamzah, A., R.I. Hapsari, and E.I. Wisnubroto. 2016. Phytoremediation of cadmium-contaminated agricultural land using indigenous plants. *International Journal of Environmental & Agriculture Research* 2: 8–14.

Han, D., H. Shen, W. Duan, and L. Chen. 2020. A review on particulate matter removal capacity by urban forests at different scales. *Urban Forestry and Urban Greening* 48: 126565.

Hansen, D., P.J. Duda, A. Zayed, and N. Terry. 1998. Selenium removal by constructed wetlands: Role of biological volatilization. *Environmental Science and Technology* 32: 591–597.

Headley, J.V., K.M. Peru, J.L. Du, N. Gurprasad, and D.W. McMartin. 2008. Evaluation of the apparent phytodegradation of pentachlorophenol by *Chlorella pyrenoidosa*. *Journal of Environmental Science and Health – Part A Toxic/Hazardous Substances and Environmental Engineering* 43: 361–364.

Jadhav, C.A., D.N. Vikhe, and R.S. Jadhav. 2020. Global and domestic market of herbal medicines: A review. *Research Journal of Science and Technology* 12 : 327–330.

Jadia, C.D. and M.H. Fulekar. 2008. Phytoremediation: The application of vermicompost to remove zinc, cadmium, copper, nickel and lead by sunflower plant. *Environmental Engineering and Management Journal* 7: 547–558.

Jadia, C.D., and M.H. Fulekar. 2009. Phytoremediation of heavy metals: Recent techniques. *African Journal of Biotechnology* 8: 921–928.

Jisha, C.K., K. Bauddh, and S.K. Shukla. 2017. Phytoremediation and Bioenergy Production Efficiency of Medicinal and Aromatic Plants. In Phytoremediation Potential of Bioenergy Plants, pp. 287–304. Springer Singapore.

Jordanovska, S., Z. Jovovic, and V. Andjelkovic. 2020. Potential of Wild Species in the Scenario of Climate Change. In Rediscovery of Genetic and Genomic Resources for Future Food Security, pp. 263–301. Springer Singapore.

Joshi, N., A. Joshi, and B. Bist. 2020. Phytomonitoring and Mitigation of Air Pollution by Plants. In Sustainable Agriculture in the Era of Climate Change, pp. 113–142. Springer International Publishing.

Kráľová, K., and E. Masarovičová. 2004. Nutrient effect on metal accumulation by some medicinal plants. *Chem Listy* 98: 710.

Kagalkar, A.N., M.U. Jadhav, V.A. Bapat, and S.P. Govindwar. 2011. Phytodegradation of the triphenylmethane dye Malachite Green mediated by cell suspension cultures of *Blumea malcolmii* Hook. *Bioresource Technology* 102: 10312–10318.

Kanwar, V.S., A. Sharma, A.L. Srivastav, and L. Rani. 2020. Phytoremediation of toxic metals present in soil and water environment: a critical review. *Environmental Science and Pollution Research* 27: 44835–44860.

Karkhanis, M., C.D. Jadia, and M.H. Fulekar. 2005. Rhizofiltration of metals from coal ash leachate. *Asian Journal of Water, Environment and Pollution* 3: 91–94.

Khellaf, N., and M. Zerdaoui. 2010. Growth response of the duckweed *Lemna gibba* L. to copper and nickel phytoaccumulation. *Ecotoxicology* 19: 1363–1368.

Koprivova, A., S. Kopriva, D. Jäger, B. Will, L. Jouanin, and H. Rennenberg. 2002. Evaluation of transgenic poplars over-expressing enzymes of glutathione synthesis for phytoremediation of cadmium. *Plant Biology* 4: 664–670.

Kshirsagar, S., and N.C. Aery. 2007. Phytostabilization of mine waste: Growth and physiological responses of *Vigna unguiculata* (L.) Walp. *Journal of Environmental Biology* 28: 651–654.

Kumar, A., and S.K. Maiti. 2015. Effect of organic manures on the growth of *Cymbopogon citratus* and *Chrysopogon zizanioides* for the phytoremediation of chromite-asbestos mine waste: A pot scale experiment. *International Journal of Phytoremediation* 17: 437–447.

Kumar, K., S.C. Gupta, S.K. Baidoo, Y. Chander, and C.J. Rosen. 2005. Antibiotic uptake by plants from soil fertilized with animal manure. *Journal of Environmental Quality* 34: 2082–2085.

Kundu, D., D. Dutta, S. Mondal, S. Haque, J.N. Bhakta, and B.B. Jana. 2017. Application of Potential Biological Agents in Green Bioremediation Technology: Case Studies. In Handbook of Research on Inventive Bioremediation Techniques, pp. 300–323.

Kvesitadze, G., G. Khatisashvili, T. Sadunishvili, and J.J. Ramsden. 2006. Biochemical Mechanisms of Detoxification in Higher Plants: Basis of Phytoremediation. Berlin: Springer.

Lahn, B. 2021. Changing climate change: The carbon budget and the modifying-work of the IPCC. *Social Studies of Science* 51 : 3–27.

Lal, K., R.K. Yadav, R. Kaur, D.S. Bundela, M.I. Khan, M. Chaudhary, R.L. Meena, S.R. Dar, and G. Singh. 2013. Productivity, essential oil yield, and heavy metal accumulation in lemon grass (*Cymbopogon flexuosus*) under varied wastewater-groundwater irrigation regimes. *Industrial Crops and Products* 45: 270–278.

Lasat, M.M. 2002. Phytoextraction of toxic metals. *Journal of Environment Quality* 31: 109–120.

Li, Y.M., R. Chaney, E. Brewer, R. Roseberg, J.S. Angle, A. Baker, R. Reeves, and J. Nelkin. 2003. Development of a technology for commercial phytoextraction of nickel: Economic and technical considerations. *Plant and Soil* 249 : 107–115.

Li, Z.S., M. Szczypka, Y.P. Lu, D.J. Thiele, and P.A. Rea. 1996. The yeast cadmium factor protein (YCF1) is a vacuolar glutathione S-conjugate pump. *Journal of Biological Chemistry* 271: 6509–6517.

Lin, C.-H., R.N. Lerch, R.J. Kremer, and H.E. Garrett. 2011. Stimulated rhizodegradation of atrazine by selected plant species. *Journal of Environmental Quality* 40: 1113–1121.

Lin, Z.-Q., R.S. Schemenauer, V. Cervinka, A. Zayed, A. Lee, and N. Terry. 2000. Selenium volatilization from a soil-plant system for the remediation of contaminated water and soil in the San Joaquin Valley. *Journal of Environmental Quality* 29: 1048–1056.

Liu, L., D. Guan, and M.R. Peart. 2012. The morphological structure of leaves and the dust-retaining capability of afforested plants in urban Guangzhou, South China. *Environmental Science and Pollution Research* 19: 3440–3449.

Lombi, E., F.J. Zhao, S.J. Dunham, and S.P. McGrath. 2001. Phytoremediation of heavy metal-contaminated soils: Natural hyperaccumulation versus chemically enhanced phytoextraction. *Journal of Environmental Quality* 30: 1919–1926.

Lomonte, C., Y. Wang, A. Doronila, D. Gregory, A.J.M. Baker, R. Siegele, and S.D. Kolev. 2014. Study of the spatial distribution of mercury in roots of vetiver grass (*Chrysopogon zizanioides*) by micro-pixe spectrometry. *International Journal of Phytoremediation* 16: 1170–1182.

Lu, H., Y. Zhang, B. Liu, J. Liu, J. Ye, and C. Yan. 2011. Rhizodegradation gradients of phenanthrene and pyrene in sediment of mangrove (*Kandelia candel* (L.) Druce). *Journal of Hazardous Materials* 196: 263–269.

Lucero, M.E., W. Mueller, J. Hubstenberger, G.C. Phillips, and M.A. O'Connell. 1999. Tolerance to nitrogenous explosives and metabolism of TNT by cell suspensions of *Datura innoxia*. *In Vitro Cellular and Developmental Biology – Plant* 35: 480–486.

Kummerová, M., D. Baráková, E. Masarovičová, and K. Kráľová. 2008. Chamomile Kusensitivity to Cd and Zn application. *Chemické listy* 102: 229.

Madejón, P., T. Marañón, J.M. Murillo, and B. Robinson. 2006. In defence of plants as biomonitors of soil quality. *Environmental Pollution* 143: 1–3.

Magalhães, M.O.L., N.M.B. do Amaral Sobrinho, F.S. dos Santos, and N. Mazur. 2011. Potential of two species of eucalyptus in the phytostabilization of a soil contaminated with zinc. *Revista Ciência Agronômica* 42: 805–812.

Mahdieh, M., M. Yazdani, and S. Mahdieh. 2013. The high potential of *Pelargonium roseum* plant for phytoremediation of heavy metals. *Environmental Monitoring and Assessment* 185: 7877–7881.

Manikandan, R., S.V. Sahi, and P. Venkatachalam. 2015. Impact assessment of mercury accumulation and biochemical and molecular response of *Mentha arvensis*: A potential hyperaccumulator plant. *Scientific World Journal* 2015: 10.

Mans, D.R.A., J.R. Toelsie, Z. Jagernath, K. Ramjiawan, A. Van Brussel, N. Jhanjan, S. Orie, M. Muringen, U. Elliot, S. Jurgens, R. Macnack, F. Rigters, S. Mohan, V. Chigharoe, S. Illes, and R. Bipat. 2004. Assessment of eight popularly used plant-derived preparations for their spasmolytic potential using the isolated guinea pig ileum. *Pharmaceutical Biology* 42: 422–429.

Maqbool, F., Z. Wang, Y. Xu, J. Zhao, D. Gao, Y.G. Zhao, Z.A. Bhatti, and B. Xing. 2012. Rhizodegradation of petroleum hydrocarbons by *Sesbania cannabina* in bioaugmented soil with free and immobilized consortium. *Journal of Hazardous Materials* 237–238: 262–269.

Marquard, R., and M. Schneider. 1998. Zur Cadmiumproblematik im Arzneipflanzenanbau. In Fachtagung, Arznei- und Gewürzpflanzen, pp. 9–15. Gießen: Justus-Liebeg-Universität.

Masarovičová, E., K. Kralová, M. Kummerová, and E. Kmentova. 2004. The effect of cadmium on root growth and respiration rate of two medicinal plant species. *Biologia* 59 : 211–214.

Masarovičová, E., K. Kráľová, and M. Kummerová. 2010. Principles of classification of medicinal plants as hyperaccumulators or excluders. *Acta Physiologiae Plantarum* 32: 823–829.

Mashhoor Roodi, M., M.A.B.M. Said, and H. Honari. 2012. Phytoremediation using the influence of aromatic crop on heavy-metal polluted soil, a review. *Advances in Environmental Biology* 6: 2663–2668.

McGrath, S.P., and F.J. Zhao. 2003. Phytoextraction of metals and metalloids from contaminated soils. *Current Opinion in Biotechnology* 14: 277–282.

Mehta, P., K. Bisht, and K.C. Sekar. 2020. Diversity of threatened medicinal plants of Indian Himalayan region. *Plant Biosystems* 2: 1–2.

Mikami, M., S. Toki, and M. Endo. 2015. Comparison of CRISPR/Cas9 expression constructs for efficient targeted mutagenesis in rice. *Plant Molecular Biology* 88: 561–572.

Minh, V. Van, N. Van Khanh, and L. Van Khoa. 2011. Potential of using vetiver grass to remediate soil contaminated with heavy metals. *VNU Journal of Science, Earth Sciences* 27: 146–150.

Mosa, K.A., I. Saadoun, K. Kumar, M. Helmy, and O.P. Dhankher. 2016. Potential biotechnological strategies for the cleanup of heavy metals and metalloids. *Frontiers in Plant Science* 7: 303.

Nabi, A., M. Naeem, T. Aftab, and M.M.A. Khan. 2020. Intimidating Effects of Heavy Metals on Mentha Species and their Mitigation Using Scientific Approaches. In Contaminants in Agriculture: Sources, Impacts and Management, pp. 305–325. Springer International Publishing.

Newman, L.A., S.E. Strand, N. Choe, J. Duffy, G. Ekuan, M. Ruszaj, B.B. Shurtleff, J. Wilmoth, P. Heilman, and M.P. Gordon. 1997. Uptake and biotransformation of trichloroethylene by hybrid poplars. *Environmental Science and Technology* 31: 1062–1067.

Pajuelo, E., P. Pérez-Palacios, A. Romero-Aguilar, J. Delgadillo, B. Doukkali, I.D. Rodríguez-Llorente, and M.A. Caviedes. 2016. Improving Legume-Rhizobium Symbiosis for Copper Phytostabilization through Genetic Manipulation of Both Symbionts. In Biological Nitrogen Fixation and Beneficial Plant-Microbe Interaction, pp. 183–193. Springer International Publishing.

Pandey, J., S. Chand, S. Pandey, Rajkumari, and D.D. Patra. 2015. Palmarosa [*Cymbopogon martinii* (Roxb.) Wats.] as a putative crop for phytoremediation, in tannery sludge polluted soil. *Ecotoxicology and Environmental Safety* 122: 296–302.

Pandey, J., R.K. Verma, and S. Singh. 2019. Suitability of aromatic plants for phytoremediation of heavy metal contaminated areas: A review. *International Journal of Phytoremediation* 21: 405–418.

Pandey, V.C., A. Rai, and J. Korstad. 2018. Aromatic Crops in Phytoremediation: From Contaminated to Waste Dumpsites. In Phytomanagement of Polluted Sites: Market Opportunities in Sustainable Phytoremediation, pp. 255–275. Elsevier.

Patra, D.K., C. Pradhan, and H.K. Patra. 2018. Chelate based phytoremediation study for attenuation of chromium toxicity stress using lemongrass: *Cymbopogon flexuosus* (nees ex steud.) W. Watson. *International Journal of Phytoremediation* 20: 1324–1329.

Peer, W.A., I.R. Baxter, E.L. Richards, J.L. Freeman, and A.S. Murphy. 2006. Phytoremediation and Hyperaccumulator Alants. In M. Tamas and E. Martinoia (eds). Topics in Current Genetics, pp. 299–340. Berlin: Springer.

Pérez-Palacios, P., A. Romero-Aguilar, J. Delgadillo, B. Doukkali, M.A. Caviedes, I.D. Rodríguez-Llorente, and E. Pajuelo. 2017. Double genetically modified symbiotic system for improved Cu phytostabilization in legume roots. *Environmental Science and Pollution Research* 24: 14910–14923.

Pierzynski, G.M., J.L. Schnoor, M.K. Schnoor, M.K. Banks, J.C. Tracy, L.A. Licht, and L.E. Erickson. 1994. Vegetative Remediation at Superfund Sites. In Mining and its environmental impact, pp. 49–70. Royal Society of Chemistry.

Pilon-Smits, E.A.H., M.P. Souza, G. Hong, A. Amini, R.C. Bravo, S.T. Payabyab, and N. Terry. 1999. Selenium volatilization and accumulation by twenty aquatic plant species. *Journal of Environmental Quality* 28: 1011–1018.

Pirzadah, T.B., B. Malik, and F.A. Dar. 2019. Phytoremediation potential of aromatic and medicinal plants: A way forward for green economy. *Journal of Stress Physiology & Biochemistry* 15: 62–75.

Prasad, A., S. Kumar, A. Khaliq, and A. Pandey. 2011. Heavy metals and arbuscular mycorrhizal (AM) fungi can alter the yield and chemical composition of volatile oil of sweet basil (*Ocimum basilicum* L.). *Biology and Fertility of Soils* 47: 853–861.

Prasad, A., A.K. Singh, S. Chand, C.S. Chanotiya, and D.D. Patra. 2010. Effect of chromium and lead on yield, chemical composition of essential oil, and accumulation of heavy metals of mint species. *Communications in Soil Science and Plant Analysis* 41: 2170–2186.

Prasad, M.N.V. and H.M. De Oliveira Freitas. 2003. Metal hyperaccumulation in plants – Biodiversity prospecting for phytoremediation technology. *Electronic Journal of Biotechnology* 6: 110–146.

Pratas, J., P.J.C. Favas, C. Paulo, N. Rodrigues, and M.N.V. Prasad. 2012. Uranium accumulation by aquatic plants from uranium-contaminated water in Central Portugal. *International Journal of Phytoremediation* 14: 221–234.

Rai, P.K. 2008. Phytoremediation of Hg and Cd from industrial effluents using an aquatic free floating macrophyte Azolla pinnata. *International Journal of Phytoremediation* 10: 430–439.

Rajakaruna, N., and B.A. Bohm. 2002. Serpentine and its vegetation: A preliminary study from Sri Lanka. *Journal of Applied Botany* 76: 20–28.

Ram, S.S., S. Majumder, P. Chaudhuri, S. Chanda, S.C. Santra, P.K. Maiti, M. Sudarshan, and A. Chakraborty. 2014. Plant canopies: Bio-monitor and trap for re-suspended dust particulates contaminated with heavy metals. *Mitigation and Adaptation Strategies for Global Change* 19: 499–508.

Raskin, I., and B.D. Ensley (eds). 1999. Phytoremediation of Toxic Metals: Using Plants to Clean Up the Environment. New York, NY: John Wiley & Sons.

Raskin, I., R.D. Smith, and D.E. Salt. 1997. Phytoremediation of metals: Using plants to remove pollutants from the environment. *Current Opinion in Biotechnology* 8: 221–226.

Reeves, R.D., and R.R. Brooks. 1983. European species of *Thlaspi* L. (Cruciferae) as indicators of nickel and zinc. *Journal of Geochemical Exploration* 18: 275–283.

Sá, R.A., O. Alberton, Z.C. Gazim, A. Laverde, J. Caetano, A.C. Amorin, and D.C. Dragunski. 2015. Phytoaccumulation and effect of lead on yield and chemical composition of Mentha crispa essential oil. *Desalination and Water Treatment* 53: 3007–3017.

Salido, A.L., K.L. Hasty, J.M. Lim, and D.J. Butcher. 2003. Phytoremediation of arsenic and lead in contaminated soil using Chinese Brake ferns (*Pteris vittata*) and Indian mustard (*Brassica juncea*). *International Journal of Phytoremediation* 5 : 89–103.

Saraswat, S. and J.P.N. Rai. 2009. Phytoextraction potential of six plant species grown in multimetal contaminated soil. *Chemistry and Ecology* 25: 1–11.

Sarma, H., N.F. Islam, R. Prasad, M.N.V. Prasad, L.Q. Ma, and J. Rinklebe. 2021. Enhancing phytoremediation of hazardous metal(loid)s using genome engineering CRISPR-Cas9 technology. *Journal of Hazardous Materials* 414: 125493.

Sasmaz, A., and E. Obek. 2012. The accumulation of silver and gold in *Lemna gibba* L. exposed to secondary effluents. *Chemie der Erde* 72: 149–152.

Saxena, G., D. Purchase, S.I. Mulla, G.D. Saratale, and R.N. Bharagava. 2020. Phytoremediation of heavy metal-contaminated sites: Eco-environmental concerns, field studies, sustainability issues, and future prospects. *Reviews of Environmental Contamination and Toxicology* 249: 71–131.

Saxena, P.K., S. KrishnaRaj, T. Dan, M.R. Perras, and N.N. Vettakkorumakankav. 1999. Phytoremediation of Heavy Metal Contaminated and Polluted Soils. In Heavy Metal Stress in Plants, pp. 305–329. Heidelberg: Springer Berlin.

Schneider, M., and R. Marquard. 1996. Investigations on the uptake of cadmium in *Hypercum perforatum*. L. (St. John's Wort). *Acta Horticulturae* 426: 435–441.

Scora, R.W., and A.C. Chang. 1997. Essential oil quality and heavy metal concentrations of peppermint grown on a municipal sludge-amended soil. *Journal of Environmental Quality* 26: 975–979.

Sen, S. 2020. 'Green'-Ing Kolkata: Creating a sustainable city – An overview. *International Journal of Research and Analytical Reviews* 7: 743–752. DOI: 10.1729/Journal.23607.

Sharma, S.C., R. Srivastava, and R.K. Roy. 2005. Role of bougainvilleas in mitigation of environmental pollution. *Journal of Environmental Science and Engineering* 47: 131–134.

Shetty, K.G., B.A.D. Hetrick, D.A.H. Figge, and A.P. Schwab. 1994. Effects of mycorrhizae and other soil microbes on revegetation of heavy metal contaminated mine spoil. *Environmental Pollution* 86: 181–188.

Shim, D., S. Kim, Y.I. Choi, W.Y. Song, J. Park, E.S. Youk, S.C. Jeong, E. Martinoia, E.W. Noh, and Y. Lee. 2013. Transgenic poplar trees expressing yeast cadmium factor 1 exhibit the characteristics necessary for the phytoremediation of mine tailing soil. *Chemosphere* 90: 1478–1486.

Shukla, A.C. 2020. The Herbal Drugs. In Advances in Pharmaceutical Biotechnology: Recent Progress and Future Applications, pp. 69–75. Springer Singapore.

Siddiqui, F., S.K. Krishna, P.K. Tandon, and S. Srivastava. 2013. Arsenic accumulation in Ocimum spp. and its effect on growth and oil constituents. *Acta Physiologiae Plantarum* 35 : 1071–1079.

Simard, M.J., and D.L. Benoit. 2010. Distribution and abundance of an allergenic weed, common ragweed (*Ambrosia artemisiifolia* L.), in rural settings of southern Quebec, Canada. *Canadian Journal of Plant Science* 90: 549–557.

Singh, G., U. Pankaj, P.V. Ajayakumar, and R.K. Verma. 2020. Phytoremediation of sewage sludge by *Cymbopogon martinii* (Roxb.) Wats. var. motia Burk. grown under soil amended with varying levels of sewage sludge. *International Journal of Phytoremediation* 22 : 540–550.

Singh, S.P., B. Keller, W. Gruissem, and N.K. Bhullar. 2017. Rice Nicotianamine Synthase 2 expression improves dietary iron and zinc levels in wheat. *Theoretical and Applied Genetics* 130: 283–292.

Sinha, S., R.K. Mishra, G. Sinam, S. Mallick, and A.K. Gupta. 2013. Comparative evaluation of metal phytoremediation potential of trees, grasses, and flowering plants from Tannery-wastewater-contaminated soil in relation with physicochemical properties. *Soil and Sediment Contamination* 22: 958–983.

de Souza, M.P., E.A.H. Pilon-Smits, and N. Terry. 2000. The Physiology and Biochemistry of Selenium Volatilization by Plants. In I. Raskin and B.D. Ensley (eds). Phytoremediation of Toxic Metals: Using Plants to Clean Up the Environment, pp. 171–190. New York: John Wiley & Sons.

Stancheva, I., M. Geneva, M. Boychinova, I. Mitova, and Y. Markovska. 2014. Physiological response of foliar fertilized *Matricaria recutita* L. grown on industrially polluted soil. *Journal of Plant Nutrition* 37: 1952–1964.

Sulastri, Y.S. and K. Tampubolon. 2019. Aromatic plants: Phytoremediation of cadmium heavy metal and the relationship to essential oil production. *International Journal of Scientific and Technology Research* 8: 1064–1069.

Suman, J., O. Uhlik, J. Viktorova, and T. Macek. 2018. Phytoextraction of heavy metals: A promising tool for clean-up of polluted environment? *Frontiers in Plant Science* 871: 1476.

Sun, Q., Z.H. Ye, X.R. Wang, and M.H. Wong. 2007. Cadmium hyperaccumulation leads to an increase of glutathione rather than phytochelatins in the cadmium hyperaccumulator *Sedum alfredii*. *Journal of Plant Physiology* 164: 1489–1498.

Sun, Y., Q. Zhou, and C. Diao. 2008. Effects of cadmium and arsenic on growth and metal accumulation of Cd-hyperaccumulator *Solanum nigrum* L. *Bioresource Technology* 99: 1103–1110.

Sun, Y., Q. Zhou, L. Wang, and W. Liu. 2009. Cadmium tolerance and accumulation characteristics of *Bidens pilosa* L. as a potential Cd-hyperaccumulator. *Journal of Hazardous Materials* 161 : 808–814.

Sundaramoorthy, P., A. Chidambaram, K.S. Ganesh, P. Unnikannan, and L. Baskaran. 2010. Chromium stress in paddy: (i) Nutrient status of paddy under chromium stress; (ii) Phytoremediation of chromium by aquatic and terrestrial weeds. *Comptes Rendus – Biologies* 333: 597–607.

Terry, N., and G. Banuelos. 1999. Phytoremediation of Contaminated Soil and Water. Routledge.

Tirillini, B., A. Ricci, G. Pintore, M. Chessa, and S. Sighinolfi. 2006. Induction of hypericins in *Hypericum perforatum* in response to chromium. *Fitoterapia* 77: 164–170.

Todorović, D., D. Popović, J. Ajtić, and J. Nikolić. 2013. Leaves of higher plants as biomonitors of radionuclides (137Cs, 40K, 210Pb and 7Be) in urban air. *Environmental Science and Pollution Research* 20: 525–532.

Tripathi, S., V.K. Singh, P. Srivastava, R. Singh, R.S. Devi, A. Kumar, and R. Bhadouria. 2019. Phytoremediation of Organic Pollutants: Current Status and Future Directions. In Abatement of Environmental Pollutants: Trends and Strategies, pp. 81–105. Elsevier.

Turgut, C., M. Katie Pepe, and T.J. Cutright. 2004. The effect of EDTA and citric acid on phytoremediation of Cd, Cr, and Ni from soil using *Helianthus annuus*. *Environmental Pollution* 131: 147–154.

Vallini, G., S. Di Gregorio, and S. Lampis. 2005. Rhizosphere-induced selenium precipitation for possible applications in phytoremediation of Se polluted effluents. *Zeitschrift fur Naturforschung – Section C Journal of Biosciences* 60: 349–356.

Vamerali, T., M. Bandiera, and G. Mosca. 2010. Field crops for phytoremediation of metal-contaminated land. A review. *Environmental Chemistry Letters* 8: 1–17.

Velmurugan, M., M. Anand, V. Davamani, K. Rajamani, and L. Pugalendhi. 2020. Landscape value of trees. *Biotica Research Today* 2: 693–695.

Voyslavov, T., S. Georgieva, S. Arpadjan, and K. Tsekova. 2013. Phytoavailability assessment of cadmium and lead in polluted soils and accumulation by *Matricaria chamomilla* (chamomile). *Biotechnology and Biotechnological Equipment* 27 : 3939–3943.

Wei, S., Q. Zhou, and U.K. Saha. 2008. Hyperaccumulative characteristics of weed species to heavy metals. *Water, Air, and Soil Pollution* 192: 173–181.

Weinstein, L.H., J.A. Laurence, R.H. Mandl, and K. Waelti. 1990. Use of Native and Cultivated Plants as Bioindicators and Biomonitors of Pollution Damage. In Plants for Toxicity Assessment, pp. 117–126. Philadelphia: American Society for Testing and Materials.

Wu, C., B. Liao, S.L. Wang, J. Zhang, and J.T. Li. 2010. Pb and Zn accumulation in a Cd-hyperaccumulator (*Viola baoshanensis*). *International Journal of Phytoremediation* 12: 574–585.

Wu, T.Y., W. Gruissem, and N.K. Bhullar. 2018. Facilitated citrate-dependent iron transloca-tion increases rice endosperm iron and zinc concentrations. *Plant Science* 270: 13–22.

Xiong, Z.T., Y.H. Li, and B. Xu. 2002. Nutrition influence on copper accumulation by *Brassica pekinensis* Rupr. *Ecotoxicology and Environmental Safety* 53: 200–205.

Yanqun, Z., L. Yuan, C. Jianjun, C. Haiyan, Q. Li, and C. Schvartz. 2005. Hyperaccumulation of Pb, Zn and Cd in herbaceous grown on lead-zinc mining area in Yunnan, China. *Environment International* 31: 755–762.

Yifru, D.D., and V.A. Nzengung. 2008. Organic carbon biostimulates rapid rhizodegradation of perchlorate. *Environmental Toxicology and Chemistry* 27: 2419–2426.

Youssef, N.A. 2021. Changes in the morphological traits and the essential oil content of sweet basil (*Ocimum basilicum* L.) as induced by cadmium and lead treatments. *International Journal of Phytoremediation* 23 : 291–299.

Zaid, A., F. Mohammad, and Q. Fariduddin. 2020. Plant growth regulators improve growth, photosynthesis, mineral nutrient and antioxidant system under cadmium stress in men-thol mint (*Mentha arvensis* L.). *Physiology and Molecular Biology of Plants* 26: 25–39.

Zhang, X., H. Xia, Z. Li, P. Zhuang, and B. Gao. 2011. Identification of a new potential Cd-hyperaccumulator *Solanum photeinocarpum* by soil seed bank-metal concentration gradient method. *Journal of Hazardous Materials* 189: 414–419.

Zurayk, R., B. Sukkariyah, R. Baalbaki, and D.A. Ghanem. 2002. Ni phytoaccumulation in *Mentha aquatica* L. and *Mentha sylvestris* L. *Water, Air, and Soil Pollution* 139: 355–364.

Index

Note: Page numbers in *italics* indicate figures and in **bold** indicate tables on the corresponding pages.